Textile Science and Clothing Technology

Series Editor

Subramanian Senthilkannan Muthu, SgT Group & API, Hong Kong, Kowloon, Hong Kong

This series aims to broadly cover all the aspects related to textiles science and technology and clothing science and technology. Below are the areas fall under the aims and scope of this series, but not limited to: Production and properties of various natural and synthetic fibres; Production and properties of different yarns, fabrics and apparels; Manufacturing aspects of textiles and clothing; Modelling and Simulation aspects related to textiles and clothing; Production and properties of Nonwovens; Evaluation/testing of various properties of textiles and clothing products; Supply chain management of textiles and clothing; Aspects related to Clothing Science such as comfort; Functional aspects and evaluation of textiles; Textile biomaterials and bioengineering; Nano, micro, smart, sport and intelligent textiles; Various aspects of industrial and technical applications of textiles and clothing; Apparel manufacturing and engineering; New developments and applications pertaining to textiles and clothing materials and their manufacturing methods; Textile design aspects; Sustainable fashion and textiles; Green Textiles and Eco-Fashion; Sustainability aspects of textiles and clothing; Environmental assessments of textiles and clothing supply chain; Green Composites; Sustainable Luxury and Sustainable Consumption; Waste Management in Textiles; Sustainability Standards and Green labels; Social and Economic Sustainability of Textiles and Clothing.

More information about this series at https://link.springer.com/bookseries/13111

Rajkishore Nayak
Editor

Lean Supply Chain Management in Fashion and Textile Industry

Editor
Rajkishore Nayak
School of Communication and Design
RMIT University Vietnam
Ho Chi Minh, Vietnam

ISSN 2197-9863 ISSN 2197-9871 (electronic)
Textile Science and Clothing Technology
ISBN 978-981-19-2110-0 ISBN 978-981-19-2108-7 (eBook)
https://doi.org/10.1007/978-981-19-2108-7

© The Editor(s) (if applicable) and The Author(s), under exclusive license to Springer Nature Singapore Pte Ltd. 2022
This work is subject to copyright. All rights are solely and exclusively licensed by the Publisher, whether the whole or part of the material is concerned, specifically the rights of translation, reprinting, reuse of illustrations, recitation, broadcasting, reproduction on microfilms or in any other physical way, and transmission or information storage and retrieval, electronic adaptation, computer software, or by similar or dissimilar methodology now known or hereafter developed.
The use of general descriptive names, registered names, trademarks, service marks, etc. in this publication does not imply, even in the absence of a specific statement, that such names are exempt from the relevant protective laws and regulations and therefore free for general use.
The publisher, the authors, and the editors are safe to assume that the advice and information in this book are believed to be true and accurate at the date of publication. Neither the publisher nor the authors or the editors give a warranty, expressed or implied, with respect to the material contained herein or for any errors or omissions that may have been made. The publisher remains neutral with regard to jurisdictional claims in published maps and institutional affiliations.

This Springer imprint is published by the registered company Springer Nature Singapore Pte Ltd.
The registered company address is: 152 Beach Road, #21-01/04 Gateway East, Singapore 189721, Singapore

Preface

The concept of lean manufacturing was coined in Japanese manufacturing sector, Toyota Production System (TPS) mainly dealing with automotive components. The major focus of lean manufacturing was to eliminate wastes from the operation cycle. Wastes can be defined as the process or operation, which does not add value to the product and the consumer is not willing to pay. The objectives of lean manufacturing were to: (a) eliminate wastes (by avoiding waiting, overproduction, reducing inventory etc.) in the manufacturing process; (b) improve productivity and efficiency; (c) optimize the utilization of space and equipment; and (d) reduce lead time. Although, there have been some technological advancements that are applied in lean manufacturing, the fundamentals of lean manufacturing have remained the same.

In recent years, the manufacturing of fashion and textiles has completely been shifted to developing countries such as Bangladesh, Vietnam, and Cambodia. The developed countries, due to high labor and utility charges are focusing to manufacture high value added clothing or technical textiles. The manufacturing of fashion and textiles is labor intensive in many developing countries despite of the technological developments and automation. Many of the fashion brands aim to buy the fashion and textile products at cheaper prices, which lead to stiff competition among the global manufactures. Further, the labor charges, raw materials prices are in continuous rise, which makes the manufacturers to face many challenges in manufacturing the garments in competitive prices. In the competitive global market, the fashion and textile manufactures should improve product quality, improve efficiency, and eliminate waste, which can be achieved by the implementation of lean manufacturing. Lean manufacturing is an integrative approach, which can be successful by the involvement of the whole organization as a team.

The fashion and textile industries are facing a huge competition due to globalization and challenges of sustainability. Fashion and textile industries are considered to be the second largest polluters after the oil sector to due generation of a large quantities of wastes, hence, the lean manufacturing concept can be utilized to reduce waste. As the global focus is changing towards achieving or producing more with limited resources, the resource-based fashion and textile manufacturing needs to be changed to a new manufacturing system. The new manufacturing system should focus on the waste reduction, energy saving, use of advanced technologies and information systems. These approaches can help the organizations to achieve excellence and provide competitive advantage, which can be achieved by the implementation of lean manufacturing. However, the transition towards a new manufacturing system requires the highest skillsets and excellent organizational leadership.

The concept of lean manufacturing is gaining increased popularity in many fashion and textile industries due its ability to manage the process and material wastes. The fundamental principle of lean manufacturing is based on continuously eliminating waste from the manufacturing processes. Implementation of lean manufacturing focuses on areas such as: increase productivity by reducing overproduction, waiting time, over processing, inventory, defects, unnecessary transportation, and unnecessary movement, which are considered the seven wastes in lean manufacturing. Lean manufacturing reduces the cycle time, lead time, and changeover time, hence, improves productivity, efficiency and profitability. Lean can help to keep the manufacturing cost low in the rising labor and material price in the global fashion and textile manufacturing sector.

Various latest technologies such as Radio Frequency Identification (RFID), Blockchain, Internet of Things (IoT), Industry 4.0, Artificial Intelligence (AI), and Automation are essential to achieve the objectives of lean manufacturing. Therefore, many organizations are replacing the traditional technologies with the advanced technologies to achieve the objectives. For example, several global retailers and manufacturers are adopting the RFID technology in their supply chain to realize the benefits, which includes accuracy of information, prevention of loss, product visibility, easy traceability, reduced labor cost and reduced lead time. Hence, RFID technology can be considered one of the leading technologies for successful operation of lean manufacturing. The use of other technologies has also been essential part of lean manufacturing as these technologies convert the lean manufacturing to digital lean, which improves the accuracy and speed in addition to timely availability of information.

This book will discuss the concepts of lean manufacturing, which have been widely used recently in the fashion and textile supply chain to eliminate waste, improve productivity and efficiency. Lean manufacturing tries to apply various fundamental tools such as Kaizen, Kanban, Jidoka, 5S and Six Sigma, to reduce various wastes for efficiency and productivity, which will be discussed in this book. In addition, the technological changes such as RFID, IoT and Industry 4.0 that are used in digital lean concepts will be also discussed in this book. The findings from the research on the use of digital technologies such as RFID that the author and some of the contributors performed in Vietnamese fashion and textile sector, will be discussed in this book. The book consists of a total of 12 chapters, contributed by the subject experts from various universities and academic institutes. The brief outline of the chapters of the book has been given below.

"Traditional Fashion and Textile Supply Chain: Concept to Consumer" provides an introduction of the traditional processes used in fashion and textile supply chain. The process sequence and summary of each process has been discussed while converting raw materials from concept to consumer. Various causes of material and process waste generated in these processes (in general) are also being discussed. There has been a brief discussion on the history of lean manufacturing and its chronological developments. Further, the fundamental principles of lean manufacturing and various types of tools used in lean manufacturing has also been discussed. Various tools used in lean manufacturing are also discussed in this chapter. Finally, the advantages, disadvantages and future directions of lean manufacturing are covered in this chapter.

"Challenges in the Traditional Fashion and Textile Supply Chain" discusses various challenges of traditional manufacturing methods. Various strategic challenges such as buyer driven supply chain, high SKU proliferation, sustainability challenges and educational challenges are discussed. Various operational challenges such as lead time and sampling, limited use of technology, and limited use of technology have been discussed in this chapter. Some of the causes of low productivity and low efficiency from fashion and textile perspective has been included in this chapter.

"Fundamental Concepts of Lean and Agile Manufacturing" focusses on the evolution of lean manufacturing-definition, objectives, and principles of lean manufacturing. The goal of lean and agile manufacturing to satisfy customer demands, has been discussed in this chapter. This chapter focusses on the fundamentals of lean and agile manufacturing concepts, in addition to the objectives, and principles. There have been discussions on the history, chronological developments, applications in the manufacturing sector, especially fashion and textile industry, and benefits of lean and agile manufacturing. There have been discussions about the implementation of lean concepts in fashion and textile manufacturing as well as how it can help to improve the productivity and efficiency. Different wastes and their control through lean manufacturing. The chapter also discusses the differences between the two concepts: lean and agile manufacturing.

"Lean Concept in Fashion and Textile Manufacturing" covers the overview of lean manufacturing. How the lean manufacturing was developed in the Toyota Production System (TPS) and Henry Ford Production System has been discussed in this chapter. Furthermore, a conceptual framework for lean concepts has been provided in this chapter. The development of lean for fashion and textile manufacturing has also been discussed in this chapter with a focus on the past works in lean manufacturing with the chronological developments. Various wastes in fashion manufacturing have been discussed with emphasis on their cause and elimination processes. Lean applications in garment manufacturing have been discussed in detail and a brief description has been given on the lean application in the textile industries. The lean application to achieve sustainability is also discussed in this chapter.

"Standardized Work in Fashion Industry" covers the processes such as time and motion study generally used in garment industries to establish standard time. The fundamental principles and the process of conducting time and motion study has been covered in this chapter. This chapter also discusses about the maximum utilization of machine, labor, cycle time, changeover time, work balancing, time measurement, tools and jigs for time management, economy of motion and time management. Various causes of idle time and bottle necking are also covered in this cheaper.

"5S and Its Implications in Fashion and Textile Industry", discusses the five pillars of 5S system (such as Sort, Set, Shine, Standardize & Sustain) in the context of fashion and textile lean manufacturing. The detailed processes for implementation of 5S technology such as observation & preparation, planning, implementing and assessment has also been discussed. Various phases and facilitating factors (such as human factors, management and leadership, staff compliance, collaborative tools) of 5S tool have been discussed. Enablers and barriers in the implementation of 5S are covered.

"Kaizen Applications in Fashion and Textile Industries" focuses on the fundamental concepts of Kaizen, in addition to the systematic developments, and application in fashion and textile industries. A brief history of Kaizen has also been included to discuss the chronological developments. Various phases of Kaizen implementation, starting from planning to execution stages, for businesses to be benefitted from waste reduction, increased productivity, and optimization of operation, are discussed in this chapter. Indeed, the application of Kaizen concept has been an effective solution for the fashion and textile industries with actual evidence from Bangladesh, Peru, and India, which are also discussed in the chapter. There have been discussions how Kaizen has resolved problems of fabric fault, maintained ergonomic standards and workers comfort to increase the productivity. Further, this chapter highlights its challenges faced for the implementation of Kaizen, advantages and disadvantages in the fashion and textile industry.

"Kanban Applications in Fashion and Textile Industries" provides a thorough explanation of the Kanban concept. Various Kanban methods as well as its implementation in the fashion and textile industries from different perspectives are covered in this chapter. Kanban system was derived from a Japanese notion, which means "visible sign", originated with the efforts of the Toyota automotive company in applying the concept to the manufacturing process. Kanban has eventually become a popular methodology that is used widely used in the automobile industry as well as in other sectors such as fashion and textiles. It has been a successful tool in the Lean Manufacturing and Just-in-time concepts with the purpose of maximizing productivity and minimizing workplace waste, which has been discussed in this chapter. The use of various signals in the Kanban system help to make the workflow as agile and efficient as possible. There has been discussion on the advantages and disadvantages of the implementation of Kanban system in the fashion and textile industries to obtain higher productivity and efficiency. This chapter also highlights some case studies focusing on the fashion brands and manufacturers using Kanban. How the implementation of Kanban tool in fashion and textile industries can help to increase the productivity, control inventory, ensure supplier and employee participation are also discussed in this chapter.

"Other Lean Tools in Fashion and Textile Manufacturing", explains how lean manufacturing tools such as Muda, Six Sigma and statistical process control, and ergonomics can help in waste reduction, process efficiency and safety in fashion and textile industries. These tools focus on reducing the seven wastes as listed for lean manufacturing process to make the process more efficient. The four Muda techniques to perform waste identification which includes: building lean thinking and cultures, hybrid approach of combining multiple decision-making methods, integrating lean and green management, and value stream mapping to prioritize responses to wastes are discussed in this chapter. The use of statistical process control (SPC) in process control in fashion and textile manufacturing and enhance the future production predictability are elucidated. The use of Six Sigma as a process-based strategy with five stages (DMAIC) to eliminate defects in manufacturing process are covered. Finally, the use of ergonomics to optimize human well-being and overall system performance, in lean manufacturing to improve quality and well-being of workers in fashion and textile industries also discussed.

"Digital Technologies for Lean Manufacturing" aims to offer thorough insights into the digital technologies that are designed to complement the operation of lean manufacturing. This chapter aims to highlight the digital technologies that are designed to complement the operation of lean manufacturing. Firstly, Industry 4.0 has been explained that helps to complement Lean Manufacturing to gain continuous improvement, better customer expectation and operational process. The results of combining digital technologies with lean manufacturing yield the concept of "Lean Industry 4.0". Secondly, blockchain occurs as a disruptive innovation to resolve the problems of lack of an integrated lean management system across the supply chain network, which is also discussed. Thirdly, the Radio Frequency Identification (RFID) system is analyzed, and its ability to offer high levels of accurate, real-time information, decreased time-consuming activities and labor cost while increasing product visibility and operation speed is covered. Fourth, Artificial Intelligence (AI) and robotics are also discussed with the ability to deal with complexity, increase in productivity and efficiency with the automatic system, and decrease production costs. Finally, other non-common yet useful tools are mentioned to give a comprehensive view of the application of digital technologies on Lean Manufacturing, including automated guided vehicles (AGVs), virtual stimulation (VS), and cybersecurity. To consolidate our findings, two case studies are presented to give realistic viewpoints of digital technology adoption from two giant firms in the textile and apparel industry, namely Uniqlo and H&M. The findings of a survey based on Vietnam's fashion and textile industries on the use of technology such as RFID is also included in this chapter.

"Lean Manufacturing: Case Studies from Fashion and Textile Industries" focuses on various case studies from global fashion and textile brands. This chapter highlights the causes relating to the failure and success of various global industries in the fashion and textile manufacturing sector. The earlier chapters have clearly mentioned that the implementation of various lean tools such as Kaizen, Muda, 5S and Kanban can help the traditional fashion and textile industries to improve their productivity and efficiency by reducing waste. This chapter will highlight the direct cases of how the implementation of lean manufacturing improves the productivity and efficiency of fashion and textile industries. Some case studies focusing on the failure of some of the industries from Africa and India due to not implementation of lean manufacturing tools has been discussed in the first section. The case studies focusing on the success of some of the fashion and textile industries by adopting lean manufacturing tools is discussed in the second section. Finally, the third and last section will focus on the future directions and conclusions of this chapter.

"Benefits, Drawbacks, and Future Directions of Lean on the Fashion and Textile Industry" is the final chapter of the book, which will cover benefits, draw backs and future directions of lean manufacturing in fashion and textile industry. The previous chapters have shown that the lean supply chain principles can be implemented in Fashion and Textile (F&T) industries to make them more competitive and sustainable. Further, the implementation of lean manufacturing can help to overcome some unexpected challenges. For example, the F&T industry has encountered the COVID-19 pandemic with many supply chain disruptions. Lean concepts have supported this sector to create a balanced flow of textile and garment products through lean with agility, supply chain visibility, and green supply chain practices. Further, this chapter also recommends the future development in lean and agility, sourcing flexibility and green supply chain practices that can help the industry to recharge and adjust during such a difficult time. This chapter focuses on structural design and supply chain network that facilitates both lean and green strategies in the fashion and textile industry.

Looking at the huge protentional of lean manufacturing in the fashion and textile industries, the author was inspired to pen down the book on Lean manufacturing. The initial background research on the subject area in 2019, resulted some published books with different publishers focusing on lean supply chain management. The author was especially interested in his area of work in fashion and textiles, manufacturing and supply chain. A detailed search in the existing body of knowledge resulted some journal publication in lean manufacturing. However, there were no commercial publications available on the application of lean manufacturing in the field of fashion and textiles. This prominent gap to gather concise information in one place led the author to publish the book. Hence, this book intensively focuses on the application of lean manufacturing in the supply chain process of fashion and textiles starting from concept to consumer. Special emphasis has been given on the garment manufacturing, which is one of the major areas of implementing lean in the fashion and textiles.

The book is a result of author's intensive work in fashion and textiles, which took about a year to complete. Author and contributors collected the information from various peer-reviewed journals; trade magazines; websites of various companies including fashion and textiles; in addition to the information collected from various research organizations; and fashion distributor's websites. Some of the chapters include the research findings from the project the author completed to know the recent status of technology in lean manufacturing.

Author has tried to make this book informative by covering all the aspects of lean manufacturing applications in the fashion and textile supply chain. The book is written in simple language so that the people with minimal knowledge in the lean manufacturing can easily understand the subject matter. It is anticipated that the book with a range of scientific knowledge, fundamental principles and application areas will dissipate knowledge to the students, academicians, research communities and manufacturers working in the fashion and textiles.

The author extends his sincere thanks to all the experts who have helped to complete all the chapters successfully. His special thanks to the authors and publishers for providing approval to reuse the images, tables and other information in the book. The author is also thankful to Springer publication team for their support for making this book a reality. The author hopes this book will help the readers in the field of fashion and textiles in acquiring relevant information on lean manufacturing.

Dr. Rajkishore Nayak
Associate Professor
Fashion Merchandising, School
of Communication and Design
RMIT University
Ho Chi Minh, Vietnam

Contents

Traditional Fashion and Textile Supply Chain: Concept to Consumer 1
Rajkishore Nayak, Majo George, and Irfan Ulhaq

Challenges in the Traditional Fashion and Textile Supply Chain 31
Prabir Jana

Fundamental Concepts of Lean and Agile Manufacturing 47
Mohammadreza Akbari, Kevin Nguyen, Kristof Van Houdt, and Seng Kiat Kok

Lean Concept in Fashion and Textile Manufacturing 67
Majo George, Lam Canh Nguyen, Hung Manh Nguyen, and Mohammadreza Akbari

Standardized Work in Fashion Industry 95
Ashish Bhardwaj

5S and Its Implications in Fashion and Textile Industry 125
Irfan Ulhaq, Majo George, and Rajkishore Nayak

Kaizen Applications in Fashion and Textile Industries 145
Majo George, Vuong Nguyen Dang Tung, Le Phan Thanh Truc, Nguyen Minh Ngoc, and Le Khac Yen Nhi

Kanban Applications in Fashion and Textile Industries 177
Majo George, Le Phan Thanh Truc, Vuong Nguyen Dang Tung, Le Khac Yen Nhi, Nguyen Minh Ngoc, and Rajkishore Nayak

Other Lean Tools in Fashion and Textile Manufacturing 199
Hiep Cong Pham, Irfan Ulhaq, Paul Yeow, and Mohammadreza Akbari

Digital Technologies for Lean Manufacturing 219
Majo George, Le Khac Yen Nhi, Nguyen Minh Ngoc,
Vuong Nguyen Dang Tung, Le Phan Thanh Truc, and Rajkishore Nayak

**Lean Manufacturing: Case Studies from Fashion and Textile
Industries** 269
Majo George, Nguyen Minh Ngoc, Le Khac Yen Nhi,
Vuong Nguyen Dang Tung, Le Phan Thanh Truc, and Rajkishore Nayak

**Benefits, Drawbacks, and Future Directions of Lean on the Fashion
and Textile Industry** 291
Hung Manh Nguyen, Scott McDonald, Bill Au,
and Mohammadreza Akbari

Traditional Fashion and Textile Supply Chain: Concept to Consumer

Rajkishore Nayak, Majo George, and Irfan Ulhaq

Abstract Fashion and textile products evolve through several processes that involve the conversion of fibre to yarn, yarn to fabric and fabric to garment. During these processes, there are numerous wastes generated, which reduces the productivity and efficiency. Many of the manufacturing facilities in developing countries still follow the traditional technologies for manufacturing, which leads to low productivity, more energy consumption, higher waste, and low product quality. The global marketplace is getting more and more competitive day by day with demand on high value-added products at cheaper prices. Hence, the fashion and textile manufacturing industries need to fulfil the demand for quality and price by improving their performance. Further, with the growing global interest on sustainability and increased consumer awareness, fashion and textile industries bring sustainable solutions in their production lines by reducing wastes. This can be achieved by the implementation of lean manufacturing in the manufacturing industries, which is the focus of this book. This chapter is the introduction to the fundamentals and concepts of lean manufacturing. A brief introduction has been given on the fashion supply chain starting from concept to consumer, which is essential to understand the problems associated with the manufacturing industries. Further, the principles and tools of lean manufacturing are also discussed including thee seven wastes in manufacturing industries for fashion and textiles. The steps to implement lean manufacturing, which includes PDCA (Plan-Do-Check-Act) cycle; advantages and disadvantages; and future directions are discussed in this book.

Keywords Lean manufacturing · Fashion supply chain · Fashion sustainability · Push and pull · Waste eliminations · Lean tools · Lean principles

---

R. Nayak (✉)
School of Communication and Design, RMIT University Vietnam, 702 Nguyen Van Linh, District 7, Ho Chi Minh, Vietnam
e-mail: rajkishore.nayak@rmit.edu.vn

M. George · I. Ulhaq
School of Business & Management, RMIT University Vietnam, 702 Nguyen Van Linh, District 7, Ho Chi Minh, Vietnam

© The Author(s), under exclusive license to Springer Nature Singapore Pte Ltd. 2022
R. Nayak (ed.), *Lean Supply Chain Management in Fashion and Textile Industry*, Textile Science and Clothing Technology, https://doi.org/10.1007/978-981-19-2108-7_1

1 Introduction

In many developing countries, the manufacturing of fashion and textiles is labor intensive using the traditional technologies. In spite of the availability of automation, many industries still operate with the traditional machineries, which are at odds to achieve the objectives of good quality and high productivity. In addition to using higher energy, the traditional technologies generate more wastage, and greenhouse gases (Niinimäki et al. 2020; Nayak et al. 2021a). The global demand of fashion and textiles is fulfilled by the manufacturing services of industries located in developing countries. There has been paradigm shift of manufacturing from developed countries to developing countries. The developing countries such as Vietnam, Cambodia and Bangladesh are the leading manufacturer of fashion and textiles. The manufacturing activities in Chaina and India, the past leaders are gradually declining in the drive of keeping the cost of production low.

The fashion and textile manufacturers in developing countries are operating in stiffer competition in the global marketplace. The increasing labor cost, rising utility charges, increase in the cost of materials and demand for sustainable fashion and textiles has made it difficult for the manufacturers to produce the products at cheaper price. Many of the large enterprises in these countries have well coped with the global demand due to larger volume of production and timely investment on new technologies (Nayak et al. 2020b). However, there are several manufacturers of raw materials, trims, semi-finished products, and final garments, especially small and medium sized enterprises (SMEs), who cannot upgrade themselves due to financial problems. Hence, they still operate using the traditional technologies. The use of traditional technologies may not be able to deliver the productivity and efficiency that can be achieved by the new technologies.

In addition to traditional technologies, there have been loss of productivity due to improper allocation of resources. Problems such as excessive reworking, improper line balancing, and lack of synchronization have also added to productivity loss (Aghajani et al. 2012; Nayak and Padhye 2015). Rework is a general problem in many industries due to lack of intermediate inspection during manufacturing. The products are fed to the next operation sequence without inspecting any defects form a process. Hence, a small defect in any process is translated into a larger defect in the final product, which then fixed after the final inspection of the finished product. Inspection of the final product is too late to fix the fault to improve the product quality. This requires reworking from the step when the defect occurred. Rework can be mainly related to stitching defects (skipped stitch, and broken stitch), seaming defects (open seam and seam puckering), fabric defects and other sewing related problems. These defects are fixed by rework, which lowers the productivity and sometimes rejection of garments. Rework impacts the regular productivity and have an impact on the manufacturing economy. Improper line balancing and lack of synchronization is the problem of mismatching of supply can demand among various sections in the manufacturing process. These non-productive activities need to be reduced, for which the implementation of lean manufacturing plays an important role.

The garment supply chain can be characterized by high volatility, small production volumes, wide range of styles, short production lifecycle and low predictability (Ren et al. 2017; Mason-Jones et al. 2000). Furthermore, there have been a decrease in volume of production for specific products and increase in the style variation, which creates challenges to source a range of raw materials such as fabrics, trims, and accessories. Earlier, when there were large order volumes, the manufacturers were ascertained the manufacturing will continue for several weeks or months. However, decrease in order quantities has led to frequent change of product lines, increased sourcing activities influencing the profitability. In order to cope with these changes, the garment manufactures need to focus on the implementation of lean manufacturing. The concept of lean manufacturing is based on reducing waste, reducing inventory, increasing productivity and efficiency by reducing the amount of rework.

Majority of the clothing manufacturers in developing countries follow the progressive bundle system (PBS), which aims at mass production of apparels (Desai et al. 2012). In the PBS systems, the work related to a specific garment is performed progressively one after the other. Machines and equipment are arranged sequentially depending on the type of the garment forming an assembly line. Multiple assembly lines can be used for a specific garment style. Garment components after cutting are supplied to different assembly lines in bundles, which moves across the assembly line in trolleys by the workers. In PBS approach, each operator only does a specific type of job repeatedly and pass to the next operator. The PBS system is labor intensive, which involves high amount of non-value-added activities and larger lead time due to longer waiting time. If one operator in the PBS line is not manufacturing at the expected level, the whole line is impacted due to slow productivity. The PBS system can be classified as batch manufacturing, which is always limited in terms of productivity. The machine arrangements in a PBS system is has been shown in Fig. 1.

To overcome these limitations, the PBS systems can be replaced with the cellular manufacturing systems or unit production systems (UPS). In UPS system, the cut pieces hanged in hangers are automatically transported by overhead conveyor system. One hanger carries all the components for a garment, which is transported automatically. Hence, the manual operations are replaced with automatic transport system. A garment is manufactured through the different steps of production and the final garment is hanged back in the hanger system. The implementation of lean manufacturing can produce better results in UPS manufacturing process compared to the PBS concept. However, the UPS systems cost a lot more and needs highly skilled operator for successful operation.

This chapter discusses the basics of fashion and textile supply chain with a special emphasis on lean manufacturing. The problems of traditional manufacturing and necessity of lean manufacturing has been discussed. The fundamental principles of lean manufacturing have been discussed with the seven wastes in the fashion and textile manufacturing. Various tools used in lean manufacturing and steps followed for the implementation are discussed. This chapter also highlights various approaches to become sustainable by the implementation of lean in the fashion and textile supply

Fig. 1 PBS system in a garment manufacturing industry

chain processes. Furthermore, the advantages, disadvantages of lean manufacturing and the future directions are discussed in this chapter.

1.1 Fashion and Textile Supply Chain

Supply chain management (SCM) deals with the management of the supply or flow of goods and services. SCM involves the procurement, storage and flow of raw materials, semi-finished and finished goods from the point of origin (manufacturing) to the point of usage (consumers) (Ross 2015; Stock and Boyer 2009). A network of upstream and downstream players is interconnected to produce the products and services required by the consumers. SCM is the management of flow of goods and services that is required to convert raw materials to finished goods for the consumers. The main objective of SCM is streamlining supply of goods and services to maximize customer satisfaction with right type of products or services (Christopher 2016). SCM can be defined as the "design, planning, execution, control and monitoring of supply-chain activities with the objective of creating net value, building a competitive infrastructure, leveraging worldwide logistics, synchronizing supply with demand and measuring performance globally (Tiwari and Jain 2013)". SCM can relate to the distribution activates in many sectors such as automotive, medicine, construction, hotel, hospital, engineering, information technology, plastics, fashion and textiles.

The supply chain activities in fashion and textiles are rather complex and involves many role players as shown in Fig. 2, which starts from the fiber producers. Fibres are known as the building block for fashion and textile manufacturing. Textile fibers can be classified into two groups: (a) natural and (b) synthetic. As per the name natural fibers are derived from natural resources, which includes cotton, wool, silk, hemp and jute (Nayak et al. 2012). On the other hand, synthetic fibers such as nylon and polyester are derived from the petroleum resources, whereas synthetic fibers such as viscose, modal and lyocell are derived from plant resources. Therefore, the synthetic fibers derived from plant resources are known as semi-synthetic fibers or regenerated fibers.

The next level of mid-stream players includes processes such as spinning and weaving industries, who produce yarns and fabrics, respectively. The yarn manufacturing process uses various fibers as the input material and produces different types of natural and synthetic yarns. Yarn manufacturing process can use different spinning systems such as ring, rotor and airjet to produce a range of yarn counts with different properties (Goyal and Nayak 2020). Ring spinning is the versatile system and produces the strongest yarn among all the systems. However, the energy consumption is the maximum for the ring spinning systems. Other spinning systems (such as rotor and airjet) consume less energy, however, there are some limitations in producing a range of yarn counts and the yarn strength. When productivity is considered, ring spinning may be considered at odds due to lowest production and highest energy consumption. However, one must consider factors such as end-use

Fig. 2 Sequence of operations for garment manufacturing

application, cost, type of product and downstream processes while selecting a yarn from various spinning systems.

Fabric manufacturing follows the spinning process and can be broadly categorized as: weaving, knitting and nonwoven (Patnaik and Patnaik 2019). Weaving is the most widely used process for manufacturing fabrics, which involves the interlacing of two sets of threads perpendicular to each other. The next widely used process is knitting, which involves the intermeshing of consecutive rows of connected loops with the next and previous rows. Nonwoven fabrics are rather weaker structures directly produced from the textile fibers (staple and filament) bonding them together by means such as mechanical, thermal, chemical, or solvent treatment. Woven fabrics vary in their structure and can be classified as plain, twill, satin, sateen, and many other structures. Knitted fabrics can be classified as single jersey, purl, rib and interlock, which have different structure, hence, vary in their properties. Nonwoven fabrics can be of various types such as spun-laced, dry-spun, melt-blown, air-laid and heat-bonded nonwoven fabrics, which also differ in their structure and properties.

Each type of fabrics has its merits and demerits, and the selection of a fabric depends on the type of product and end use applications. For example, woven fabrics are used for apparel clothing (such as shirts, trousers, ladies dress and activewear), home decorations (such as curtains, bedsheets, draperies, bed towels and table cloths), uniforms (such as security, police and army) and industrial applications; knitted fabrics are used for apparel clothing (such as Tees, jackets, coats, jumpers and pants), innerwear (such as camisoles, undies, girdles, sleepwear and vests), sportswear (such as sock, sports shoes, track suits and vests), home furnishings (such as cleaning wipes, laundry bags and mosquito nets) and automotive textiles (such as headrest lining, cushioning material and lining for motorbike helmets); nonwovens are used for medical textiles (such as surgical masks, surgical gowns, gloves, wipes, caps, plasters and wound dressing), filtration (such as vacuum bags, filter membranes for gas and liquid, and allergen membranes), geotextiles (such as erosion control, canal liners, soil stabilizer and drainage systems) (Sinclair 2014).

The fabrics manufactured by various processes (as discussed above) are in griege state and not ready to be converted into garment. Greige fabrics are in unfinished state directly coming out of a loom or a knitting machine, which contains impurities such as wax, oil, gums, soluble impurities, and dirt in addition to the sizing materials. These impurities make hard for textile coloration by dyeing and printing. Hence, these impurities need to be removed by various chemical processing steps before they are ready for garment making. The chemical processing steps include desizing, scouring, bleaching, dyeing, printing, and finishing. Among all the upstream and downstream processes of garment construction, the textile chemical processing is the most polluting step as it produces the maximum amount of waste water (Muthu 2018). Therefore, any drive to reduce wastes (especially wastewater) during fashion supply chain should emphasize on the textile chemical processing.

Fabric chemical processing makes the fabrics ready to be converted into garments. The next process is garment manufacturing, where the two-dimensional fabrics (2D fabrics) are converted into three dimensional garments (3D garments). The garment manufacturing process involves activities such as fabric inspection, spreading, cutting, bundling, fusing, sewing, ironing, inspection and packaging (Nayak and Padhye 2015). The global trend in garment manufacturing has undergone many changes from the beginning of the twenty-first century. The global demand for apparel textiles is being fulfilled by only some developing countries such as Bangladesh, Cambodia, China, Laos, India, Vietnam and some countries in the European Union (EU). The manufacturing of garments at cheaper prices in developed countries such as the US or Australia or Canada is not viable due to high labor cost and energy charges. Therefore, they rely on the developing countries to fulfill their demand for fashion and textiles.

In developing countries, the fashion and textile industries provide employment to a large section of the population. This sector also contributes largely towards the gross domestic product (GDP) of the countries (Nayak et al. 2019). Earlier, the manufactures in developing countries were getting larger orders in bulk quantities, which is easier to control. Once, the production line is set for bulk orders, it would be running for some months or even some weeks. However, the order quantities are shrinking now a days, the garment manufactures have to change the line setting in some days or few weeks. Further, the raw materials and trims are sourced in smaller volumes, which reduces the flexibility and profit margin. The volatility and shrinking volume can be better managed by the implementation of lean manufacturing in the production floor.

The competitive global marketplace is creating a lot of challenges for the fashion and textile manufactures. The demand for low cost by fashion brands has necessitated to increase the efficiency and productivity in the manufacturing process (Čiarnienė and Vienažindienė 2014). The traditional manufacturing processes are at odds if high productivity and efficiency are claimed. Hence, the manufactures need to invest in the new technologies and automation to replace the labor-intensive processes. The automation technologies can be implemented at any stage of the manufacturing processes of fashion and textiles starting from fibre production to garment manufacturing (Nayak and Padhye 2017, 2018b). Furthermore, the investment in the automation and new technologies need to be appropriately operated to reduce the wastes from the manufacturing processes, which can be achieved by the implementation of lean manufacturing. Hence, it can be concluded that lean manufacturing will become an indispensable part for the fashion and textile manufactures in the future.

1.2 Fundamentals of Lean Manufacturing

The business environment is rapidly changing for the global fashion and textile enterprises due to reduced profit margin, demand for high quality product, high product varieties and increasing pressure to reduce the lead time. In order to stay competitive and survive in this environment, strategies such as enterprise resource planning (ERP), quality management (ISO 9000), total quality management (TQM), just-in-time (JIT) or quick response, business process reengineering, and lean manufacturing are essential (Schonberger 2014; Fink 2003). Among these, the most significant for the manufacturing industries such as fashion and textiles is the lean manufacturing. The lean manufacturing approach has been widely adopted by many fashion and textile enterprises in order to enhance productivity, efficiency and to achieve the desired quality (Carmignani 2016).

The operational costs in garment production can be reduced by adopting the concept of "Lean manufacturing", which focuses on eliminating the process waste, improving productivity, empowering people with greater communication, and converting the organization into a learning organization (Bruce et al. 2004). The process waste can be reduced by avoiding overproduction, wastes, waiting time, unnecessary motion, unnecessary transportation, improper inventory management, and overprocessing. Continual improvement (generally known as the Japanese word 'Kaizen') is the major principle of lean manufacturing. 'Kaizen' promotes continuous and necessary changes (big or small) towards the achievement of a desired goal.

The fundamental thrust of lean manufacturing is to produce a high quality product at lower cost by reducing or eliminating the seven cardinal wastes such waiting, inventory, overproduction, repair, inappropriate processing, excess motion, and transportation from the value stream through continuous improvement and to deliver the value to the customer (Benders and Van Bijsterveld 2000). The goal of the lean manufacturing is to create an integrated system using multi-dimensional approach that includes adoption of management practices such as pull strategy, just in time philosophy, total quality management, cellular manufacturing, electronic data interchange (EDI), Kanban and co-design. Lean production techniques create a sustainable and positive work environment by emphasizing on empowering the workers and adopting the tools which enhance the operational efficiency by cycle reduction, cellular manufacturing, working in teams and stabilizing workflow etc. Yang et al. (2010) adds that when the different forms of wastes in lean culture are reduced, will in turn useful in managing the environmental waste by enhancing environmental performance.

1.3 Concept of Lean Manufacturing

The concept of lean manufacturing was developed by Toyota in 1950s to eliminate waste and maximize the process efficiency so that they can compete with the automanufactures in the US. Tools such as Kanban, Kaizen 5S, value stream mapping (VSM) was used to achieve the objectives and tangible results were achieved by the Toyota Production System (TPS) (Prasad et al. 2020; Jana and Tiwari 2021). The concept was then applied to other continuous manufacturing sectors and descent results were achieved. During that time, good success was achieved in the industries with continuous process compared to the discontinuous manufacturing sectors. Industries such as textiles, were not in the favor of implementing lean concepts at that time due to lack of knowledge and understanding of the strategies. Further, the management was not in favor of changing the existing production methods as they assumed that the manufacturing is done effectively. It was also assumed that the lean concepts work well for continuous processes or discontinuous processes with large volume and low variety of products. Furthermore, it was assumed that switching to new technologies from traditional technologies may lead to disaster and adversely affect the productivity.

Lean manufacturing concept refers to the process from raw materials manufacturing to the delivery of the finished goods to the end-user, with tactful management of resources resulting in monetary and time savings. It is sometimes referred to as just-in-time (JIT) systems due to their highly coordinated activities and delivery of goods that occur just as they need (Stevenson 2020). Lean manufacturing minimizes waste and in turn reduces costs, from which the technical term "lean" became popular ever since. Lean Supply Chain (LSC) is the smart way of linking a set of organizations by managing the up and downstream flows of funds, information, services, and products satisfying the end-user by the reduction of cost and wastage (Tortorella et al. 2017). Lean manufacturing or JIT and supply chain management can be classified into upstream activities like procurement and supply, and downstream activities like distribution, which includes warehousing, transportation, retailing, etc., and the integration of Logistics and Supply Chain activities.

2 Principles and Tools of Lean Manufacturing

The fundamental concept of lean manufacturing is to eliminate the waste from the entire operations of an organization. In fashion and textiles, the lean manufacturing principles can be implemented in product design, sourcing, manufacturing, inventory management, quality assurance and packaging (Ciano et al. 2019; Marodin et al. 2017). Further, it can also be implemented in the maintaining good relations with the suppliers and customers. Lean manufacturing aims at producing good quality products with less inventory, less error, less waiting time, and less human effort (Wickramasinghe and Wickramasinghe 2017). The focus of lean manufacturing is to eliminate the activities that are not adding any value to the product and for which the consumers are not willing to pay. There are various tools that help the lean manufacturing to be successfully implemented in the manufacturing process. This section discusses the fundamental principles and tools used in lean manufacturing process.

2.1 Principles of Lean Manufacturing

The fundamental principle of lean manufacturing lies on defining the value of a product from consumer or final customer's perspective. Once, the value stream has been identified, the focus then is to identify the existing and future value streams. Then various wastes in the existing value streams are identified and plans are made how to reduce and eliminate these wastes. Waste elimination helps to improve the productivity and efficiency while adding value in the manufacturing process (Cherrafi et al. 2018). The five major principles of lean manufacturing are:

1. **Principle 1**: Specify values from customer or consumer perspective for products and services,
2. **Principle 2**: Identify the existing and future value stream followed by identifying the wastes and nonvalue-added activities along the value stream. The wastes or nonvalue adding activities are reduced and gradually eliminated from the existing value stream for future improvement.
3. **Principle 3**: Once the wastes are eliminated, the products and services flow across the value stream without any interruption.
4. **Principle 4**: The products and services are the manufactured and delivered to the customers following the principles of pull approach.
5. **Principle 5**: The process of identifying and eliminating the wastes is repeated until approaching to the perfection in the value stream.

As mentioned above, lean manufacturing focuses on the elimination of waste from the product and service value stream. In lean concept, any nonvalue-added activity or process is called waste, which needs to improve and eliminated continuously (Borikar et al. 2018). Waste in a manufacturing process or in a service can be defined as the process that is not benefiting the customer, therefore, the customer is not willing to pay for this. Hence, all the wastes should be removed. When the wastes are eliminated, the value-added processes are remaining which helps to create the product more efficiently. Various wastes in lean manufacturing are listed in Table 1.

Out of the seven wastes mentioned above, most of the garment industries suffer from problems of overproduction, excessive inventory and waiting. In several instances, the garments are produced in excess quantities, which must be stored for finishing. Similarly, excessive inventory (such as fabrics and trims) needs warehouse space, labor, material handling equipment and inventory carrying cost, which leads to increased cost of manufacturing. The garment manufacturing industries suffer from the problem of synchronization, i.e., products waiting for long time to be processed, which leads to the accumulation of semi-finished components (Fig. 3). This is a major problem in many garment industries, which can occur at any time mainly due to improper line balancing and breakdown of machines. The long waiting time can be reduced through proper line balancing, reducing the changeover time, changing the plant layout and regular preventive maintenance. All these waste problems in garment industries can be overcome by the implementation of lean manufacturing.

Unnecessary movement can be the other major waste in many garment industries, which can be attributed to wrong layout of various departments. For example, some industries house the warehouse section in a separate building than the manufacturing unit, which is far from the spreading and cutting department. Hence, the fabrics and trims need to move from the warehouse to the cutting department. Similarly finished garments are transported from the manufacturing units again to the warehousing building. However, some industries follow linear approach in the warehousing and manufacturing, which reduces the overall cost due to the elimination of unnecessary transportation (Fig. 4).

Table 1 Seven wastes of lean manufacturing

Seven wastes	Explanation	Meaning/examples in fashion and textile context
1 Overproduction	Producing goods in excess that are not in demand or not needed	Many fashion brands overproduce clothes that they cannot sell. This overproduction leads to climate change and exploitation of resources
2 Inventory	Involves raw-materials, parts, semi-finished and finished products for which capital is invested	Excessive inventory such as fabrics, trims sourced for production sits in the warehouse for longtime without use. Othe example, cutting racks or finishing trolleys are idle for longer times than they are used
3 Overprocessing	Unnecessary use of technology, oversized equipment, and resources	Adding features to the product that are not needed by the customers. Example, adding a range of care labels (5–7) in different languages
4 Waiting	Time spent in waiting in between various processes, which is nonvalue-added time	Example includes sewing operators waiting for cut components, merchandisers waiting for buyer approvals, designers waiting for fabrics and trims
5 Defects	Manufacturing products that do not meet the quality standards	Producing garments with defects and reworking to fix them is a common trend in garment manufacturing
6 Movement	Movement of material, labor or machinery in excess that does not add any value to the product	Excess motion during sewing or fusing is a common problem, which arises due to poor training of operators. Industrial engineering department conducts work and motion study to avoid this problem
7 Transport	Moving materials between different locations that does not add any value to the product	Example includes moving cut fabric bundles from cutting to sewing floor, moving finished garments from finishing to quality assurance (QA) department

The above-mentioned wastes (in Table 1) can be present in any of the processes such as souring, designing manufacturing or even if in policies and procedures. All the wastes consume time and resources, but do not add any value to the products. Lean manufacturing is an integrated set of tools and practices, designed to analyze the root causes of problems in order to find solutions and eliminate these wastes. Lean manufacturing focuses on the elimination of the nonvalue added activities from the entire process of designing to dispatching. Value can be defined as the process or activity in a manufacturing plant for which the customers or the consumers are willing to pay. In garment manufacturing process there are many activities which are not adding any value to the product. When lean manufacturing is installed, these

Fig. 3 Accumulation of products: **a** semi-finished garments, and **b** finished garments in a manufacturing industry for the next processing

processes are identified and eliminated or replaced with value-added activities. As wastes are continuously eliminated in lean manufacturing, the benefits are obtained in terms of reduced lead time, reduced cost, increased productivity, improved quality, and overall profitability in a manufacturing industry.

The main objective of lean manufacturing is to add maximum value to the product with the use of the least use of resources (Kumar et al. 2018). Lean focuses on creating the best system from the concept stage till consumer stage so that maximum value is provided to the customers. In lean approach, the industries strive for continuous improvement by applying the steps as shown in Fig. 5 to redesign their value stream. The main area of focus is to identify, define and map the value stream, in order to identify the nonvalue-added processes. Once, identified, the nonvalue-added activities are redesigned or eliminated so that improvements can be made. Pull-based approach is followed to reduce the inventory and continuous effort has been made to achieve perfection (Solaimani and Sedighi 2020).

2.2 Tools in Lean Manufacturing

Various tools used in lean manufacturing are designed to eliminate wastes as explained in the preceding section. Waste is known as Muda in Japanese language, which relates to uselessness. Hence, lean tools are designed to eliminate Muda from various processes or services in an organization to improve quality. In other words, eliminating Muda or the non-value adding activities from the processes is the key applications of lean tools. There are many tools that are implemented in lean manufacturing to eliminate wastes to bring the most value to the customers (Leksic et al. 2020). Various tools used in lean manufacturing has been discussed in Table 2. However,

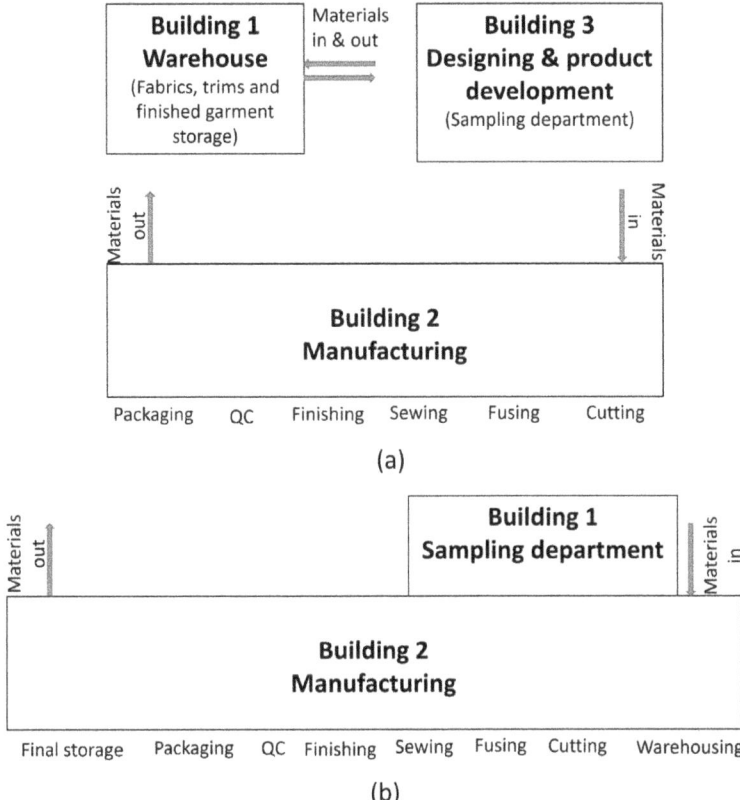

Fig. 4 Plant layouts from unnecessary movement viewpoint: **a** Excessive unnecessary movement, and **b** reduced or very low unnecessary movement. (*Note* These layouts are just indicative purposes only as there may be some other layouts possible for reducing unnecessary movement)

some of the most important tools in lean manufacturing are Kanban, Kaizen, 5S, six sigma, just in time (JIT) and Jidoka, which are discussed in this book and shown in Fig. 6. The lean manufacturing tools can be utilized in a wide range of industries starting from manufacturing to engineering to finance.

2.3 Implementation of Lean Manufacturing

The lean manufacturing concept is applied in the garment industry following the steps of PDCA (Plan-Do-Check-Act (Chong and Perumal 2019; Ghatorha et al. 2022). The PDCA cycle is an iterative model for improvement in processes and labor. As indicated by the name, there are the four seps in PDCA cycle which are: (1) Plan (planning to achieve the desired results), (2) Do (implementing the plan),

Fig. 5 Basic lean manufacturing steps

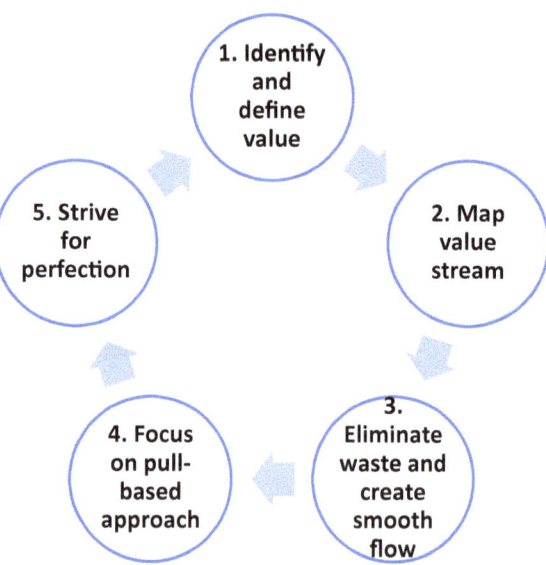

(3) Check (verifying the results achieved with the set standard), and (4) Act (review, evaluate and do it again). In the plan stage a proper plan is prepared considering the problem to be solved and resources available. The plan is reviewed and revises so that the failure change is small. Next, the actions are taken as per the plan in the previous stage. Unexpected problems may arise during this step, which can be avoided by testing the plan in a small section using controlled condition. In check stage inspection is done to find if the plan is successful or not. If the plan was not executed smoothly, find the root cause of the problem. Finally, if all the three steps worked, in the final stage the plan is executed to meet the objectives. The PDCA cycle (shown in Fig. 7) is an essential tool to solve problems and reduce waste in any sector of the organization.

2.3.1 Analysis of Methods

The first in lean manufacturing in fashion and textile industries is the scientific analysis of existing processes. The analysis involves understanding the methods, process flow, material flow, time, costs and wastes in the manufacturing industry. Based on the analysis, some processes where the waste is the maximum with higher production volumes are selected for improvement. Tools such as pareto chart is used to find the slow processes and the root cause of the problems. The status of the orders running are analyzed and the causes of delay, if any, are investigated. Plans are prepared how to improve process efficiency and productivity by eliminating wastes from the processes.

Table 2 Various tools used in lean manufacturing

Tool name	Description
Kanban	Kanban is a Japanese word which means instruction card. A Kanban tool uses visual indication approaches to identify Muda. Kanban tools regulate the flow of goods both within and outside the manufacturing facilities. Signal cards are used to indicate when goods are needed for automatic replenishment
Kaizen	A Japanese concept, which means continuous improvement in the workplace. Kaizen involves small and gradual improvements for more productive, efficient, and safer workplace
5S	5S is a Japanese method for systematically organizing and standardizing the workplace, for completing the work effectively. 5S involves cleanliness of workplace and arranging equipment where it belongs, which avoids the risk of injury. The English letters for 5S are: sort, set in order, shine, standardize and sustain
Six sigma	Six sigma (6-σ) is a statistical technique that was developed by Motorola to improve the quality of e products. Six sigma is used to identify the causes of variation to remove defects from produced in manufacturing process
Just in time (JIT)	JIT technique involves producing products exactly as per the customer demand. JIT follows a pull approach to produce the right product with the right quality, in right quantity, in the right place, at the right time
Poka Yoke	This is an approach to design error detection and prevention to achieve zero defects in the manufacturing process. Poke Yoke ensures operators are not making any error during the manufacturing process
Jidoka	In Jidoka the focus is to design equipment which can make the process partially or fully automatic. Partial automation can help to detect the fault or stop the machine when needed and improve the quality
Heijunka	Focuses on manufacturing in much smaller batches by mixing various product designs within the same process. This helps to reduce the lead times and lower inventory is needed due to smaller batch size

(continued)

2.3.2 Improvement Plan

The value stream mapping (VSM) is performed on the processes selected above and wastes are identified. In this step, plans made in the earlier step are implemented based on the data obtained from the analysis. Th tools of lean manufacturing are implemented, and efforts are made to eliminate wastes form the processes to reduce

Table 2 (continued)

Tool name	Description
Total productive maintenance (TPM)	TPM is a maintenance schedule that focuses on proactive and preventative maintenance to minimize the breakdown of tools and equipment, which is important for smooth running with least disruption. TPM helps in improving the productivity by reducing cycle times, breakdown maintenance time and eliminate defects
Cellular manufacturing (CM)	CM is based on the group technology, where the products are manufactures in groups
PDCA (Plan-Do-Check-Act)	PDCA is an iterative approach for improvement by following the four seps: (1) Plan (planning to achieve the desired results), (2) Do (implementing the plan), (3) Check (verifying the results achieved with the set standard), and (4) Act (review, evaluate and do it again)
Pareto chart	Pareto chart is a graphical representation of the problems that arise in the manufacturing process, ranked from the most occurring to the least occurring, arranged in descending order from left to right. Pareto chart is used to find the root cause of problems and their solutions

Fig. 6 Tools used in lean manufacturing

lead time, improve efficiency and productivity. In order to improve the performance in the manufacturing process and reduce delay, time study and motion study are carried out to establish standard time by reducing movement (motion study) and calculate the productivity (Takt time). Industrial engineering department is helpful

Fig. 7 The PDCA cycle

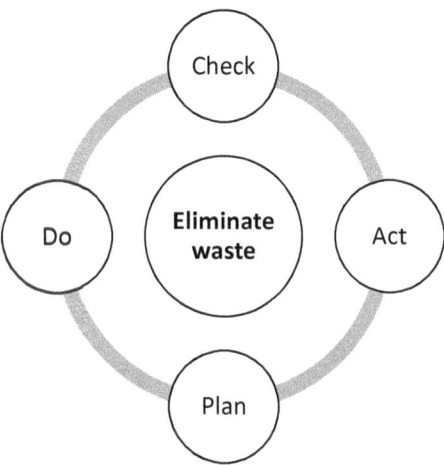

to establish the motion study and time study. New processes established by the work study are implemented.

2.3.3 Evaluate Efficiency

In this step, various approaches are used to calculate the efficiency of the new processes. The efficiency of processes is calculated by dividing the time that adds value and the total time (time that adds value & nonvalue added time).

$$Process\ efficiency = \frac{Time\ that\ adds\ value}{Total\ time\ (vale\ added + nonvalue\ added)} * 100$$

Similarly, the other parameters, labor productivity, machine productivity and Takt time are calculated by the formulas using below.

$$Labor\ productivity = \frac{Total\ units\ produced}{Total\ labor\ hours * number\ of\ workers} * 100$$

$$Machine\ productivity = \frac{Total\ units\ produced}{Total\ machine\ hours * number\ of\ machines} * 100$$

$$Takt\ time = \frac{Total\ production\ time\ available\ for\ a\ product\ line}{Total\ customer\ demand}$$

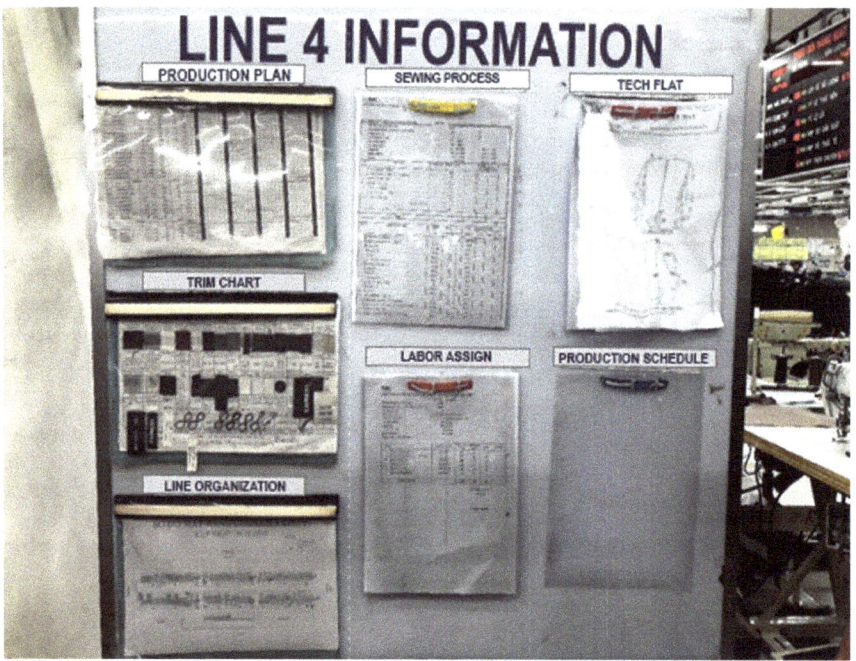

Fig. 8 Wall mounted printed documents such as product tech packs, trims used and quality charts for a specific garment style. (*Courtsey* Ca Nha Tran, Fashion Enterprise, RMIT University)

2.3.4 Seek Continuous Improvement

From the data obtained for efficiency, productivity and takt time above; the bottlenecks are identified and removed. Then, these parameters are calculated again, and the steps are repeated until the target results are achieved. Once, the targets are achieved, the values and processes for doing the work is standardized.

Some visual tools are used as guidelines to improve the productivity and efficiency as shown in Figs. 8, 9 and 10. Figure 8 indicates the wall mounted printed documents such as product tech packs, trims used and quality charts for a specific garment style. Figure 9 indicates the automation of stitching garment components without labor involvement. Figure 10 indicates the digital display of error rates in a specific garment industry. (Authors acknowledge and thanks to the industry for allowing to take the pictures).

2.4 Traditional Manufacturing versus Lean Manufacturing

Many of the fashion and textile industries in developing countries are still following the traditional approach for manufacturing (Nayak 2020). The traditional approach

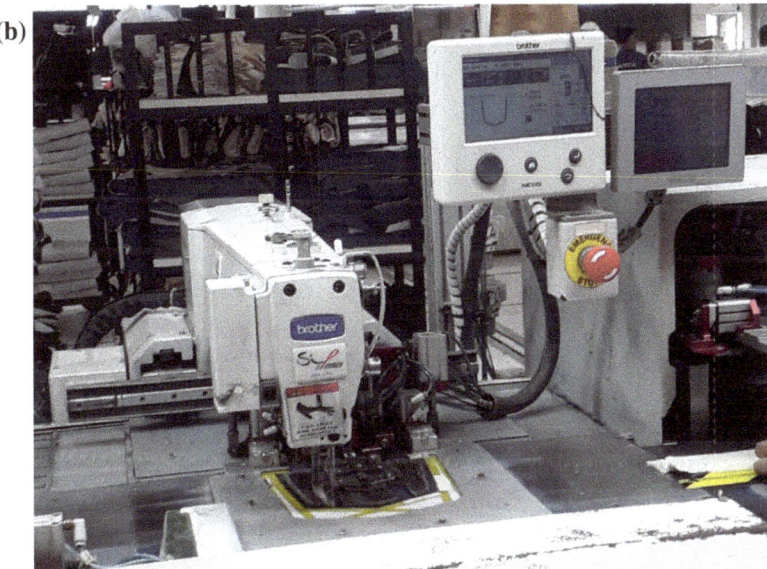

Fig. 9 Automation in: **a** cutting department and **b** stitching department (without labor involvement) of a garment manufacturing industry

Fig. 10 Digital display of error rates in a specific garment industry using lean manufacturing. (*Courtsey* Ca Nha Tran, Fashion Enterprise, RMIT University)

is characterized by large volume of production without the proper forecasting of the demand which is known as a push approach. Hence, there are large amounts of wastes of unsold products remaining. Normally several of the slow fashion brands following the traditional approach face this problem. Some fast fashion brands not following the fashion forecasting or buying in large quantities due to cheaper product also face this problem. On the other hand, in the lean manufacturing the products are manufactured following the pull approach, where the number of products is exactly as per the consumer demand. Hence, there are not much waste at the end of season or unsold clothes.

In the traditional system, products are manufactured considering the economies of scale, whereas in the lean manufacturing customer demand is given priority for manufacturing (Gupta and Jain 2013). The mass manufactured products are not sold in the traditional processes, whereas the products are enough to meet the consumer demand in the lean approach. In traditional process, there are large number of nonvalue-added activities, but in the lean manufacturing the nonvalue-added activities are known as wastes, which are eliminated to improve productivity and efficiency. In the traditional approach, the leadership operates as per executive command, but in lean they work in a team by vison and involvement. Traditional manufacturing is charactered by large volume production, large inventory, functional layout and long runtimes, whereas the lean manufacturing is charactered by small volume of production, small

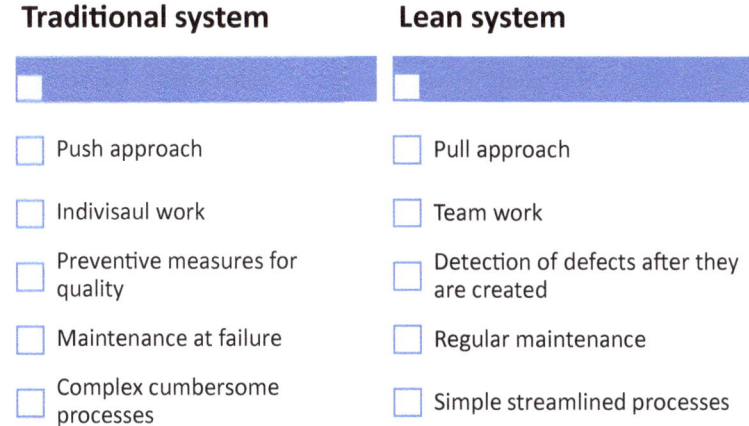

Fig. 11 Traditional system for manufacturing versus Lean manufacturing

inventory, cellular layout and small runtimes (Lyonnet and Toscano 2014). Some of the fundamental difference of lean manufacturing has been shown in Fig. 11.

3 Lean Manufacturing and Sustainability

As discussed in this chapter, the major focus of lean manufacturing is elimination of wastes. The implementation of lean manufacturing in fashion and textile industries can help the industries to achieve the objectives of sustainability. Sustainability is a main concern in most of the manufacturing sectors and sustainable practices means commitment beyond the company and customers while the community and environment are important. In recent years sustainable practices have been gaining impetus in fashion and textile manufacturing due to increased consumer awareness and stricter global legislations (Nayak 2020; Fletcher 2012; Nayak et al. 2020a). The term sustainability was coined in 1987 in Brundtland report, which means "satisfying the current needs without compromising the future generation's needs" (Keeble et al. 2003). The sustainable practices are viewed from three perspectives, which are environmental, economic and social also known as the "Triple Bottom Line (TBL)" of sustainability (Hacking and Guthrie 2008; Nayak et al. 2020c). Almost all the three TBL of sustainability are neglected during the supply chain activities relating to the fashion and textile industries. There has been growing interests in the environmental aspects of sustainability, which is the major concern among the TBL of sustainability.

The fashion and textile manufacturing industries around the world are struggling with varying degrees of environmental problems (Roberts 2003). The inherent nature of the production processes required for garments, largely impacts the environment due to a large amount of energy usage and water consumption; greenhouse gas (GHG) emission; hazardous waste generation; and discharge of toxic effluent containing dyes, finishes and auxiliaries to the eco-system (Gardetti and Torres 2013; Niinimäki and Hassi 2011). It has been shown that about 20% of all water pollutions are caused by the textile treatments such as dyeing, which greatly impacts the environment. Although the carbon emission intensity in garment production (between 1990 and 2005) has been decreased for grey cloth, jute goods and polyester goods by 1.90%, 2.07% and 0.72%, respectively; it has not achieved the desired results to save the planet earth (Anonymous 2014). Cotton yarn production has shown the highest increase in the emission intensity by 7.37%, which indicates that cotton yarn continued to be produced without caring for the environment, due to the consumption of harmful pesticides and fertilizers, which releases hazardous wastes into the nearby land and water systems. Hence, there is a need for green practices in the production of sustainable garments.

There are several studies focusing on the environmental benefits of lean manufacturing by reducing the wastes (Ghobadian et al. 2020). Although there are some contradictions on the statement "lean is green (i.e., lean manufacturing can lead to sustainability)"; in fashion and textile sectors; there are tangible sustainability benefits that can be directly achieved by the implementation of lean manufacturing. The lean manufacturing approach helps to reduce the process and materials waste, which is a common problem in fashion and textile manufacturing. The implementation of lean manufacturing can help to reduce the generation of wastewater, which can reduce the water pollution problem as it is one of the major challenges these industries are facing at this stage. The elimination of excessive motion, transportation and waiting time leads to the consumption of reduced energy (Henao et al. 2019). Hence, there are indirect environmental benefits due to the implementation of lean manufacturing.

There has not been significant research in fashion and textiles on the economic and social benefits of implementation of lean manufacturing (Babalola et al. 2019). However, from the nature of fashion and textile manufacturing, it is indicative that there are social and economic benefits of lean manufacturing (Nayak et al. 2021b). For example, lean manufacturing can provide the elimination of material wastes, which helps to achieve the economic benefits due to materials savings. Similarly, improvement in the quality of the products lead to reduced rejection and reworks that provide economic benefits and saving of materials (Yadav et al. 2020). These will be increase in profits due to reduced working capital invested in inventories by inventory reduction. The improvement in productivity, efficiency; reduction in the lead time and cycle time is an indicative of saving of energy providing economic benefits. The implementation of lean manufacturing helps to change the attitude of workers as they work in a team with increased responsibility for the elimination of waste. The lean manufacturing helps to enhance the motivation as the health and safety of the workers are increased in addition to higher wages. The workers also get training and skill upgradation leading to better performance in achieving their KPI (key performance index). However, there may be negative impacts such as monotonous and standard work, which are repetitive. The employees may be more stressed due to lack of freedom and working under constant supervision and being rated at all the times. Various sustainability benefits from lean manufacturing have been shown in Fig. 12.

Fig. 12 Sustainability benefits from lean manufacturing

4 Advantages and Disadvantages of Lean Manufacturing

Lean manufacturing has been a firm foundation for enterprise in every industry, including the fashion and textile industry to achieve the success with the fastest way and the least costly way. However, there might be some advantages and disadvantages when considering the lean application, which are discussed in this section.

4.1 Advantages

- On one hand, lean manufacturing can reduce wastes or nonvalue-added activities, thus initially providing a competitive edge by cutting the cost as well as ameliorating quality and productivity (Sangwa and Sangwan 2018).
- Many authors separated the benefits of lean manufacturing into quantitative and qualitative aspects. Particularly, the application of lean manufacturing can reduce the waste of raw materials and unnecessary motion within the production line.

- Besides, lean manufacturing can reduce cycle volatility and enhance performances by optimizing the cycle time, production lead time, processing time as well as the setup time (Bhamu et al. 2020).
- Lean manufacturing also improves the effectiveness and maximize the usage of infrastructure as well as its capacity (Prasad et al. 2020). In terms of cost, this implementation can directly decrease the cost of labor, maintenance and inventory of raw materials, work-in-progress and final products thanks to the better material flow and scheduling (Paneru 2011).
- Moreover, enterprises can get rid of defects and scrap within the production. Moving to the qualitative factors, lean manufacturing direct improve employees' morale, job satisfaction and responsibility. Lean manufacturing can promote employees' participations within the production owing to the additional responsibility when they must monitor the product quality themselves instead of separated quality checkers (Paneru 2011).
- Furthermore, team communication, management and decision-making can also be enhanced by the implementation of lean manufacturing. The company can improve its standardized housekeeping as well as maximize its flexibility to adopt the changes of market customers' demands or markets trend, thus efficiently fulfilling the orders (Bhamu et al. 2020).
- The implementation of lean manufacturing provides sustainability benefits covering the triple bottom line of sustainability (Environmental, social, and economic benefits). Hence, lean manufacturing not only makes the brand and manufactures more profitable, but also make them more competitive in the global fashion market for sustainable products.

4.2 Disadvantages

- On the other hand, the issues of lean manufacturing are divided into pre-implementation stage, implementation stage, and post-implementation stage. In the pre-implementation stage, the lack of lean awareness and the scepticism surrounding the implementation and benefits of lean manufacturing at different hierarchical level within any enterprises mostly occurred. It is proven that shop floor employees tend to have various misconceptions about the key objectives of lean manufacturing, thus prevent the managers from acquiring the accurate operational data from these employees (Bhamu et al. 2020). As a result, the imprecise data leads to the higher cycle time and reduce the operational efficiency.
- Moving to the implementation stage, the disproportionate distribution of lean application amongst different departments and stages of a particular enterprise cannot help them to achieve lean manufacturing benefits. Further, in the implementation stage, the employees may be reluctant to adopt to the changes required to achieve the lean objectives. They may not be coordinating with the lean team for successful implementations.

- In the post-implementation stage, higher expenses are needed to meet the objectives of lean manufacturing. The problem that many companies observe in the post-implementation stage is that the improvement remains localized instead of ensuring continuous improvements (Bhamu et al. 2020). This is derived from the insufficient post-implementation analysis and planning. Since process review is one of the main steps of lean manufacturing implementation, companies that forego this step cannot review the entire process, thus creating no opportunities for continuous improvements.
- It is unfeasible for SMEs (small and medium-sized enterprises) to adopt lean manufacturing since it requires the reformation or reconstruction of management, production line and working station, thus incurring a tremendous cost for them. It also requires hands-on training for supervisory level and long-term commitment to attain the quality culture, which is unaffordable for firms that are operationally and financially unstable.
- Lean manufacturing can lead to the worse customer–supplier relationship due to the wrongful tools and techniques, which cannot induce on-time delivery (Bhamu et al. 2020).
- Furthermore, for some firms, the positive results of implementing lean manufacturing may come late, thus it requires patience.

5 Future Directions in Lean Manufacturing

Many researchers have proposed various ideas for the application of lean manufacturing, especially in the fashion and textile industries, which can utilize the advancement of technology in the era of Industry 4.0 (Pagliosa et al. 2019). There will be several technologies such as artificial intelligence (AI), robotics, virtual prototyping, augmented reality and virtual reality (AR & VR), radio frequency identification (RFID), internet of things (IoT), and blockchain (Buer et al. 2018; Nayak and Padhye 2018a). Specifically, Buer et al. (2018) introduced a conceptual framework regarding the relationship between lean manufacturing and Industry 4.0 technologies (Fig. 13). Accordingly, lean manufacturing can maximize its practices and exert

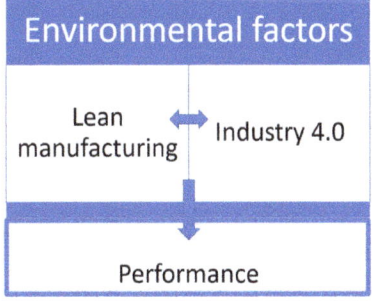

Fig. 13 Relationships between lean manufacturing, industry 4.0, performance, and environmental factors

facilitating effect on the adoption of the technologies in Industry 4.0. Also, the environmental factor plays as a main catalyst for this integration, resulting in a huge influence on different performance levels of the system.

However, the real effects of lean integration are still controversial. Buer et al. (2018) pointed out some key assessment to conclude the success of this relationship. Moreover, the study of Pagliosa et al. (2019) suggested for more assessment of the possible synergy levels between lean manufacturing and the industry 4.0. It requires more empirical validation for such synergy relied on the systematic literature review as well as quantifying its intensity levels. This can be achieved by conducting cross-sector survey-based research which can offer more robust arguments and support the proposition for this relationship. Besides, the researchers proposed to categorize the industry 4.0 technologies and lean manufacturing in different value stream levels. Accordingly, the potential benefit when considering the relationship between lean manufacturing and Industry 4.0 is still ambiguous, which requires companies to initiate more research and pilot test for application to enhance its understanding. With such understanding, the manufacturers can create a positive impact on their organization at all dimensions of value stream, such as intracellular, macro-processes, or supply chain levels.

For instance, Braglia et al. (2020) suggested a maturity model that considers the systematic assessment of fashion-luxury manufacturing firms in the context of the industry 4.0 vision. With the scientific purpose, this model aims to acquire the solid data about fashion-luxury manufacturing firms' current state and their strategies related to the adaptation to industry 4.0, thus extracting the potential success elements. In terms of practical purpose, companies that implement lean manufacturing can intensively evaluate the industry 4.0 maturity and reflect its suitability to the current operational technologies. These evaluations and assessment can help define an ad hoc supply chain cyber-physical systems (CPS) for the fashion and textile industry. As a result, the companies can increase its supply chain reliability thanks to the dual characteristic of the model, which digitally operates within a cloud platform and utilize the integrated knowledge from both data-driven analytical algorithms and physical capability to simulate its health condition.

6 Conclusions

This chapter has discussed the fundamental concepts of lean manufacturing in the fashion and textile industries. As the demand for better quality products; higher efficiency and productivity; lower lead time; cheaper cost; and sustainable products are in continuous demand; fashion and textile industries need to meet these demands to survive in the global marketplace. Further, the fashion product volumes are shrinking; the market is becoming more volatile; the manufacturing sector is being hit by pandemics such as Covid-19; the consumer choice is changing; the cost of labor and materials is continuously increasing; which creates more challenges for the manufacturers and retailers in the global marketplace. Hence, the manufacturing industries should switch to new tools and concepts to survive in the global marketplace. This chapter has also discussed the fashion and textile supply chain starting from concept to consumer in order to understand the existing trend in the global marketplace and the challenges faced in the current scenario.

Further, the seven wastes relating to process and material in lean manufacturing, which are generally found in the fashion and textile industries has been discussed in this chapter. Various approaches to eliminate these wastes to improve the productivity and efficiency are also discussed. The steps used to implement lean manufacturing, which includes PDCA (Plan-Do-Check-Act) cycle is also discussed in this chapter. It is essential to understand these steps to successfully implement the lean manufacturing in fashion and textile industries. Fashion and textile manufacturing industries in developing countries are facing several challenges. The major challenge, which makes the lean manufacturing difficult to implement is the traditional manufacturing industries with low productivity, higher waste generation, and high effluent generation. Some of the industries are switching to the newer technologies to reduce the wastes. However, several industries are unable to change due to the high cost involved in the newer technologies and the cost of implementing the implementation of the new technologies. These industries should realize the benefits received from the implementing the new technologies to achieve the objectives of lean manufacturing.

References

Aghajani M, Keramati A, Javadi B (2012) Determination of number of kanban in a cellular manufacturing system with considering rework process. Int J Adv Manuf Technol 63:1177–1189

Anonymous (2014) Understanding industrial energy use: Physical energy intensity changes in Indian manufacturing sector. Accessed 20 December. https://vdocuments.site/understanding-industrial-energy-use-physical-energy-intensity-changes-in-indian.html

Babalola O, Ibem EO, Ezema IC (2019) Implementation of lean practices in the construction industry: a systematic review. Build Environ 148:34–43

Benders J, Van Bijsterveld M (2000) Leaning on lean: the reception of a management fashion in Germany. N Technol Work Employ 15:50–64

Bhamu J, Bhadu J, Sangwan KS (2020) Lean manufacturing implementation in ceramic industry: a case study. In: Enhancing future skills and entrepreneurship (Springer, Cham)

Borikar S, Purohit R, Hulle A (2018) Waste elimination in textile industry. Man-Made Text India 46

Braglia M, Marrazzini L, Padellini L, Rinaldi R (2020) Managerial and Industry 4.0 solutions for fashion supply chains. J Fash Market Manage: An Int J

Bruce M, Daly L, Towers N (2004) Lean or agile: a solution for supply chain management in the textiles and clothing industry? Int J Oper Prod Manag 24:151–170

Buer S-V, Strandhagen JO, Chan FTS (2018) The link between Industry 4.0 and lean manufacturing: mapping current research and establishing a research agenda. Int J Prod Res 56:2924–2940

Carmignani G (2016) Lean supply chain model and application in an Italian fashion luxury company. In: Understanding the lean enterprise (Springer)

Cherrafi A, Garza-Reyes JA, Kumar V, Mishra N, Ghobadian A, Elfezazi S (2018) Lean, green practices and process innovation: a model for green supply chain performance. Int J Prod Econ 206:79–92

Chong JY, Puvanasvaran AP (2019) Conceptual framework for lean manufacturing implementation in SMEs with PDCA approach. In: Symposium on intelligent manufacturing and mechatronics, pp 410–18. Springer

Christopher M (2016) Logistics & supply chain management (Pearson UK)

Ciano MP, Pozzi R, Rossi T, Strozzi F (2019) How IJPR has addressed 'lean': a literature review using bibliometric tools. Int J Prod Res 57:5284–5317

Čiarnienė R, Vienažindienė M (2014) Management of contemporary fashion industry: characteristics and challenges. Procedia Soc Behav Sci 156:63–68

Desai A, Nassar N, Chertow M (2012) American seams: an exploration of hybrid fast fashion and domestic manufacturing models in relocalised apparel production. J Corp Citizsh 53–78

Fink D (2003) A life cycle approach to management fashion: an investigation of management concepts in the context of competitive strategy. Schmalenbach Bus Rev 55:46–59

Fletcher K (2012) Durability, fashion, sustainability: the processes and practices of use. Fash Pract 4:221–238

Gardetti MA, Torres AT (2013) Sustainability in fashion and textiles: values, design, production and consumption (Greenleaf Publishing)

Ghatorha KS, Sharma R, Singh G (2022) Lean manufacturing through PDCA: a case study of a press manufacturing industry. In: Proceedings of the international conference on industrial and manufacturing systems (CIMS-2020), pp 167–87. Springer

Ghobadian A, Talavera I, Bhattacharya A, Kumar V, Garza-Reyes JA, O'regan N (2020) Examining legitimatisation of additive manufacturing in the interplay between innovation, lean manufacturing and sustainability. Int J Prod Econ 219:457–468

Goyal A, Nayak R (2020) Sustainability in yarn manufacturing. In: Sustainable technologies for fashion and textiles (Elsevier)

Gupta S, Jain SK (2013) A literature review of lean manufacturing. Int J Manage Sci Eng Manage 8:241–249

Hacking T, Guthrie P (2008) A framework for clarifying the meaning of triple bottom-line, integrated, and sustainability assessment. Environ Impact Assess Rev 28:73–89

Henao R, Sarache W, Gómez I (2019) Lean manufacturing and sustainable performance: trends and future challenges. J Clean Prod 208:99–116

Jana P, Tiwari M (2021) Lean management in apparel manufacturing. Lean Tools Apparel Manuf (Elsevier)

Keeble JJ, Topiol S, Berkeley S (2003) Using indicators to measure sustainability performance at a corporate and project level. J Bus Ethics 44:149–158

Kumar S, Dhingra A, Singh B (2018) Lean-Kaizen implementation: a roadmap for identifying continuous improvement opportunities in Indian small and medium sized enterprise. J Eng Des Technol

Leksic I, Stefanic N, Veza I (2020) The impact of using different lean manufacturing tools on waste reduction. Adv Prod Eng Manage 15

Lyonnet B, Toscano R (2014) Towards an adapted lean system–a push-pull manufacturing strategy. Prod Plan Control 25:346–354

Marodin GA, Tortorella GL, Frank AG, Filho MG (2017) The moderating effect of Lean supply chain management on the impact of lean shop floor practices on quality and inventory. Supply Chain Manage Int J

Mason-Jones R, Naylor B, Towill DR (2000) Lean, agile or leagile? Match Supply Chain Marketplace Int J Prod Res 38:4061–4070

Muthu SS (2018) Sustainable innovations in textile chemical processes (Springer)

Nayak R, Padhye R (2015) Garment manufacturing technology. Woodhead Publishing, Cambridge, UK

Nayak R (2020) Supply chain management and logistics in the global fashion sector: the sustainability challenge (Routledge)

Nayak R, Akbari M, Far SM (2019) Recent sustainable trends in Vietnam's fashion supply chain. J Clean Prod 225:291–303

Nayak R, Nguyen L, Patnaik A, Khandual A (2021a) Fashion waste management problem and sustainability: a developing country perspective. In: Waste management in the fashion and textile industries. Netherlands, Elsevier

Nayak R, Van Thang LN, Nguyen T, Gaimster J, Morris R, George M (2021b) Sustainable developments and corporate social responsibility in Vietnamese fashion enterprises. J Fash Market Manage An Int J

Nayak R, Padhye R (2017) Automation in garment manufacturing (Woodhead Publishing)

Nayak R, Padhye R (2018a) Artificial intelligence and its application in the apparel industry. In: Automation in garment manufacturing (Elsevier)

Nayak R, Padhye R (2018b) Introduction to automation in garment manufacturing. In: Automation in garment manufacturing (Elsevier)

Nayak R, Nguyen LTV, Panwar T, George M, Ulhaq I (2020a) Sustainable supply chain management. Supply Chain Manage Logist Glob Fash Sect: Sustain Chall 1

Nayak R, Nguyen LTV, Panwar T, George M, Ulhaq I (2020b) Sustainable supply chain management: challenges from a fashion perspective. Supply Chain Manage Logist Glob Fash Sect 3–32

Nayak R, Panwar T, Nguyen LVT (2020c) Sustainability in fashion and textiles: a survey from developing country. Sustain Technol Fashion Text 3–30

Nayak RK, Padhye R, Fergusson S (2012) Identification of natural textile fibres. In: Handbook of natural fibres (Elsevier)

Niinimäki K, Hassi L (2011) Emerging design strategies in sustainable production and consumption of textiles and clothing. J Clean Prod 19:1876–1883

Niinimäki K, Peters G, Dahlbo H, Perry P, Rissanen T, Gwilt A (2020) The environmental price of fast fashion. Nat Rev Earth Environ 1:189–200

Pagliosa M, Tortorella G, Espindola Ferreira JC (2019) Industry 4.0 and lean manufacturing: a systematic literature review and future research directions. J Manuf Technol Manage

Paneru N (2011) Implementation of lean manufacturing tools in garment manufacturing process focusing sewing section of Men's Shirt

Patnaik A, Sweta P (2019) Fibres to smart textiles: advances in manufacturing, technologies, and applications (CRC Press)

Prasad MM, Dhiyaneswari JM, Ridzwanul Jamaan J, Mythreyan S, Sutharsan SM (2020) A framework for lean manufacturing implementation in Indian textile industry. Mater Today: Proceed 33:2986–2995

Ren S, Chan H-L, Ram P (2017) A comparative study on fashion demand forecasting models with multiple sources of uncertainty. Ann Oper Res 257:335–355

Roberts S (2003) Supply chain specific? understanding the patchy success of ethical sourcing initiatives. J Bus Ethics 44:159–170

Ross DF (2015) Distribution planning and control: managing in the era of supply chain management (springer)

Sangwa NR, Kuldip SS (2018) Development of an integrated performance measurement framework for lean organizations. J Manuf Technol Manage

Schonberger R (2014) Quality management and lean: a symbiotic relationship. Qual Manag J 21:6–10

Sinclair R (2014) Textiles and fashion: materials, design and technology (Elsevier)

Solaimani S, Mohamad S (2020) Toward a holistic view on lean sustainable construction: a literature review. J Clean Prod 248, 119213

Stevenson WJ (2020) Operations management (McGraw HIll)

Stock JR, Stefanie LB (2009) Developing a consensus definition of supply chain management: a qualitative study. Int J Phys Distrib Logist Manage

Tiwari A, Jain M (2013) Analysis of supply chain management in cloud computing. Int J Innov Technol Explor. Eng 3:152–155

Tortorella GL, Miorando R, Marodin G (2017) Lean supply chain management: empirical research on practices, contexts and performance. Int J Prod Econ 193:98–112

Wickramasinghe GLD, Vathsala W (2017) Implementation of lean production practices and manufacturing performance: the role of lean duration. J Manuf Technol Manag

Yadav G, Sunil L, Donald H, Sachin KM, Balkrishna EN, Yang L (2020) Development of a lean manufacturing framework to enhance its adoption within manufacturing companies in developing economies. J Clean Prod 245:118726

Yang C-L, Lin S-P, Chan Y-H, Sheu C (2010) Mediated effect of environmental management on manufacturing competitiveness: an empirical study. Int J Prod Econ 123:210–220

Challenges in the Traditional Fashion and Textile Supply Chain

Prabir Jana

Abstract The fashion and textile supply chain is being hailed as the highest employment generation industry and engine of growth for early industrialisation in least developed economies. The challenges in traditional fashion & textile supply chain are strategic as well as operational. While strategic challenges are structural and relationship related, generic across the world and stereotyped; some of these challenges are buyer dominance in the supply chain, diverging SKU in downstream network, sustainability and circularity issues due to fast fashion induced over consumption and lack of dependency on formal education. The operational challenges are more of execution oriented, may vary across geography and market. Some of the operational challenges are long lead time, multiple iteration during product development, low penetration of technology and lower rate of digitisation in manufacturing, over dependence on management philosophies, fringe use of scientific methods and confused perception of artisanal or industrial production. Moreover, the workplace often is ergonomically vulnerable. The chapter discusses each challenge in detail.

Keywords Strategic challenge · Operational challenge · Technology · Education · Lean

1 Introduction

Traditional fashion and textile supply chain is probably the most globalised and distributed supply chain for any products ever conceived. For economic reasons, it has been a standard practice in fashion and textile trade for yarn produced in one country to be woven into fabrics in another and cut components or partially sewn garments are shipped to yet another country for final finishing (Pogany et al. 1985). There are certain characteristics that are typical to the traditional fashion and textile supply chain, and specifically the manufacturing of apparel; i.e. migratory in nature

---

P. Jana (✉)
Department of Fashion Technology, Shahi Chair Professor, Industry 4.0, National Institute of Fashion Technology, Delhi, India
e-mail: prabir.jana@nift.ac.in

© The Author(s), under exclusive license to Springer Nature Singapore Pte Ltd. 2022
R. Nayak (ed.), *Lean Supply Chain Management in Fashion and Textile Industry*, Textile Science and Clothing Technology, https://doi.org/10.1007/978-981-19-2108-7_2

(Dickerson 1991), ergonomically vulnerable workforce (Jana 2008c), high labour orientation, less automation (Jana 2018a), higher participation of women workforce (Lopez and Robertson 2016), less capital intensive and comparatively lesser threshold of tech knowledge requirement (Jana 2018a) etc. While some of the characteristics are positive traits, rests built up the challenges faced by the sector.

These challenges can be divided into strategic and operational in nature. Strategic challenges like buyer-driven supply chain (Gereffi and Memedovic 2003), high SKU proliferation (Jana and Tiwari 2021), environmental impact (Gardetti and Muthu 2015) low-tech adoption (Gereffi and Memedovic 2003) etc. Operational challenges are Long lead time and iteration during garment sampling (Jana 2010), limited use of science and technology (Jana 2018a), sewing operation's vulnerability to RSI/CTD (Jana 2008c), etc.

2 Strategic Challenges

2.1 Buyer-Driven Supply Chain

A supply chain involves activities in the design, production and marketing of a product, and there is a critical distinction between buyer-driven and producer-driven value chains (Gereffi and Memedovic 2003). Producer-driven supply chains are large transnational manufacturers, who play the important role in coordinating production networks (including their backward and forward linkages). Typically, capital and technology intensive industries such as automobiles, aircraft, computers, semiconductors and heavy machinery follow producer-driven network. While apparel industry is a buyer-driven value chain that contains three types of lead firms: retailers, marketers and branded manufacturers (Gereffi and Memedovic 2003).

If we trace back the history of some of the oldest textile raw material manufacturers, we see that they are centuries old; iconic fibre manufacturer DuPont was established in 1920 as Fibersilk Company, Courtauld was founded in 1794 as a silk, crepe and textile business and lately Reliance Textile was established in 1960 and went on to become one of world's largest polyester fibre manufacturer. Traditionally these fibre giants used to dictate the fashion trends with their new developments showcased annually through some trend forecasting exhibitions like Premiere Vision (France), Interstoff (Germany), to name a few. The brands, retailers, manufacturers used to take inspiration from such exhibitions to develop their next season's collections.

In comparison the fashion retailers are comparatively new; Walmart, the American discount department store chain, was founded in 1962, H&M, the Swedish multinational fast-fashion clothing company was founded in 1947, Inditex, the Spanish Fast fashion was founded in 1960s. By the turn of this century these retailers amassed important consumer insight (due to ready availability of Point-of-sale (POS) data and through different loyalty programs) and started using that as power to dictate the contour of next season's development. The fashion and textile network has become a buyer-driven network. A major hypothesis of the global value chains approach is that national development requires linking up with the most significant lead firms in an industry (Gereffi and Memedovic 2003). Lead firms in fashion and textile supply chain are not necessarily the traditional vertically integrated manufacturers, nor are they necessarily involved in making finished products, rather fashion designers or brands or private label retailers, located upstream or downstream from manufacturing, and involved in the supply of critical components or information. What distinguishes lead firms from non-lead firms is that they control access to major resources (such as product design, new technologies, brand names or consumer demand) that generate the most profitable returns.

Although the raw material, manpower and logistics cost increased significantly over last two decades, being a buyer driven network, the retailers did not allow the retail price to be increased significantly (to win over the consumers loyalty) even sometimes reduced! For example, a pair of Levi's 501 jeans cost $50 in 1998, and in 2008 it is cheaper at $46. A Lacoste Polo shirt went for $95 in 1998, the retail price of same in 2008 is $75 (Wilson 2008). Factoring for inflation, each of these examples is resulting undue pressure on manufacturers to reduce cost of production so that the retailer's margin remains unchanged.

Being a buyer-driven network, consolidation and infusion of technology happened at retail and warehouse domain, manufacturing left behind ignored. Secondly scouring and migrating to cheap labour countries were motto for all brands and retailers, leaving very less scope for manufacturers to adopt technology. The recent re-shoring, near shoring initiatives is likely to infuse technology at the manufacturing domain.

2.2 High SKU Proliferation

Unlike automotive supply network, apparel and textile supply chain take the shape of matrix and not tiers (Massey 2000). The supply chain networks in context to textile and apparel may be completely disconnected, partially overlapping or completely overlapping. While disconnected networks works very similar to vertically integrated organizations, the completely overlapping networks in contrast works on "globalization" concept, very common for fashion merchandise sourcing from contract manufacturers in South East Asia. Figure 1 shows three different types of Supply Chain networks that are common in fashion and textile industries.

Historically, companies 'managed' linked supply chain operations by ownership, i.e., vertical integration (Harrigan 1983), but rare to find now. Courtaulds had fibre manufacturing to retail functions in the U.K. (Knox and Newton 1998). Some of the other recent examples of vertically integrated organisations are Luthai Textiles, Shahi Exports, Pratibha Syntex, to name a few.

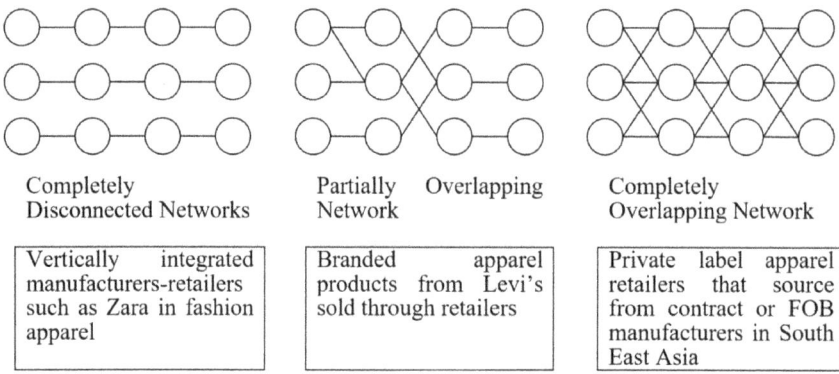

Fig. 1 Different Supply Chain networks. *Source* (Jana 2010)

Luthai Textiles, a fully vertically integrated company in China with $1.3 billion turnover manages an end-to-end supply chain that include cotton planting, spinning, dyeing, weaving, finishing and shirt sewing. With 80% of business is textile production and remaining 20% is Original Design manufacturing (ODM) of shirts for other brands like Brooks Brothers, Marks & Spencer, Burberry, Kohl's and Lacoste. Pratibha Syntex Ltd. is a vertically integrated, sustainability oriented, knitted textile apparel company in India with $ 75 million turnover that operates from farm to the finished product. It works with a total of 25,000 farmers, employs 6500 workers and works with renowned global apparel brands across 20 countries like Patagonia, Pact, Timberland, Gstar, Zara, H&M, & many more.

Shahi Exports, the largest apparel manufacturer from India with $ 1.0 billion turnover involves in spinning, dyeing, weaving, finishing and garmenting and operates 65 state-of-the-art manufacturing facilities across India with a diverse workforce of 120,000 people. Approximately, 70% of the fabric utilized is from their vertically integrated setup. The company manufactures for brands like Tommy Hilfiger, Gap Inc, Decathlon, Kohls, Walmart, H&M, Uniqlo, Calvin Klein, A&F, Marks & Spencer, Zara, Nike, and many other.

Supply chain of aerospace, home appliance, etc. follow a converging network or an A-shape network, that has numerous raw material (SKU) at upstream and limited variety of finished products at downstream, whereas textile and apparel supply chain follow diverging supply chain network or a V-shape network that starts with limited raw material (SKU) at upstream, but SKU proliferate as it moves downstream (Jana and Tiwari 2020). For example cotton fibre bale with a given staple length and micronaire is single SKU, but when spun into three different counts of yarn, becomes three SKU; now those yarns can be woven into twelve different specifications of fabric (two weave types, two ends per cm × picks per cm combinations), becomes twelve SKU; once these twelve different types of fabrics are dyed and printed in two colourways each, becomes forty eight SKU; finally once these forty eight types of fabrics are cut/sew/finished in garmenting facility say in three different styles and five sizes each, becomes $48 \times 3 \times 5 = 720$ SKUs.

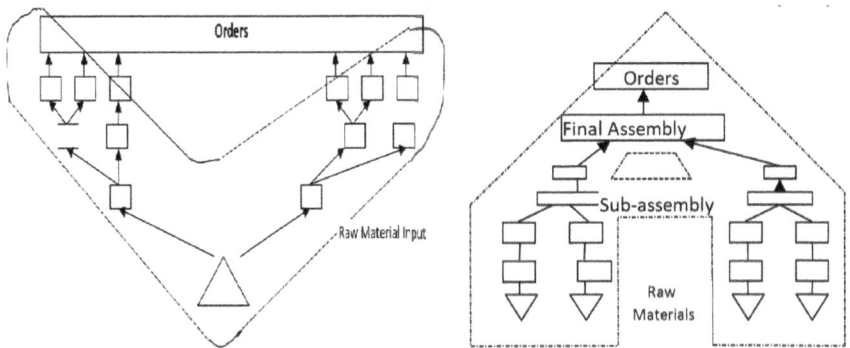

Fig. 2 V-A-supply chain structure. *Source* (Jana 2021)

Figure 2 shows schematic diagram of 'V' and 'A' type supply network. As most of the industries follow converging supply chain network, the generic "industry neutral" ERP systems or other standardized workflow systems (which were developed keeping converging network in mind) faced challenges during implementation in textile and apparel organizations. This can be attributed to the reason behind limited success of SAP implementation in apparel manufacturing organisations globally.

Partnership is one of the key pillars behind both V-shape and A-shape network. One of the primary objectives of supply chain optimisation is trust between partners; basis of partner selection should be collaborative/technology/core competency-based and not price-based (adversarial) (Jana and Tiwari 2020). In textile and apparel supply chains, retailers should partner with the apparel manufacturers, who in turn would partner with fabric and accessories suppliers. Marks and Spencer's erstwhile partnership with UK manufacturers (up to late twentieth century) speaks about the unmatchable trust and loyalty achieved in those years (Bevan 2007).

However, the perishability of fashion merchandise creates a perceived sense of secrecy among the partners in the supply chain, which acts as a deterrent for trust based supply chain network to operate. Take for example two competing fast fashion retailer H&M and A&F both are prototyping their upcoming collection with a leading apparel manufacturer in Southeast Asia. While, it is common for the apparel manufacturer to have separate merchandisers for both the retailers, but it will be very difficult for the merchandisers to shield information of the styles like fabric/colour/silhouette from each other. Even if we assume that H&M and A&F will not source from same apparel manufacturer and working with two different manufacturers ABC Exports and XYZ Exim respectively, there is every possibility that for fabric development both apparel manufacturer still approaches one textile mill. As it is common for production executives from brands & manufacturers to visit textile mills during development, there is every possibility that competing brands/retailers will come to know about each other's development much before the styles hit the shop.

If the product is diaper or toothpaste or even white classic shirt, sharing product information by competing brands across common contract manufacturers upstream the supply chain will not have any perceived problem, however for fashion products this is still not acceptable. The same logic applies for sharing sales information also, resulting poor visibility, high inventory, and heavy markdown in textile and apparel supply chain.

2.3 Sustainability and Circularity

Among many definitions, "Sustainability" in the context of fashion and textiles may be defined as an activity that can be continued indefinitely without causing harm; and meeting a current generation's needs without compromising those of future generations (Fletcher 2008) (Report of the World Commission on Environment and Development 1987). Once considered pollution free industry, fashion and textile industry is now touted as one of the worst polluters in the world. The challenges are multi-fold; first, globalisation related carbon emission; second, solid waste generation; third, water consumption and pollution; and last but not the least the polyester fibre itself.

Due to globalised sourcing model, raw material and manpower need to travel across geographical boundaries resulting travel related carbon emission. In 2018 global fashion industry produced around 2.1 billion tonnes of GHG emissions in 2018, equalling 4% of the global total (Mckinsey & Company 2020). From the Fig. 3 it can be seen that around 71% of the emissions come from upstream production, while energy-intensive raw material production takes up highest 38%, and apparel

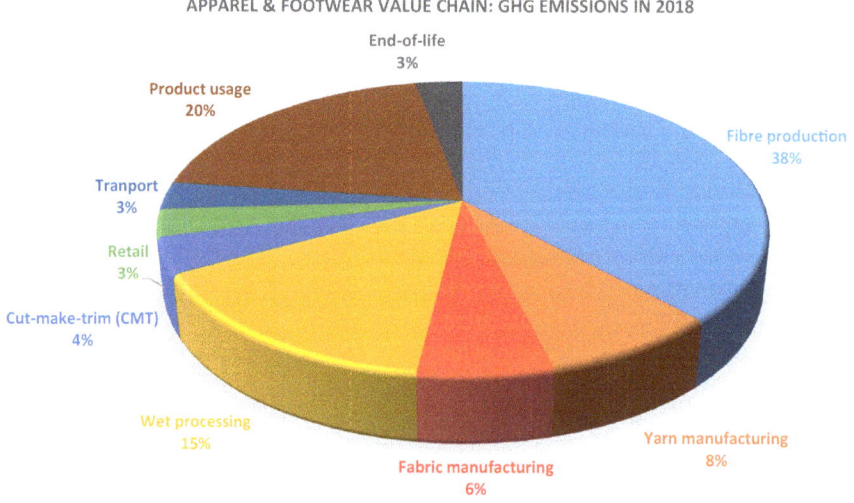

Fig. 3 Fashion and textile CHG emissions. *Source* Reproduced from Mckinsey & Company (2020)

production takes up the least 4%. The remaining 29% are generated by downstream activities such as transport, packaging, retail operations, usage and end-of-use.

Almost 150 billion garments are produced every year by fashion and textile industry, which is almost 20 times the world population. While 30% of clothes produced are never sold, another third only leaves the shops with a discount. Not only are brands producing more, but customers are also using large amounts of clothing for a shorter time. According to Plenitude: The New Economics of True Wealth (Schor 2010), Americans for instance, consume three times as much as their ancestors did fifty years ago, and they buy twice as many items of clothing as they did twenty years ago. An average American buys 70 apparel items a year, the average closet of a UK citizen contains 152 items, lifetime of an apparel item in developed countries is less than three years and more than 50% of fast fashion produced is disposed in under one year. Such scale of overproduction and over consumption has a huge impact on the environment, while about 12.8 million tons of clothing is sent to landfills annually, another 92 million tons of solid waste is produced annually using 98 million tons of natural resources (Rudenko 2018).

Every year the fashion industry uses 93 billion cubic metres of freshwater, enough to meet the consumption needs of five million people. Every garment produced consume huge amount of freshwater; for example, 7,000 L of water is needed to produce one pair of jeans and for a t-shirt, it is 2,700 L (The World Bank 2019). The processing and finishing of leather also have a high water impact; a pair of bovine leather shoes is estimated to use up to 8,000 L to manufacture (Chan 2020). The industry not only consumes freshwater, but also spews out dirty water aplenty. Around 20% of wastewater worldwide comes from fabric dyeing and treatment. And the chemicals used in tanneries are also a concern when it comes to water pollution. The different chemical processes required to manage and convert the polyester can have significant pollution impacts, harmful chemicals such as cobalt, sodium bromide and antimony oxide entering waterways if not managed properly (Chan 2020).

Polyester and cotton combined controls approximately 76% of the fashion and textile supply chain. Polyester production results in more emissions since it is produced from fossil fuels such as crude oil. According to estimates (Pandey 2018), 262% more CO_2 is emitted to produce a single polyester T-shirt than a cotton shirt. But substituting polyester with its recyclable counterpart offers up to a 90% reduction of toxic substances, a 60% reduction in energy usage, and up to a 40% dip in emissions. Every year a half a million tons of plastic microfibers are dumped into the ocean, the equivalent of 50 billion plastic bottles. These microfibers cannot be extracted from the water, and they can spread throughout the food chain (The World Bank 2019).

2.4 Educational Challenge

During 1940s, fashion and apparel industry in US were faced with a dwindling number of qualified people to help them run and carry on their businesses. The next generation wanted to be doctors and lawyers—not tailors. A group of industry members, led by Mortimer C. Ritter, an educator with an interest in programs for young working people, and Max Meyer, a retired menswear manufacturer, set about organizing a school to ensure the vitality of their businesses. That's the beginning of FIT, New York in 1944.

Apparel manufacturing was never considered a tech education and started as a polytechnic worldwide. Most of the world's best fashion institutes started their journey not as a degree giving institutes, rather as polytechnics; Nottingham Trent University, UK had started as "Trent polytechnic" in 1970; Manchester Metropolitan University, UK traces its origins to the Manchester Mechanics Institute and the Manchester School of Design, which formed Manchester Polytechnic in 1970.

The Hong Kong Polytechnic University also started its journey as "Hong Kong Polytechnic" in 1972. India's National Institute of Fashion Technology started in 1986 to offer certificate and diploma programs to train manpower for India's burgeoning apparel export, only to be converted to a statutory institute by an Act of the Indian Parliament in 2006 to award degree in bachelors, masters and doctorate level. In comparison, textile education is centuries old; for example Heriot-Watt University, Edinburgh (UK) was established in 1821 and one of India's oldest textile college "College of Textile Technology Serampore" was established in 1908 in West Bengal.

While the erstwhile polytechnics justified their existence till 1990s by supplying the trained manpower to apparel manufacturing industry, the conversion of those polytechnics to universities resulted more focus in knowledge domain and gradually eroded the skill domain. While the technology improved across industry, apparel industry is skill labour oriented. The same was explained very nicely by John Holusha in New York Times (Holusha 1990) "*Many apparel companies operate today about the way they did at the turn of the century. Rows of workers apply the same stitches, piece after piece, to bundles of partially finished garments. Then they pass the bundles on to others for the next step. In other industries, this sort of repetitious handwork has all but given way to automation and robotics*".

Apparel manufacturing is still dependent on skills of sewing operator. For example, science of assembling of garments (the subject is popularly called garment construction in academic institutes across world) is a skill based subject and a key expertise in achieving right quality in garment. Ironically that is grossly ignored across fashion institutions today due to lack of qualified teacher. No teacher wanted to develop themselves in this field due to perceived low prestige among peers, limited career progression and lure of ever-dominant and over-glamourized retail sector.

Traditionally textile education was categorised under "Engineering & Technology" discipline and fashion education was categorised under "Art and Design" discipline. With fashion education gained prominence globally some of the textile colleges started courses on fashion. With migration of industry from Europe and USA to Asia, it is interesting to see the gradual dilution of the apparel manufacturing related courses in European and US institutes and simultaneously proliferation of apparel manufacturing related courses in Asian institutes.

While countries like Germany, Italy, Japan and Korea still recognizes vocational skill as prestigious profession and equal socio-economic status, countries like India and other South East Asian countries are obsessed with knowledge oriented career and low perceived value to vocational and skill based career. India's recent push towards skilling India mission and re-introduction of Vocational education by MHRD is a positive step towards creating more skilled manpower. Indian Government introduced the National Skill Qualification Framework (NSQF) in 2013 through a Cabinet notification, as a single unified framework for all qualifications. Ministry of Human Resource and Development (MHRD), Government of India, launched the B.Voc. program in 2014 to offer NSQF-aligned vocational courses as part of college or university education.

It is yet to establish the co-relation of availability of professional education and progress of apparel industry in any particular country. Some of the interesting yet intriguing facts are: India's first fashion institute was established in 1986 and yet in 2020 with around 70 plus institutes offering professional degree programs in fashion, India's share of global exports did not improve much. Bangladesh and Vietnam are World's number two and number three producer of apparel after China without significance presence of professional fashion education in their countries.

Another interesting fact is the non-glamourised nature of apparel manufacturing; designations of some of the important manufacturing executives are like "pattern master", "fit technician", "line supervisor", etc. that are colonial and non-attractive for the new generations. In echoing with the educational challenge, finding young professionals for such job functions are becoming extremely difficult. While every young talent aspires to become a fashion designer, hardly anyone would like to become a pattern maker.

3 Operational Challenges

3.1 Lead Time and Sampling

Fashion and textile supply chain traditionally operated with two clearly defined seasons; Spring/Summer and Autumn/Winter (Jackson and Shaw 2001). While globalised multi-country sourcing practice requires longer lead time (shipment time for material), two seasons in a year supported a theoretical lead time of 28 weeks for conceptualising, designing, producing and shipping the merchandise. A typical critical path (Fig. 4) for an up-market UK fashion retailer buying tailoring from the United Kingdom using Italian fabric is shows 15–30 weeks (Jackson and Shaw 2001).

Fast fashion can be defined as low-cost clothing collections that mimic current luxury fashion trends and helps sate deeply held desires among young consumers in the industrialized world for luxury fashion, even as it embodies unsustainability (Joy et al. 2012). The unprecedented growth of fast fashion promoted by H&M and Zara

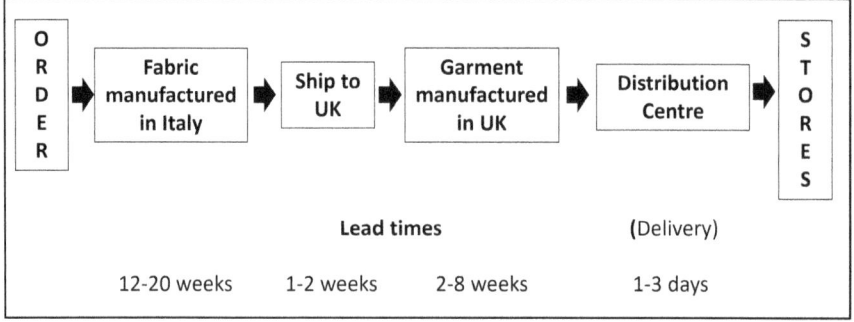

Fig. 4 Critical path for an up-market UK fashion retailer

compressed the standard turnaround time from catwalk to consumer of 28 weeks to a matter of mere weeks. While the long lead time provided enough cushion to manufacturers to make iterations in different stages of supply chain, the shorter lead time popularised by fast fashion retailers enforces undue pressure to do "right first time".

In traditional sourcing model 70–74% of the manufacturing lead time is consumed by pre-production, 40–52% of the pre-production time is spent on approval and 30% of the approval time is spent on iteration, i.e. approximately 12% of the total lead time is wasted on iteration (Jana 2010). The challenge is that the time-cushion in the supply chain has vanished; be it for sourcing of raw material, during sample development or for bulk production. Manufacturers are resorting to innovative technology and management technique for addressing these challenges.

The virtual sampling is being adopted globally by manufacturers and retailers to fast-track the garment sampling process and keep the iterative process entirely digital. Digital iterative process is less time consuming as well as less resource consuming. Virtual sampling process involves virtual dress forms (avatars), three-dimensional simulation of fabric drape over the virtual dress forms, which may be static or dynamic (walking, jumping, dancing, etc.) and as last step the virtual sampling can also quantify the tightness or looseness of the garment by calculating the space between the virtual body surface and virtual garment surface and friction properties of fabric.

Virtual sampling technology is more or less established and some of the popular commercial brands offering the solution are CLO, Tukatech, Lectra, etc. Human body is soft and compressible with different level of softness (or hardness) at different places; whereas the dress forms (either physical or digital) are made out of hard non-compressible material. Therefore, "live fitting session" is considered mandatory activity in sample approval process. To overcome such challenge, Tukatech and few other brands even came out with soft dress forms mimicking human body for realistic dress form fitting process, resulting higher rate of success.

3.2 Limited Use of Science and Technology

Traditionally apparel manufacturing (cut-sew-finish) is considered by many as artisanal rather industrial process, resulting low use of scientific processes and low use of high-tech machinery and equipment. Operations Research (OR) is interdisciplinary branch of mathematics which use methods like mathematical modelling, statistics, and algorithms to help decision makers arrive at optimal or good decisions in complex problems which are concerned with optimizing the maxima (profit, faster assembly line, greater yield, etc.) or minima (cost loss, lowering of risk, etc.) of some objective function (Jana 2011). Although there are many such requirements in apparel manufacturing, however hardly any OR tools are used in reality. Cut planning software use multi-priority optimisation technique, but a handful of organisation in the world uses same.

Pre-production activities in any organisation are characterized by people-oriented functions, interdependent activities are synchronised between succeeding and preceding activities to make a non-linear process network, while Gantt chart can scientifically manage the supply chain, critical path method (CPM) can optimise the lead time (Jana et al. 2005); however neither Gantt chart nor CPM tools are commonly used in apparel manufacturing. Managers are supposed to achieve multiple objectives of the organisation subject to the numerous constraints of the operating environment. These constraints can be resources, time, labour, energy, materials, money, or they can be restrictive guidelines, such as recipe for a garment wash, technical specifications of a style, etc. Linear programming (LP) is a mathematical method in OR for determining the way to achieve the best outcome (such as maximum profit or lowest cost) subject to such resource constraints in such scenarios; but currently everyone depends just on experience and gut feel.

Dimensionally unstable fabrics are the biggest challenge towards any automation in apparel manufacturing. Though cutting & finishing process was automated to a large extent, automation in sewing had limited success due to the very nature of raw material, i.e. the limp, dimensionally unstable fabric. Initially it was thought that by automating the entire sewing operation it will be possible to eliminate (minimise!) dependency on human operators (Jana 2015). Another challenge for robotic sewing technology was inability of traditional end effectors to pick-up single ply from stack of porous fabrics, especially with count and constructions varying from style to style. The electro-adhesion fabric pickup technology lately showing great promise (Jana 2018b). The vision sewing technology with advanced vision capture and increased computing power can now imitate the human hand eye coordination, additionally with of advancement of prosthetics, it is a matter of time when a pair of hands without brain will overpower the last challenge towards robotic sewing and cyborgs dominate the sewing floor (Jana 2018b).

Limited digitization is another challenge in apparel manufacturing organisations. Initial processes like receiving of raw material to inspection and issuance to cutting department is digitized. And the end functions, shipment data to invoicing, warehousing and retailing are again digitized. The very important cut-sew-finish process is largely non-digitized. Although daily reports are regularly collected and archived in almost every manufacturing organisations, most of the data is not usable for any further analysis or extracting meaningful insight. The primary reason behind this is the mistrust between the stakeholders; there are varying expectations of stakeholders when talking about digitalisation (Jana 2021); while management is trying to use digitalisation for performance improvement, the worker is feeling threatened that his/her efficiency is being watched. While top management is looking for transparent production status from the shopfloor, production executives may not like to share the actual production status. Unless all the stakeholders are convinced and agreed on a common goal of digitization, the usable data can't be collected.

3.3 Abundance of Management Philosophies

Traditionally apparel manufacturers favoured management solutions over technology solutions. The reasons behind management solutions gaining prominence, is the increasing inter-industry movement of top-level executives, zero capital investment involved, ease of comprehension by top management and ready availability of consultants. On the other hand, the technology solutions getting marginalized due to requirement of additional capital expenditure and lack of knowledgeable experts in technology domain (Jana 2016).

Different management philosophies became popular among apparel manufacturers during different time period; for example, ISO 9000 was popular during 1990s, Statistical Process Control (SPC) was popularised by Liz Claiborne during 1997–2005s, Lean manufacturing started gaining importance (in Apparel Manufacturing in Asia) since 2010 and became the ultimate weapon for many brands and retailers to discipline their suppliers. Six Sigma became fad for sometime during 2010 but weaned way as reducing manufacturing defects to no more than 3.4 occurrences per million units or events is neither logical nor economical in apparel manufacturing activity.

Among all the management philosophies used in apparel manufacturing, lean manufacturing probably has the longest run, wider coverage and maximum impact. According to one Harvard Business Review article Zara had partnered with Toyota to apply the Lean and JIT principles to its production facilities way back in 1990. Inditex is continuing their lean project under the wider umbrella of sustainability and the project's aim is to help supplier factories develop a production management system based on Lean philosophy, with a view to improving the working conditions of workers owing to improved production management systems (Inditex 2015).

Lean Manufacturing efficiency programme promoted by the Adidas Group is to improve overall factory performance in terms of delivery, quality and cost. The Adidas Social and Environmental Affairs (SEA)-Lean Project identify the Lean benchmarks that can measure working conditions of suppliers and to increase Standard of Engagement (SOE) transparency points for measuring working conditions (Adidas 2005). Nike lean supplier capability program involves identification of core value stream and production orientation around same, balancing production processes using takt-time, eliminating waste through reduction of inventory buffers and work-in-progress, increasing operator participation in quality control and problem-solving for continuous improvement and improving operational stability with standardised work and visual management techniques (International Labour Organisation nd). Decathlon runs their operation excellence (OPEX) programme based on lean manufacturing philosophies for capacity building of their manufacturers globally. While the use of lean manufacturing is becoming norms in fashion and apparel industry, the misuse of the term is also rampant in annual report, vision and mission statement, and all promotional documents.

4 Conclusion

While some of the strategic challenges may not have any immediate solutions, like transparency among partners in a supply network producing fashion merchandise, challenges like buyer-driven supply network, sustainability and lack of skilled manpower may see some reversal in near future. There are already some tech-startups like (www.groyyo.com), Fashinza (https://fashinza.com/) zetwerk (https://www.zetwerk.com), Zilingo (https://zilingo.com/en/) who are trying to break the streotype business model. While some trying to infuse technology, some are trying to address the financial woes and some are trying to organise the supplier network like never before.

The pressure from all front (political, social, and environmental) will ultimately force the fashion supply chain to adopt zero-impact circular model. While the sustainability push is now reached saturation point in terms of stakeholder awareness, it is time now to push for developing and democratizing the technology for wider adoption. While some of the technologies like "less water" or "zero water" textile processing is becoming popular, zero-discharge effluent treatment is being adopted, commercialization and democratization of fibre recycling (chemical) technology should be the next mission. Adoption of Industry 4.0 in apparel manufacturing and Government's push towards vocation education may add some muscle to apparel manufacturing.

As the continuous search for manufacturing industry to find lower labour cost countries is bottoming out, the onus to reduce cost of manufacturing is becoming increasingly important. And eliminating or reducing all types of wastes from the process would ultimately reduce the cost of the merchandise. The recent interest towards re-shoring, on-shoring and demand activated manufacturing is shifting part of the fashion and textile supply chain from a cost centric to time-centric sourcing model, and those lean tools aiming to levelling of production flow can play a very important role in removing "time-waste" from the supply chain. Lean tools will remain as one of the most important tools for addressing operation challenges due to low capital investment, ease of implementation, albeit slightly longer gestation time. Last but not the least ergonomic vulnerability of sewing workforce is a chronic challenge, which were least addressed in South East Asian countries. While we have addressed the occupational safety till date, closer attention to occupational health is the need of the hour. Scientific assessment of ergonomic vulnerability, workplace ergo-engineering and use of exoskeletons in apparel manufacturing is the future.

References

Adidas (2005) https://www.adidas-group.com. Accessed November 21, 2021. https://www.adidas-group.com/media/filer_public/2013/07/31/adidas_sustainability_website_content_march2006_en.pdf

Bevan J (2007) The rise & fall of marks & spencer . . . and how it rose again. Profile books limited, London

Chan E (2020) Vogue. March 22. Accessed September 3, 2021. https://www.vogue.in/fashion/content/the-fashion-industry-is-using-up-too-much-water-heres-how-you-can-reduce-your-h2o-footprint

Closa S (2015) HBS digital initiative. December 8. Accessed November 21, 2021. https://digital.hbs.edu/platform-rctom/submission/zara-disrupting-the-fashion-industry/

Dickerson K (1991) Textiles and apparel in the international economy. Macmillan Publishing Company, New York

Fletcher K (2008) Sustainable fashion & textiles: design journeys. Earthscan Publications Ltd., Oxford

Gardetti MA, Muthu SS (2015) Handbook of sustainable luxury textiles and fashion. Springer, London

Gereffi G, Memedovic O (2003) The global apparel value chain: what prospects for upgrading by developing countries. United Nations Ind Dev Organ, Vienna

Harrigan KR (1983) Strategies for vertical integration. Lexington books, Lexington, MA

Holusha J (1990) New York Times, September 9

Inditex (2015) Inditex Annual Report 2015. Inditex

International Labour Organisation (nd) https://www.ilo.org. Accessed November 21, 2021. https://www.ilo.org/wcmsp5/groups/public/---asia/---ro-bangkok/---ilo-jakarta/documents/presentation/wcms_565091.pdf

Jackson T, Shaw D (2001) Mastering fashion buying and merchandising management. Palgrave Macmillan, London

Jana P (2010) An investigation into Indian apparel and textile supply chain networks. The Nottingham Trent University, Nottingham

Jana P (2011) Apparelresources. Accessed November 14, 2021. https://apparelresources.com/business-news/manufacturing/pertcpm-operation-research-application-apparel-industry/

Jana P (2015) Sewing equipment and work aids. In: Nayak R, Padhye R (eds) Garment manufacturing technology, pp 275–315. Woodhead Publishing, Cambridge

Jana P (2016) Apparelresources. April 12. Accessed November 12, 2021. https://in.apparelresources.com/business-news/manufacturing/are-we-neglecting-technology-experts-speak/

Jana P (2018a) Automation in sewing technology. In: Nayak R, Padhye R (eds) Automation in garment manufacturing, p 199. Woodhead Publishing, Duxford

Jana P (2018b) Industry 4.0: possibilities in apparel manufacturing. StitchWorld 20–22

Jana P (2008c) Ergonomics in apparel manufacturing-I, importance and impact of ergonomics. Stitch World 6:42–47

Jana P, Tiwari M (2021) Lean tools in apparel manufacturing. Woodhead Publishing, Duxford

Jana P (2021) Digital Fashion 2: overcoming the information asymmetry. Apparel OnLine, May 27. Accessed November 21, 2021. https://apparelresources.com/technology-news/manufacturing-tech/digital-fashion-2-overcoming-information-asymmetry/

Jana P, Gupta S, Joshi C, Knox A (2005) Sypply chain dynamics in Indian apparel export manufacturing. www.techexchange.com

Joy A, John F Jr, Sherry AV, Wang J, Chan R (2012) Fast fashion, sustainability, and the ethical appeal of luxury brands. Fash Theory 16(3):273–296

Knox A, Newton E (1998) Clothing industry supply chain management & technology. In: International conference on apparel manufacturing: future scenario. Natl Inst Fash Technol, New Delhi

Lopez G, Robertson R (2016) Stitches to riches? apparel employment, trade, and economic development in South Asia. World Bank Group, Washington, DC

Macbeth DK, Ferguson N (1994) Partnership sourcing: an integrated supply chain approach. Pitman Publishing

Massey L (2000) an investigation into apparel and textile supply chain developments. The Nottingham Trent University, Nottingham

Mckinsey & Company (2020) Fashion on climate. Mckinsey & Company

Pandey K (2018) Down to earth. Accessed November 12, 2021. https://www.downtoearth.org.in/news/environment/fashion-industry-may-use-quarter-of-world-s-carbon-budget-by-2050-61183

Pogany P, Tomlinson K, Whisler J, Myers R (1985) Trade between the US and the non-market economy countries during 1984. United States International Trade Commission, Quarterly, Washington DC

Report of the World Commission on Environment and Development (1987) Our common future. Oxford University Press

Rice JB, Hoppe RM (2001) Supply chain versus supply chain: the hype and the reality. Supply Chain Manage Rev

Rudenko O (2018) Share cloth. Accessed September 23, 2021. https://sharecloth.com/blog/reports/apparel-overproduction

Lindsey R, Whiteaker C, Townsend M, Bhasin K (2018) Bloomberg. February 5. Accessed November 21, 2021. https://www.bloomberg.com/graphics/2018-death-of-clothing/

Schor J (2010) Plenitude: the new economics of true wealth, 1st edn. The Penguin Press, New York

Thackray J (1986) America's vertical cutback. Management Today

The World Bank (2019) Accessed 2021. https://www.worldbank.org/en/news/feature/2019/09/23/costo-moda-medio-ambiente

The World Bank (2019) The World Bank. September 23. Accessed July 23, 2021. https://www.worldbank.org/en/news/feature/2019/09/23/costo-moda-medio-ambiente

Wilson E (2008) Dress for less and less. The New York Times, May 29

Fundamental Concepts of Lean and Agile Manufacturing

Mohammadreza Akbari, Kevin Nguyen, Kristof Van Houdt, and Seng Kiat Kok

Abstract To stay competitive in a world of heightened uncertainty, manufacturers are continually designing and redesigning their production, supply chain networks, and logistics systems. Lean and agile manufacturing play a particularly key role in this by focusing on the reduction in cycle-time, lowering manufacturing costs and reducing inventory. The goal of lean and nimble manufacturing is to create a better product value to satisfy customer demands, while at the same time optimizing the efficiency of the entire supply chain. This chapter focusses on the evolution of lean and agile manufacturing concepts, objectives, and principles. The history and chronological developments and applications in the manufacturing sector, especially fashion industry, will be covered. The chapter will also discuss the differences between the two concepts: lean and agile manufacturing.

Keywords Lean · Agile manufacturing · Supply chain · Logistics · Chronological developments · Added value · Customer satisfaction

1 Introduction

The concept of lean manufacturing or practices was introduced for the first time by Toyota in Japan as the part of Toyota Production System (TPS) (Gupta and Jain 2013). Lean manufacturing supports the production enhancement by focusing on getting the right processes to the right place, in the right quantity, at the right time to achieve efficient and flexible workflow (Singh et al. 2010). The aim is to reduce inventories and lower any waste throughout the production process. Therefore, understanding these differences is very crucial for any supply chains moving towards lean practices. (Lee-Mortimer 2006). As the business environment changes quickly, understanding

---

M. Akbari (✉)
James Cook University, Townsville, Australia
e-mail: reza.akbari@rmit.edu.vn

K. Nguyen · K. Van Houdt · S. K. Kok
RMIT University Vietnam, Ho Chi Minh City, Vietnam

© The Author(s), under exclusive license to Springer Nature Singapore Pte Ltd. 2022
R. Nayak (ed.), *Lean Supply Chain Management in Fashion and Textile Industry*,
Textile Science and Clothing Technology,
https://doi.org/10.1007/978-981-19-2108-7_3

these shifts are crucial to improve the design of manufacturing systems (Akbari et al. 2017; Gupta and Jain 2013). Globalization adds to the growth and complexity of supply chains (Akbari 2018; Akbari and Hopkins 2019). To aid industries, lean and agile manufacturing can play important roles to their overall operational success (Chopra and Meindl 2015; Christopher 2016).

Implementation of lean practices in any supply chain, directly impacts manufacturing. Lean focuses on the elimination of any waste processes. A waste is defined as a non-value-added process or activity. According to Seth and Gupta (2005), a successful implementation of lean can lead to an increase in productivity and product outputs, as well as lower inventories and work-in-progress (WIP) goods. The main goal of lean practices is to produce goods of higher quality and speed at the lowest possible cost by eliminating all waste processes (Dennis 2015).

The Toyota production system (TPS) was developed by the Toyota Motor Corporation to achieve the highest quality product by lowering manufacturing costs and increasing the lead time through eliminating waste in all processes (Kehr and Proctor 2016). This approach originated from the supermarket chain Piggy Wiggly and the early mass production system implemented by Ford Motor Company (Liker 2004; Fujimoto 1999). TPS represents lean manufacturing as two pillars of Jidoka (automation with human touch) and Just-in-Time (JIT) product delivery (Fig. 1).

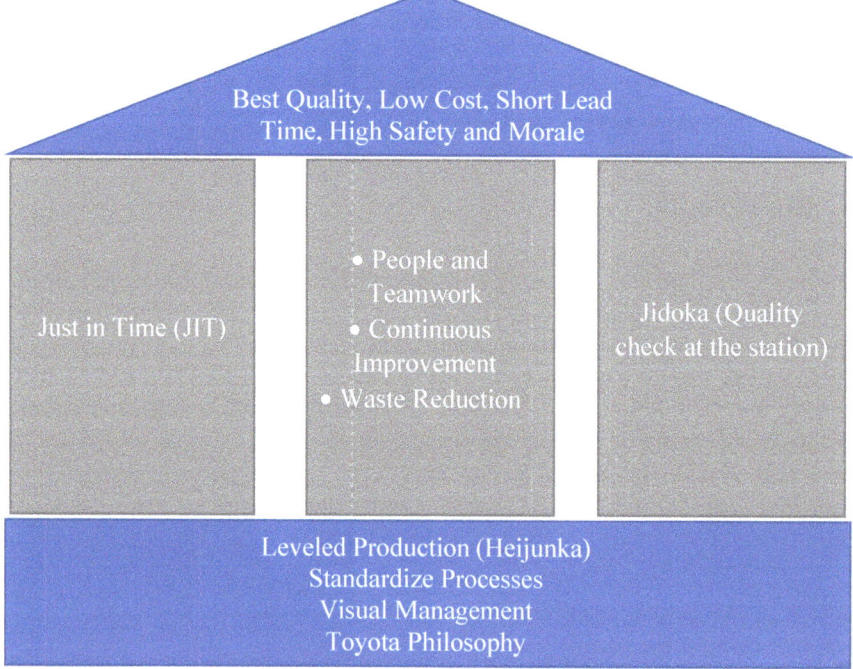

Fig. 1 TPS house diagram. *Source* Reproduced from Kehr and Proctor (2016)

Lean manufacturing is re-structuring and re-engineering its processes to be able to respond to the rapid changes in customer demands and upcoming challenges. Agile manufacturing is a concept in manufacturing which focuses on improving the competitiveness of companies (Gunasekaran 1999a, b). Agility tackles new ways of operating business processes to overcome challenges. In a fast-changing environment, organizations should be able to alleviate the changes and be more flexible and responsive (Akbari and McClelland 2020; Gould 1997).

The successful implementation of agile manufacturing is dependent on a variety of changes ranging from governmental regulations, business cooperation, information technology, re-engineering, to employee flexibility (Hormozi 2001). From time to time, implementing agile manufacturing requires modification of government regulations to support the working relationships among different firms (e.g., creation of virtual corporation). Since agile manufacturing relies on short-term relationships, many businesses might fear sharing information and losing competitiveness. Additionally, to facilitate agile implementation, robust information technology and infrastructure is needed. Furthermore, business process re-engineering is needed to notice the inefficiency and redesign of the entire process. Finally, it is human nature to resist these changes, but agile manufacturing requires employees to be more creative and constantly accept challenges in current procedures.

For many businesses, the term Lean manufacturing and Agile manufacturing may seem to be the same, but there are some major differences. According to Sanchez and Nagi (2001), Lean responds to the competitive pressures from the market with restricted resources, while Agile is more focused on responding to the complexity brought about by the constant global changes. Lean is a collective use of operational techniques to utilize productive resources, and Agile is more of an overall strategy to address unpredictable environments.

The primary purpose of this chapter is to focus on the evolution of lean and agile manufacturing, objectives, and principles of lean and agile manufacturing. The history and chronological developments and applications in manufacturing will be covered. The difference between them will be also discussed.

2 Lean and Agile Manufacturing

2.1 Lean Manufacturing

2.1.1 Pre-World War II

Although lean manufacturing is a term coined only about 40 years ago, forms of the process have been in existence for centuries. In 1104, the Venice Arsenale was founded by the local government to build ships for the navy. At its production heights, nearly 20,000 workers were employed, and the company manufactured one ship daily

(Giove et al. 2008). This ship builder introduced the idea of mass production lines to increase the assembly speed and cut down on wood waste (Charron et al. 2015).

Fast forward 700 years and the enormous changes influenced by the industrial revolution in the 1800s required manufacturers to improve their production efficiency. In 1816, the American clock maker Eli Terry focused on the mass production of his products but encountered issues with the cost of running an assembly line. By adding an adjustable escapement to the clock and thus altering the product design, he was able to lower production costs significantly (Hoke 1985).

One century later with the invention of the automobile, assembly lines really came to the forefront in twentieth century in America (Womack et al. 1990). The American industrialist Henry Ford incorporated a system to advance the manufacturing process. Ford's view was that productivity could become more efficient by reducing waste in any form throughout the manufacturing process (Zarbo and D'Angelo 2006). This resulted in an increased product output and decreased the assembly cost. Ford claimed that there was "nothing incompatible between quality manufacturing and mass production (Ford and Crowther 1926)."

To achieve this, workers were key to the entire process and therefore deserved the utmost respect for their contributions. Additionally, standardized equipment was incorporated, and manufacturing facilities were designed in such a way to minimize transportation costs. All these improvements lead to an increased efficiency and JIT manufacturing (Ford and Crowther 1926).

2.1.2 Post-World War II

After 1945 and a period of reconstruction, Japan flourished economically. The country's real GDP (gross domestic product) grew 9.1% annually from 1953 to 1973. A manufacturing concept that inspired Japanese industrial organizations during this period was the Toyota Production System or Toyotism. This approach would become known as lean production (Valli 2017).

Taiichi Ohno is credited with being the founding father of this manufacturing method. In collaboration with Shigeo Shingo (Charron et al. 2015), he directed the Toyota car company to a stock of inventory that would focus only on what the customer pulled out of stock. Subsequently, Toyota moved away from a push system and towards a pull system or Kanban (Roser 2017). Some crucial prerequisites for this Kanban approach to work effectively were short changeover times, level production schedules, and reducing production variability in the form of equipment downtime to just name a few. Issues within these elements could lead to difficulties achieving the JIT goal (Inman 1999).

In the 1980s, another major corporation that put a heavy focus on quality improvement was IBM. The goal was to implement a system in which all essential business processes would have integrated control methods. This meant that specific employees would become responsible for business operations such as inventory control, supplier relations, payroll etc. The aim was to improve the efficiency of the organization (Charron et al. 2015).

2.1.3 Lean Six Sigma

Bill Smith introduced the six-sigma approach in 1986 to help maximize the profit of the mobile phone provider Motorola (Costa et al. 2018). As a result, manufacturing costs were reduced by $1.4 billion and share values increased four-fold (Antony and Banuelas 2002). Using data, the aim was to minimize product deficiencies as much as possible to enhance customer satisfaction (Ikumapayi et al. 2020).

To measure the variability of a process, statisticians use the letter Sigma, σ, from the Greek alphabet (Pyzdek and Keller 2010). The term six sigma refers to a target of only having 3.4 defects per million opportunities (Linderman et al. 2003). Consumer satisfaction heavily relies on eliminating these mistakes in the business process (Snee 2000). In order to solve problems efficiently, six sigma uses the DMAIC (Define-Measure-Analyze-Improve-Control) method (de Mast and Lokkerbol 2012). Relevant DMAIC tools aid stakeholders to complete projects and gain valuable problem-solving skills (Arumugam et al. 2014).

Industries tended to implement lean and six sigma separately, but practical experience and research showed that each method had its limitations (George 2002). According to Antony et al. (2017), the drawback of lean is its inability to solve and analyze more complex problems. While on the other hand, not all production obstacles require a detailed analysis as is done in six sigma. This has resulted in an increased combined approach of lean with six sigma by businesses over the last decade (Rodgers et al. 2019).

2.1.4 Benefits of Lean Manufacturing

The past few decades have seen various lean manufacturing initiatives with an emphasis on waste reduction and a gain in efficiency to reduce costs. However, the competitive nature of the market continues to pose a challenge for many industries. Manufacturing higher number of products and diversifying a product line has an influence on overall costs (Durakovic et al. 2018).

In order to achieve success in the lean manufacturing process, Womack and Jones (2003) suggested five critical principles. Firstly, customers need to perceive a product and/or service as a good value. It needs to meet the demands of the consumer both from a price and production time perspective. The only way to achieve this is by having dedicated production teams who communicate with and try to understand the customer. Ignoring this process will negatively influence the value of the product.

Identifying the value stream is a second important component of lean manufacturing. Womack and Jones (2003) mentioned some essential elements in this phase. Products go through a significant manufacturing process that starts from conception to actual design, and finally the production launch. Running this problem-solving step effectively is key to business success. Additionally, carefully managing orders to meet scheduled deliveries or needs will lead to an efficient transformation of the raw materials to the finished product.

Thirdly, a challenge of many managers is to create a flow within the value chain, especially related to time (de Treville et al. 2014). Time is a valuable and finite resource that cannot be replaced or recovered (Wikner 2018). Employees and departments need to think about how they fit in the product development process flow. The management's task is to get staff onboard and positively contribute to the value stream. It is imperative that workers see the flow as part of their best interest (Womack and Jones 2003).

Fourthly, once the flow has been established, customers are able to pull value from the next upstream activity (Pinto et al. 2018). The pull production system is an effective approach to reduce inventory, and thus waste. Products are only manufactured once the customer has made the order (Achanga 2006). Myerson (2012) points out that a critical factor in this pull process is the partnership between companies and their suppliers. Close collaboration is imperative to ensure essential materials are delivered to ensure JIT production.

Achieving value specification, a value stream, a continuous production flow, and pull value leads to the final component of the lean manufacturing process: perfection. The interaction between the first four principles creates a vicious circle which allows organizations to reflect on their production process. Hidden costs, mistakes, waste and lost time suddenly get exposed which allows the production process to improve even more (Womack and Jones 2003).

For lean manufacturing to be a success, the implementation of these 5 principles is a high priority. Wyrwicka and Mrugalska (2017) have noticed that human, organizational and technical elements influence the introduction of lean production. Change can negatively impact staff engagement or perhaps management lacks the skills to push through reforms. Organizational culture or department silos can also restrict the introduction of this approach. Lastly, an underdeveloped technical infrastructure will obviously impact production.

Making the necessary changes can have profound benefits for companies. A company can increase its competitive edge by cutting costs and improving the quality of its products. Ultimately, this will lead to satisfied customers who are able to purchase the best product and service at the lowest possible price (Bhamu and Singh Sangwan 2014).

2.1.5 Potential for the Fashion and Textile Industry

The fashion and textile industry have seen an enormous growth in the last few decades (Nayak et al. 2019). Nonetheless, Fletcher (2010) argued that manufacturing has taken on a fast fashion approach. Fashion items are mass produced in a standardized form and then quickly disposed to the market. Consequently, the consumption of cheap and easy to replace garments has put a significant burden on environmental and social sustainability (Todeschini et al. 2017). Saha et al. (2019) describe three key terms: sustainable fashion, ethical fashion and green fashion. Sustainable fashion includes the green and ethical aspects, while green fashion focuses on the sale of biodegradable attires or clothes that were manufactured from recycled materials (Shen et al. 2013; Nayak 2019).

The enormous production and disposal of clothes has resulted in the textile and apparel industry becoming the second largest polluter globally (Nayak et al. 2020a, b). The use of pesticides to cultivate cotton and use of chemicals in the production process all contribute to this pollution. The carbon footprint of the sector is no less than 10% of the total emissions being generated worldwide (Pensupa 2020). This raises the importance for the fashion industry to reevaluate its short fast cycle attributes and start innovating its manufacturing processes (Khanzada et al. 2020). Consumer awareness about environmental issues is increasing and legislations are also becoming stricter on a global scale (Fletcher 2012; de Brito et al. 2008; Marcuccio and Steccolini 2005).

As markets are changing and the power is shifting towards the consumer (Itani et al. 2019), the value of a product needs to be clear and meet the demands of the buyers. Competition between businesses has led to a focus on engaging customers with the products. Kumar and Pansari (2016) claim that businesses can get a sustainable competitive advantage from customer engagement and push value from the customer to the business (Kumar et al. 2010). Lean manufacturing can help bridge the gap between the consumer and the fashion industry.

Ultimately though, implementing sustainable environmental goals in the lean manufacturing process depend on decisions made by management. It must be a conscious decision from the management top to achieve these objectives (Conti et al. 2006; de Treville and Antonakis 2006). Once approval has been given by the major stakeholders, Longoni and Cagliano (2015) suggest a range of lean manufacturing practices that can lead to environmental sustainability.

Lowering inventories reduces energy consumed in warehouses to store the products, thus reducing pollution and emissions (King and Lenox 2001). Rothenberg et al. (2001) add that by reducing setup time, defects are detected much earlier and result in less rework and loss of resource consumption. This attention on quality will automatically lead to a reduction of scrap materials. Nevertheless, the key to creating an environmental consciousness in an organization's culture falls with management. Training and encouraging an environmentally friendly attitude amongst employees are paramount (Longoni et al. 2014). Promoting a proactive view on the creation of environmentally sound products empowers workers to come up with new ideas and support the organization on its mission to become environmentally sustainable (Beard and Rees 2000).

2.2 Agile Manufacturing

The business environment has changed dramatically over the last two decades due to fierce competition in global and local areas, fast advancement in technology, and turbulent environmental conditions around the world. Maskell (2001) discussed four common challenges that businesses face in the twenty-first century, including quick and unpredictable changes in marketplace, individualism and customization trends in customer demand, short life-cycle products and services, and market requirements for low volume, high quality, and specific products. Recent research also identified similar pressures of volatile customer requirements, fast changes in technology, and unforeseen market conditions that businesses must deal with to be competitive and remain sustainable (Devadasan et al. 2012; Dubey and Gunasekaran 2015; Gunasekaran et al. 2019; Nabass and Abdallah 2018). To address these challenges and pressures, manufacturing companies have adapted their operations and invested resources in improving their agility capabilities (Leite and Braz 2016).

Several strategies and techniques such as total quality management (TQM), JIT, lean manufacturing, and computer-based management systems have been implemented over the years by manufacturing companies (Nandakumar et al. 2011). However, the engineering techniques of TQM, JIT or lean manufacturing focus on saving resources, cutting costs, and improving internal production processes; the things that companies can control. These techniques have limitation in dealing with continuously changing and unpredictable market conditions (Gunasekaran et al. 2019; Maskell 2001). Other methods including mass customization, optimizing supply chain network, automating manufacturing processes, empowering employees have also been developed and implemented to gain competitiveness, flexibility and agility (Carvalho et al. 2012). These methods set a foundation for a new manufacturing paradigm, which is referred by the academic community as agile manufacturing. Agile manufacturing has evolved as an alternative production method, which focuses on fast responsiveness and flexibility to address the anticipated and unanticipated changes in manufacturing conditions (Potdar et al. 2017).

2.2.1 Definition of Agile Manufacturing

Since its introduction in the early of 1990s, agile manufacturing has been well-researched by the academic community and adopted by many industries regardless of their line of business (Potdar et al. 2017). Customer-sensitive industries such as electronic, entertainment, food and beverages, fashion, and footwear, have shifted their focus toward agile manufacturing (Potdar et al. 2017). There are various definitions of agile manufacturing depending on the research industry, business context, and the study's focus. Looking at the competition perspective, agile manufacturing can be referred as "the capability of surviving and prospering in a competitive environment of continuous and unpredictable change by reacting quickly and electively to changing markets, driven by customer-designed products and services" (Gunasekaran 1999a, b). In the internal business operation perspective, agile manufacturing is defined as an organizational ability to adjust internal resources such as business processes and activities, strategies, technological tools, and personnel to respond to market uncertainty quickly and effectively (Gligor and Holcomb 2012; Tseng and Lin 2011). In the operational strategic perspective, agile manufacturing is a business mindset that focuses on utilizing structures, infrastructures, local and global competencies as a strategy to gain effective and efficient responsiveness towards rapidly changed customer demand (Gunasekaran et al. 2019). In business functional perspective, agile manufacturing is considered as the company's capability to design and produce a broad range of high quality but low-cost products and services with short lead times, in various sizes and meeting customer specific demand (Gunasekaran et al. 2019; Vázquez-Bustelo et al. 2007a, b).

In general, agile manufacturing refers to a company's agility to provide flexibility and responsiveness to compete in a global fierce competition, meet the market ever-changing conditions, and satisfy customer demand for highly customized products and services. A manufacturer is considered to have production flexibility when it can utilize resources to develop and manufacture products desired by the customer, while responsiveness is defined as the company's ability to transform the customer requirements and align its operational processes to create products and services in the shortest time with the lowest possible cost (Sanchez and Nagi 2001). By applying agile manufacturing strategy, companies will shift away from mass production and emphasize on timely delivery of highly customized and innovative products to the market ahead of competitors (Dubey and Gunasekaran 2015). Low cost and high quality are not the only business objectives in agile manufacturing. Important competitive and business objectives of agile manufacturing may include flexibility, technology leadership, product customization, customer loyalty, strategic partner relationship, and operation efficiency (Hasani et al. 2012; Wu et al. 2017).

2.2.2 Historical Background

Since the early 1980s, under the pressures of rapid market changes, fast technology advancements and shifting customer demands, manufacturing companies has put efforts to gain greater flexibility, reduce unnecessary inventory, shorten development and production lead times, and increase quality of products and customer services. These efforts led to a new manufacturing paradigm called agile manufacturing (Sanchez and Nagi 2001). The concept of agile manufacturing was first introduced in the Agile Manufacturing Enterprise Forum (AMEF) in 1991. The AMEF, organized by the Iacocca Institute at Lehigh University, was a research initiative of the US government, industry executives, and academics, to map a long-term strategy for US manufacturers to compete in the global competition. The agile forum identified the key strategy was agility in manufacturing, which would assist US manufacturers become more flexible and responsive to continuous market changes (Sanchez and Nagi 2001; Thilak et al. 2015).

The initial report from the 1991 agile forum pointed out several drivers for manufacturing agility such as fiercer global competition, increasing level of market fragmentation, stronger strategic supply chain relationship, and rapid changes in customer requirements of customized products and services (Jin-Hai et al. 2003; Maskell 2001). Following the initial report, a range of publications outlined critical areas that companies need to emphasize to achieve agility, including customer demand, competitor, technology adoption, internal resources and external network (Jin-Hai et al. 2003). These articles also proposed different ways to define agile manufacturing in various business contexts such as competition, business performance, supply chain management, and political involvement (Gunasekaran 1999a, b; Hormozi 2001; Leite and Braz 2016; Potdar et al. 2017; Thilak et al. 2015). For example, Gunasekaran et al. (2019) described production agility as a system utilizing internal resource competencies, both tangible and intangible assets, to meet the customer requirements with speed and flexibility. These assets include technological tools to facilitate information and communication management, highly motivated workforce with proper skills and knowledge, flexible production capability, and dynamic business environment to nurture the development of agile manufacturing (Gunasekaran et al. 2019; Jin-Hai et al. 2003).

Agile manufacturing is currently a well-researched paradigm. In their systematic review of literature about agile manufacturing, Potdar et al. (2017) identified 300 scholarly articles published from 1993 to 2016 by different researchers and industrial practitioners. The articles cover a wide range of industries seeking to leverage the advantages of agile manufacturing, i.e. manufacturing industry (Almahamid et al. 2010; Vázquez-Bustelo et al. 2007a, b), textile and clothing industry (Bruce et al. 2004; Zerenler 2007), fast-moving-consumer-goods (FMCG) industry (Agarwal et al. 2006), and electronic industry (Bottani 2009; Ismail et al. 2011). Agile manufacturing is widely accepted as a new production approach which is based on knowledge and practices of lean manufacturing and flexible manufacturing. While lean manufacturing focuses on utilizing limited resources and applying various operational techniques to reduce production costs, agile manufacturing emphasizes on being most flexible and responsive to meet customer dynamic demands (Potdar et al. 2017). Therefore, lean manufacturing is a producer-centered approach and agile manufacturing is a customer-centered approach. Agile manufacturing is an organizational-level strategy to achieve excellence in an unpredictable business environment and leading position ahead of competitors (Sanchez and Nagi 2001).

2.2.3 Benefits of Agile Manufacturing

There is an extensive literature on benefits of agile manufacturing, which can be grouped into three main categories of satisfying customer dynamic demands, leading position in the global competition, and better business performance (Dowlatshahi and Cao 2006; Dubey and Gunasekaran 2015; Gligor and Holcomb 2012; Gunasekaran et al. 2019; Hasani et al. 2012; Leite and Braz 2016; Nabass and Abdallah 2018; Potdar et al. 2017; Wu et al. 2017).

Achieving customer satisfaction through customized, innovative, and high-quality products and services is one of the obvious benefits of agile manufacturing brought to businesses. Maskell (2001) pointed out that there was an increased customer segmentation in many markets worldwide. Customers demand to be treated as an individual; products and services need to be customized to meet their requirements, delivered in small quantities for better uniqueness, and priced reasonably (Maskell 2001). By emphasizing on agility, agile manufacturing enables producers to have greater flexibility to quickly response to customer dynamic requirements. With optimized operation processes and high-skilled workforce, agile manufacturing adopters are proactive in offering products and services with customer centric design, mass customization, shortened lead time, and delivery at cost of mass production (Leite and Braz 2016). Such products would exceed customer expectations and gain competitive advantages (Gunasekaran et al. 2019).

Gaining a leading position in the global competition is another important benefit of agile manufacturing. Agile manufacturing helps companies to become more competitive in ever-changing business environments (Dowlatshahi and Cao 2006). Agile manufacturing requires companies to take competition into context with optimized operation processes, low cost and highly customized products, continuous addition of new customer-desired product features, and strong supply chain network (Hasani et al. 2012).

Nabass and Abdallah (2018) conducted a study using survey data collected from 282 manufacturing companies from various industries in Jordan. The study's results concluded that agile manufacturing positively and significantly impacted business performance in three measurements of flexibility, quality, and delivery. The study also revealed other benefits in business performance such as clearer strategic objectives, more flexible and responsive capabilities, stronger internal and external collaboration, and better utilization of resources by applying information technology, having knowledgeable and high-skilled workforce, and optimizing business processes. These benefits are consistent with previous research (Hallgren and Olhager 2009; Inman et al. 2011; Leite and Braz 2016; Narasimhan et al. 2006; Vázquez-Bustelo et al. 2007a, b). Gunasekaran et al. (2019) discussed that agile manufacturing addresses a wide range of competitive and business objectives, including mass customization, cost leadership, flexibility and responsiveness, technology front runner, and customer satisfaction and loyalty. Agile manufacturing strongly promotes strategic collaboration among suppliers, customers, and cross-functional teams (Sanchez and Nagi 2001). The strategic collaboration could be implemented through virtual enterprise approach, which leverages the competencies of internal resources and external stakeholders, such as partners, suppliers, customers, and even competitors (Gunasekaran et al. 2019; Potdar et al. 2017). Technology adoption is also critical in agile manufacturing. Information systems is the foundation to facilitate information sharing, business management, monitoring and control, and build flexibility and responsiveness capabilities for companies (Dubey and Gunasekaran 2015; Theorin et al. 2017).

2.2.4 Agile Manufacturing in Fashion and Textile Industry

The fashion and textile industry conducts business in a highly competitive environment due to intensive globalization, customer demanding expectations, shortened product life cycles, and high volatile markets (Chen et al. 2013; Nayak 2020a, b). Several studies explained different characteristics of the fashion industry, including market volatility, short product life cycle, large product variety, low demand predictability, trendy and disposable products, and complex supply chain network (Cachon and Swinney 2011; Lee and Rhee 2008; Moon et al. 2017). Fashion customers continuously demand more variety of trendy and seasonal styles, colors, sizes, quality fabrics, and fast delivery; however, their purchase behaviors are highly impulsive (Lemieux et al. 2012; Nishat Faisal et al. 2006).

To meet the aforementioned customer demands and expectations, fashion companies have to constantly introduce new trendy styles, clothing materials, and colors by seasons. To attract more customers and satisfy their desire for new apparel, fashion companies are proactively creating fashion trends, increasing the number of seasons, offering a large number of collections, various product ranges and styles (Lemieux et al. 2012). According to Carugati et al. (2008), fashion manufacturers often introduce four to six collections per year. Several major fashion companies such as Zara, Uniqlo and H&M even present new collection in a weekly basis.

In order to offer that constant new fashion, at affordable prices, and fit the trends, fashion companies need to develop agile manufacturing capability to response quickly to market changes and customer shifting requirements and preferences (Moon et al. 2017). Agile manufacturing enables fashion companies to satisfy rising customer demand and stay competitive by building flexibility and responsiveness capabilities, optimized operation processes, low cost and highly customized products, continuous addition of new customer-desired product features, and strong supply chain network (Hasani et al. 2012).

3 Differences Between Lean and Agile Manufacturing

While lean is seen the application of approaches and practices that is focused upon the reduction of waste (Ghobakhloo and Tang 2014) and enhancement of efficiency, agile manufacturing remit is to ensure appropriate responses and adaptability in an ever-changing environment. Lean processes are focused upon streamlining activities and operations towards optimization with reduction or removal of non-value-added elements (Kamble et al. 2019). This keen focus is on ensuring the firm maximizes its operational capacities and with it, reduces the costs associated with inefficiency.

In contrast, agile manufacturing, applies an ideology that focuses upon flexibility, adaptability and the capability to change to uncertainty (Vázquez-Bustelo et al. 2007a, b; Hallgren and Olhager 2009). Indeed, in some circles, agile manufacturing has been defined as programs to enhance performance in areas such as responsiveness, shortened lag times towards change and even efficiency in knowledge management (Narasimhan et al. 2006; Iqbal et al. 2020). Iqbal et al. (2020) go as far as viewing agile manufacturing as firms 'flexing muscles' which are highly adept and can change to altering demands.

It is important to note that there are some complimentary aspects and philosophies between lean and agile manufacturing, namely in the drive towards efficiency via quality and responsiveness, where cost/waste is the dominant focus in the former, with flexibility a major tenet in the latter (Iqbal et al. 2020). Often the complimentary aspects of lean and agile ideologies, are both applied simultaneously for the benefit of the organization and the business it is undertaking. Their focus on enhancing effectiveness, efficiency and economy are all vital aspects of business strategy and ideology (Qamar et al. 2018).

Nonetheless, while the approaches are often applied in tandem in business operations and considerations, they are known as competing (Hallgren and Olhager 2009) and mutually exclusive paradigms (Iqbal et al. 2020). Several previous research has discussed numerous dimensions where lean and agile differ not only in their principles, but also in the core operational focus between the concepts and in how each develops a competitive advantage for the business. For instance, where lean is focused on continuous improvement in resource and process usage, agile is focused upon the manipulation of activities to quickly meet ever-changing customer demands (Yusuf and Adeleye 2002).

Similarly, the focus, objectives and the strategic direction and decisions of the firm often drives the application of either methods. These decisions, survival of the firm, profitability, market share or market entry, can be at considerable contrasts and as a product of this, may equate to usage or non-usage of lean or agile manufacturing ideologies. Hallgren and Olhager (2009) argue that a cost-leadership approach is more closely aligned to lean management, focusing on cost effectiveness, where flexibility and a strategy of differentiation reflects the ideologies of agile. Narasimhan et al. (2006) and Vázquez-Bustelo et al. (2007a, b) contend that the state of the external environment can provide similar contrasting use of either ideology, with lean being the dominant focus in times of economic stability and agile, in times of uncertainty. Booth and Hammer's (1995) earlier work, suggests a similar line of argument, where lean ideologies improve effectiveness and efficiency while reducing waste, it lacks the ability to absorb shocks in turbulent and ever-changing environments.

Harrison (1997) and Inman et al. (2011) suggest that usage of one approach requires trade-offs to be made in the other, with this notion indicating the mutual exclusivity of both approaches, rather than its complimentary coexistence. Booth and Hammer's (1995) view contends that lean advances all aspects of the mass production process, where agile manufacturing is more focused on leaving this behind and managing continuous change in requirements as a routine function. This view is shared by Hofer et al. (2012), where agile systems are able to manage variances and derive advantages through flexibility, where as lean's advantage is to maximize efficiency and with it reduce such variability (Table 1).

4 Conclusion

The chapter has provided both an overview and deeper insights into the notions of lean and agile manufacturing, its inherent approaches, benefits and the rationale for its use in the fashion industry. Through understanding its emergence and the etymology of both, it provides readers with clearer knowledge of the origins of lean and agile manufacturing and the various dimensions for their successful application.

Table 1 Differences between lean and agile

Dimensions	Lean	Agile
Principles	Elimination of waste	Customer enrichment through flexibility
	Perfect first-time quality	Change to enhance competitiveness
	Kaizen	Leverage the impact of turbulence and uncertainty
Production quantity	Enhancement of mass production	Break with mass production; emphasis on mass customization
Market conditions	Fairly stable market, suitable for sequential customization of product families	Turbulent market, most suitable for parallel customization as market demands vary randomly
Competitive advantage	Productive efficiency through continuous improvement in resource and process usage	Customer enrichment through timely mobilization of enterprise-wide competencies
Core operational focus	Constant systems check and updates towards flexible machines for JIT deliveries	Manipulate intelligent machines to quickly replicate custom solutions

Source Adapted from Yusuf and Adeleye (2002) and AL-Tahat and Bataineh (2012)

With its humble beginnings dating back to pre-World War II, lean manufacturing fully gained prominence in Japan as part of the Toyota company's approach to manufacturing. Mirrored in the US by IBM, lean manufacturing ideologies was further enhanced through greater controls via Lean Six Sigma. Indeed, the approaches espoused by lean have gained more widespread usage within industry as a product of its dominant focus on reducing waste and developing efficiencies, processes that all add value to businesses.

Agile manufacturing's dominant focus in on ensuring flexibility and with it adaptability to not only customer desires but also to the external environment. Instead of focusing on enhancing efficiencies in standardization, agile manufacturing prioritizes innovation, creativity and systems that can deal with constant flux, allowing quick pivoting to be both proactive and reactive to change. Indeed, adaptive capabilities support the ability to weather uncertain environmental events and provide businesses with the opportunity to survive and at times thrive (Alonso et al. 2020). Born out of a new to be constantly competitive in an ever-changing environment, agile manufacturing promotes an ideology that relishes change and is readily prepared for it.

The chapter has also identified that both approaches have a vital role in supporting the fashion and textile industry, in ensuring appropriate quality thresholds without wastage and also developing capacities to innovate as well as weather uncertainty. In essence, while both frameworks differ in their approaches and in a number of dimensions of focus, their complimentary use also affords considerable potential for success. Given the nature of FMCG and the seasonality of fashion, agile manufacturing provides a means to build a competitive advantage for the future. Likewise, careful waste management and efficient use of resources through being lean would ensure stability in the business enterprise. As such, while there are differing reasons for either the use of lean or agile manufacturing, appropriate application of both can be highly beneficial in not only ensuring cost-effective solutions but manufacturing that befits the turbulent demands of customers. These notions are fundamental concepts of lean and agile manufacturing and if applied appropriately afford multiple benefits to not only the fashion industry, but to other domains as well.

References

Achanga P (2006) Critical success factors for lean implementation within SMEs. J Manuf Technol Manag 17(4):460–471. https://doi.org/10.1108/17410380610662889

Agarwal A, Shankar R, Tiwari MK (2006) Modeling the metrics of lean, agile and leagile supply chain: an ANP-based approach. Eur J Oper Res 173(1):211–225. https://doi.org/10.1016/j.ejor.2004.12.005

Akbari M (2018) Logistics outsourcing: a structured literature review. Benchmark Int J 25(5):1548–1580. https://doi.org/10.1108/BIJ-04-2017-0066

Akbari M, Hopkins J (2019) An investigation into anywhere working as a system for accelerating the transition of Ho Chi Minh city into a more livable city. J Clean Prod 209:665–679. https://doi.org/10.1016/j.jclepro.2018.10.262

Akbari M, McClelland R (2020) Corporate social responsibility and corporate citizenship in sustainable supply chain: a structured literature review. Benchmark Int J 27(6):1799–1841. https://doi.org/10.1108/BIJ-11-2019-0509

Akbari M, Clarke S, Maleki Far S (2017) Outsourcing best practice-the case of large construction firms in Iran. In: Proceedings of the informing science and information technology education conference, Ho Chi Minh City, Vietnam. http://proceedings.informingscience.org/InSITE2017/InSITE17p039-050Akbari3237.pdf

Almahamid S, Awwad A, McAdams AC (2010) Effects of organizational agility and knowledge sharing on competitive advantage: an empirical study in Jordan. Int J Manag 27(3):387–404,579

Alonso AD, Kok SK, Bressan A, O'Shea M, Sakellarios N, Koresis A, Solis MAB, Santoni LJ (2020) COVID-19, aftermath, impacts, and hospitality firms: an international perspective. Int J Hosp Manag 91:102654. https://doi.org/10.1016/j.ijhm.2020.102654

Al-Tahat MD, Bataineh KM (2012) Statistical analyses and modeling of the implementation of agile manufacturing tactics in industrial firms. Math Probl Eng 2012:1–24. https://doi.org/10.1155/2012/731092

Antony J, Banuelas R (2002) Key ingredients for the effective implementation of Six Sigma program. Measuring Bus Excellence 6(4):20–27. https://doi.org/10.1108/13683040210451679

Antony J, Rodgers B, Cudney EA (2017) Lean six sigma for public sector organizations: is it a myth or reality? Int J Qual Reliab Manag 34(9):1311–1402. https://doi.org/10.1108/IJQRM-08-2016-0127

Arumugam V, Antony J, Linderman K (2014) A multilevel framework of six sigma: a systematic review of the literature, possible extensions, and future research. Qual Manag J 21(4):36–61. https://doi.org/10.1080/10686967.2014.11918408

Beard C, Rees S (2000) Green teams and the management of environmental change in a UK county council. Environ Manag Health 11(1):27–38. https://doi.org/10.1108/09566160010314161

Bhamu J, Singh Sangwan K (2014) Lean manufacturing: literature review and research issues. Int J Oper Prod Manag 34(7):876–940. https://doi.org/10.1108/IJOPM-08-2012-0315

Booth C, Harmer M (1995) Agility, the future for ceramic manufacturing. Ceramic Eng Sci Proc 16(1):220–225. https://doi.org/10.1002/9780470314616.ch33

Bottani E (2009) On the assessment of enterprise agility: issues from two case studies. Int J Log Res Appl 12(3):213–230. https://doi.org/10.1080/13675560802395160

Bruce M, Daly L, Towers N (2004) Lean or agile: a solution for supply chain management in the textiles and clothing industry? Int J Oper Prod Manag 24(2):151–170. https://doi.org/10.1108/01443570410514867

Cachon GP, Swinney R (2011) The value of fast fashion: quick response, enhanced design, and strategic consumer behavior. Manage Sci 57(4):778–795. https://doi.org/10.1287/mnsc.1100.1303

Carugati A, Liao R, Smith P (2008) Speed-to-fashion: managing global supply chain in Zara. In: 2008 4th IEEE international conference on management of innovation and technology, pp 1494–1499. https://doi.org/10.1109/ICMIT.2008.4654593

Carvalho H, Azevedo SG, Cruz-Machado V (2012) Agile and resilient approaches to supply chain management: influence on performance and competitiveness. Logist Res 4(1):49–62. https://doi.org/10.1007/s12159-012-0064-2

Charron R, Harrington JH, Voehl F, Wiggin H (2015) The lean management systems handbook. CRC Press, Boca Raton

Chen J, Sohal AS, Prajogo DI (2013) Supply chain operational risk mitigation: a collaborative approach. Int J Prod Res 51(7):2186–2199. https://doi.org/10.1080/00207543.2012.727490

Chopra S, Meindl P (2015) Supply chain management: strategy, planning, and operation, 6th edn. Pearson, London

Christopher M (2016) Logistics & supply chain management. Pearson, UK.

Conti R, Angelis J, Cooper C, Faragher B, Gill C (2006) The effects of lean production on worker stress. Int J Oper Prod Manag 26(9):1013–1038. https://doi.org/10.1108/01443570610682616

Costa LBM, Filho MG, Fredendall LD, Paredes FJG (2018) Lean, six sigma and lean six sigma in the food industry: a systematic literature review. Trends Food Sci Technol 82(2018):122–133. https://doi.org/10.1016/j.tifs.2018.10.002

de Brito MP, Carbone V, Meunier Blanquart C (2008) Towards a sustainable fashion retail supply chain in Europe: organisation and performance. Int J Prod Econ 114(2):534–553. https://doi.org/10.1016/j.ijpe.2007.06.012

de Mast J, Lokkerbol J (2012) An analysis of the Six Sigma DMAIC method from the perspective of problem solving. Int J Prod Econ 139(2):604–614. https://doi.org/10.1016/j.ijpe.2012.05.035

de Treville S, Antonakis J (2006) Could lean production job design be intrinsically motivating? Contextual, configurational, and levels-of-analysis issues. J Oper Manag 24(2):99–123. https://doi.org/10.1016/j.jom.2005.04.001

de Treville S, Bicer I, Chavez-Demoulin V, Hagspiel V, Schürhoff N, Tasserit C, Wager S (2014) Valuing lead time. J Oper Manag 32(6):337–346. https://doi.org/10.1016/j.jom.2014.06.002

Dennis P (2015) Lean production simplified (2nd Edition): a plain-language guide to the world's most powerful production system. Productivity Press, New York

Devadasan SR, Sivakumar V, Murugesh R, Shalij PR (2012) Lean and agile manufacturing: theoretical, practical and research futurities. PHI Learning Pvt. Ltd.

Dowlatshahi S, Cao Q (2006) The relationships among virtual enterprise, information technology, and business performance in agile manufacturing: an industry perspective. Eur J Oper Res 174(2):835–860. https://doi.org/10.1016/j.ejor.2005.02.074

Dubey R, Gunasekaran A (2015) Agile manufacturing: framework and its empirical validation. Int J Adv Manuf Technol 76(9–12):2147–2157. https://doi.org/10.1007/s00170-014-6455-6

Durakovic B, Demir R, Abat K, Emek C (2018) Lean manufacturing: trends and implementation issues. Period Eng Nat Sci 6(1):130–143. https://doi.org/10.21533/pen.v6i1.45

Fletcher K (2010) Slow fashion: an invitation for systems change. J Design Creat Process Fash Ind 2(2):259–265. https://doi.org/10.2752/175693810X12774625387594

Fletcher K (2012) Durability, fashion, sustainability: the processes and practices of use. Fash Pract 4(2):221–238. https://doi.org/10.2752/175693812X13403765252389

Ford H, Crowther S (1926) Today and tomorrow. Doubleday, Page & Co., New York, NY

Fujimoto T (1999) Evolution of a manufacturing system at Toyota. Oxford University Press, Cary, NC, USA

George ML (2002) Lean six sigma: combining six sigma quality with lean production speed. McGraw-Hill, New York, NY

Ghobakhloo M, Tang SH (2014) IT investments and business performance improvement: the mediating role of lean manufacturing implementation. Int J Prod Res 52:5367–5384. https://doi.org/10.1080/00207543.2014.906761

Giove S, Rosato P, Breil M (2008) A multicriteria approach for the evaluation of the sustainability of re-use of historic buildings in Venice. Sustainability indicators and environmental valuation paper—Fondazione Eni Enrico Mattei. https://doi.org/10.22004/ag.econ.46625

Gligor DM, Holcomb MC (2012) Understanding the role of logistics capabilities in achieving supply chain agility: a systematic literature review. Supply Chain Manag Int J 17(4):438–453. https://doi.org/10.1108/13598541211246594

Gould P (1997) What is agility. Manuf Eng 76(1):28–31. https://doi.org/10.1049/me:19970113

Gunasekaran A (1999a) Agile manufacturing: a framework for research and development. Int J Prod Econ 62(1999):87–105. https://doi.org/10.1016/S0925-5273(98)00222-9

Gunasekaran A (1999b) Agile manufacturing: a framework for research and development. Int J Prod Econ 62(1):87–105. https://doi.org/10.1016/S0925-5273(98)00222-9

Gunasekaran A, Yusuf YY, Adeleye EO, Papadopoulos T, Kovvuri D, Geyi DG (2019) Agile manufacturing: an evolutionary review of practices. Int J Prod Res 57(15–16):5154–5174. https://doi.org/10.1080/00207543.2018.1530478

Gupta S, Jain SK (2013) A literature review of lean manufacturing. Int J Manag Sci Eng Manag 8(4):241–249. https://doi.org/10.1080/17509653.2013.825074

Hallgren M, Olhager J (2009) Lean and agile manufacturing: external and internal drivers and performance outcomes. Int J Oper Prod Manag 29(10):976–999. https://doi.org/10.1108/01443570910993456

Harrison A (1997) From leanness to agility. Manuf Eng 76:257–260

Hasani A, Zegordi SH, Nikbakhsh E (2012) Robust closed-loop supply chain network design for perishable goods in agile manufacturing under uncertainty. Int J Prod Res 50(16):4649–4669. https://doi.org/10.1080/00207543.2011.625051

Hofer C, Eroglu C, Hofer AR (2012) The effect of lean production on financial performance: the mediating role of inventory leanness. Int J Prod Econ 138(2):242–253. https://doi.org/10.1016/j.ijpe.2012.03.025

Hoke D (1985) Ingenious Yankees: the rise of the American system of manufactures in the private sector. Bus Econ History 14:223–235. http://www.jstor.org/stable/23702661

Hormozi AM (2001) Agile manufacturing: the next logical step. Benchmark Int J 8(2):132–143. https://doi.org/10.1108/14635770110389843

Ikumapayi OM, Akinlabi ET, Mwema FM, Ogbonna OS (2020) Six sigma versus lean manufacturing: an overview. Mater TodayProc 26(2):3275–3281. https://doi.org/10.1016/j.matpr.2020.02.986

Inman RR (1999) Are you implementing a pull system by putting the cart before the horse? Prod Invent Manag J 40(2):67–71

Inman RA, Sale RS, Green KW, Whitten D (2011) Agile manufacturing: relation to JIT, operational performance and firm performance. J Oper Manag 29(4):343–355. https://doi.org/10.1016/j.jom.2010.06.001

Ismail HS, Poolton J, Sharifi H (2011) The role of agile strategic capabilities in achieving resilience in manufacturing-based small companies. Int J Prod Res 49(18):5469–5487. https://doi.org/10.1080/00207543.2011.563833

Itani OS, Kassar AN, Loureiro SMC (2019) Value get, value give: the relationships among perceived value, relationship quality, customer engagement, and value consciousness. Int J Hosp Manag 80(2019):78–90. https://doi.org/10.1016/j.ijhm.2019.01.014

Iqbal T, Jajja MSS, Bhutta MK, Qureshi SN (2020) Lean and agile manufacturing: complementary or competing capabilities?. J Manuf Technol Manag 31(4):749–774

Jin-Hai L, Anderson AR, Harrison RT (2003) The evolution of agile manufacturing. Bus Process Manag J 9(2):170–189. https://doi.org/10.1108/14637150310468380

Kamble S, Gunasekaran A, Dhone NC (2019) Industry 4.0 and lean manufacturing practices for sustainable organisational performance in Indian manufacturing companies. Int J Prod Res 58(5):1319–1337. https://doi.org/10.1080/00207543.2019.1630772

Kehr TW, Proctor MD (2016) People pillars: re-structuring the Toyota Production System (TPS) house based on inadequacies revealed during the automotive recall crisis. Qual Reliab Eng Int 33:921–930. https://doi.org/10.1002/qre.2059

Khanzada H, Khan MQ, Kayani S (2020) Cotton based clothing. In: Wang H, Memon H (eds) Cotton science and processing technology. Textile science and clothing technology. Springer, Singapore. https://doi.org/10.1007/978-981-15-9169-3_15

King AA, Lenox MJ (2001) Lean and green? An empirical examination of the relationship between lean production and environmental performance. Prod Oper Manag 10(3):244–256. https://doi.org/10.1111/j.1937-5956.2001.tb00373.x

Kumar V, Pansari A (2016) Competitive advantage through engagement. J Mark Res 53(4):497–514. https://doi.org/10.1509/jmr.15.0044

Kumar V, Aksoy L, Donkers B, Venkatesan R, Wiesel T, Tillmanns S (2010) Undervalued or overvalued customers: capturing total customer engagement value. J Serv Res 3(3):297–310. https://doi.org/10.1177/1094670510375602

Lee CH, Rhee B-D (2008) Optimal guaranteed profit margins for both vendors and retailers in the fashion apparel industry. J Retail 84(3):325–333. https://doi.org/10.1016/j.jretai.2008.07.002

Lee-Mortimer A (2006) A lean route to manufacturing survival. Assem Autom 26(4):265–272. https://doi.org/10.1108/01445150610705155

Leite M, Braz V (2016) Agile manufacturing practices for new product development: Industrial case studies. J Manuf Technol Manag 27(4):560–576. https://doi.org/10.1108/JMTM-09-2015-0073

Lemieux A-A, Pellerin R, Lamouri S, Carbone V (2012) A new analysis framework for agility in the fashion industry. Int J Agile Syst Manag 5(2):175–197. https://doi.org/10.1504/IJASM.2012.046904

Liker J (2004) The Toyota way: 14 management principles from the world's greatest manufacturer. McGraw-Hill

Linderman K, Schroeder RG, Zaheer S, Choo AS (2003) Six Sigma: a goal-theoretic perspective. J Oper Manag 21(2):193–203. https://doi.org/10.1016/S0272-6963(02)00087-6

Longoni A, Cagliano R (2015) Cross-functional executive involvement and worker involvement in lean manufacturing and sustainability alignment. Int J Oper Prod Manag 35(9):1332–1358. https://doi.org/10.1108/IJOPM-02-2015-0113

Longoni A, Golini R, Cagliano C (2014) The role of new forms of work organisation in developing sustainability strategies in operations. Int J Prod Econ 147(1):147–160. https://doi.org/10.1016/j.ijpe.2013.09.009

Marcuccio M, Steccolini I (2005) Social and environmental reporting in local authorities. Public Manag Rev 7(2):155–176. https://doi.org/10.1080/14719030500090444

Maskell B (2001) The age of agile manufacturing. Supply Chain Manag Int J 6(1):5–11. https://doi.org/10.1108/13598540110380868

Moon K-LK, Lee J, Lai SC (2017) Key drivers of an agile, collaborative fast fashion supply chain: Dongdaemun fashion market. J Fashion Market Manag Int J 21(3):278–297. https://doi.org/10.1108/JFMM-07-2016-0060

Myerson P (2012) Lean supply chain and logistics management. McGraw-Hill, New York, NY

Nabass EH, Abdallah AB (2018) Agile manufacturing and business performance: the indirect effects of operational performance dimensions. Bus Process Manag J 25(4):647–666. https://doi.org/10.1108/BPMJ-07-2017-0202

Nandakumar MK, Ghobadian A, O'Regan N (2011) Generic strategies and performance—evidence from manufacturing firms. Int J Product Perform Manag 60(3):222–251. https://doi.org/10.1108/17410401111111970

Narasimhan R, Swink M, Kim SW (2006) Disentangling leanness and agility: an empirical investigation. J Oper Manag 24(5):440–457. https://doi.org/10.1016/j.jom.2005.11.011

Nayak R (ed) (2019) Sustainable technologies for fashion and textiles. Woodhead Publishing.

Nayak R, Akbari M, Far SM (2019) Recent sustainable trends in Vietnam's fashion supply chain. J Clean Prod 225:291–303

Nayak R, Houshyar S, Patnaik A, Nguyen LT, Shanks RA, Padhye R, Fegusson M (2020a) Sustainable reuse of fashion waste as flame-retardant mattress filing with ecofriendly chemicals. J Clean Prod 251:119620

Nayak R, Nguyen LTV, Panwar T, George M, Ulhaq I (2020b) Sustainable supply chain management: challenges from a fashion perspective. Supply Chain Manag Logist Global Fashion Sector 3–32

Nishat Faisal M, Banwet DK, Shankar R (2006) Mapping supply chains on risk and customer sensitivity dimensions. Ind Manag Data Syst 106(6):878–895. https://doi.org/10.1108/02635570610671533

Pensupa N (2020) 12—Recycling of end-of-life clothes. In: Nayak R (eds) Woodhead publishing series in textiles, sustainable technologies for fashion and textiles. Woodhead Publishing, pp 251–309. https://doi.org/10.1016/B978-0-08-102867-4.00012-8

Pinto JLQ, Matias JCO, Pimentel C, Azevedo SG, Govindan K (2018) Just in time factory: implementation through lean manufacturing tools. Springer International Publishing. https://doi.org/10.1007/978-3-319-77016-1

Potdar PK, Routroy S, Behera A (2017) Agile manufacturing: a systematic review of literature and implications for future research. Benchmark Int J 24(7):2022–2048. https://doi.org/10.1108/BIJ-06-2016-0100

Pyzdek T, Keller P (2010) The six sigma handbook. McGraw-Hill

Qamar A, Hall MA, Collinson S (2018) Lean versus agile production: flexibility trade-offs within the automotive supply chain. Int J Prod Res 56(11):3974–3993. https://doi.org/10.1080/00207543.2018.1463109

Rodgers B, Antony J, Gupta S (2019) A critical perspective on the changing patterns of Lean Six Sigma research. Int J Product Perform Manag 68(1):248–258. https://doi.org/10.1108/IJPPM-08-2017-0196

Roser C (2017) "Faster, better, cheaper" in the history of manufacturing: from the stone age to lean manufacturing and beyond. CRC Press

Rothenberg S, Pil FK, Maxwell J (2001) Lean, green, and the quest for superior environmental performance. Prod Oper Manag 10(3):228–243. https://doi.org/10.1111/j.1937-5956.2001.tb00372.x

Saha I, Bhandari U, Mathew DJ (2019) A study on consumer awareness towards green fashion in India. Res Design Connected World 134:483–494. https://doi.org/10.1007/978-981-13-5974-3_42

Sanchez LM, Nagi R (2001) A review of agile manufacturing systems. Int J Prod Res 39(16):3561–3600. https://doi.org/10.1080/00207540110068790

Seth D, Gupta V (2005) Application of value stream mapping for lean operations and cycle time reduction: an Indian case study. Prod Plan Control 16:44–59. https://doi.org/10.1080/09537280512331325281

Shen D, Richards J, Liu F (2013) Consumers' awareness of sustainable fashion. Market Manag J 23(2):134–147

Singh B, Garg SK, Sharma SK (2010) Scope of lean implementation: a survey of 127 Indian industries. Int J Rapid Manuf 1:323–333. https://doi.org/10.1504/IJRapidM.2010.034253

Snee RD (2000) Impact of six sigma on quality engineering. Qual Eng 12(3):9–14. https://doi.org/10.1080/08982110008962589

Theorin A, Bengtsson K, Provost J, Lieder M, Johnsson C, Lundholm T, Lennartson B (2017) An event-driven manufacturing information system architecture for Industry 4.0. Int J Prod Res 55(5):1297–1311. https://doi.org/10.1080/00207543.2016.1201604

Thilak VMM, Devadasan SR, Sivaram NM (2015) A literature review on the progression of agile manufacturing paradigm and its scope of application in pump industry. Sci World J 2015:e297850. https://doi.org/10.1155/2015/297850

Todeschini BV, Cortimiglia MN, Callegaro-de-Menezes D, Ghezzi A (2017) Innovative and sustainable business models in the fashion industry: entrepreneurial drivers, opportunities, and challenges. Bus Horiz 60(6):759–770. https://doi.org/10.1016/j.bushor.2017.07.003

Tseng Y-H, Lin C-T (2011) Enhancing enterprise agility by deploying agile drivers, capabilities and providers. Inf Sci 181(17):3693–3708. https://doi.org/10.1016/j.ins.2011.04.034

Valli V (2017) The economic rise of Asia: Japan. Academia University Press, Indonesia and South Korea

Vázquez-Bustelo D, Avella L, Fernández E (2007a) Agility drivers, enablers and outcomes: empirical test of an integrated agile manufacturing model. Int J Oper Prod Manag 27:1303–1332. https://doi.org/10.1108/01443570710835633

Vázquez-Bustelo D, Avella L, Fernández E (2007b) Agility drivers, enablers and outcomes: Empirical test of an integrated agile manufacturing model. Int J Oper Prod Manag 27(12):1303–1332. https://doi.org/10.1108/01443570710835633

Wikner J (2018) An ontology for flow thinking based on decoupling points—unravelling a control logic for lean thinking. Prod Manuf Res 6(1):433–469. https://doi.org/10.1080/21693277.2018.1528904

Womack JP, Jones DT (2003) Lean thinking: banish waste and create wealth in your corporation. Simon & Schuster UK Ltd., London

Womack JP, Jones DT, Roos D (1990) The machine that changed the world: based on the Massachusetts Institute of Technology 5-million dollar 5-year study on the future of the automobile. Rawson Associates, New York

Wu K-J, Tseng M-L, Chiu ASF, Lim MK (2017) Achieving competitive advantage through supply chain agility under uncertainty: a novel multi-criteria decision-making structure. Int J Prod Econ 190:96–107. https://doi.org/10.1016/j.ijpe.2016.08.027

Wyrwicka M, Mrugalska B (2017) Mirages of lean manufacturing in practice 7th international conference on engineering. Project Prod Manage Procedia Eng 182:780–785

Yusuf YY, Adeleye EO (2002) A comparative study of lean and agile manufacturing with a related survey of current practices in the UK. Int J Prod Res 40(17):4545–4562. https://doi.org/10.1080/00207540210157141

Zarbo RJ, D'Angelo R (2006) Transforming to a quality culture: the henry ford production system. Pathol Patterns Rev 126:21–29. https://doi.org/10.1309/KVT7NWVPJR73T4K6

Zerenler M (2007) Information technology and business performance in agile manufacturing: an empirical study in textile industry. In: Fourth international conference on information technology (ITNG'07), pp 543–549. https://doi.org/10.1109/ITNG.2007.111

Lean Concept in Fashion and Textile Manufacturing

Majo George, Lam Canh Nguyen, Hung Manh Nguyen, and Mohammadreza Akbari

Abstract The concept of "lean manufacturing" refers to a process of manufacturing, starting from raw materials until the delivery of the finished goods to the end-users, with prudent management of resources resulting in monetary and time savings. Lean principles have helped fashion and textile firms become more competitive and sustainable. Lean manufacturing has minimized waste and in turn reduced costs by applying reforms introduced by "Toyota" in the 1990s, and from which the technical term "lean" was derived and became popular. This chapter discusses the benefits and challenges of key lean principles in the fashion and textile industry from product, process design and supply chain perspectives using lean operations. Modular design and postponement are two major lean operations that have had a significant impact on the fashion and textile industry. The literature and real case information from the most prominent fashion retailers, Zara, Benetton and H&M, have been used to illustrate lean applications and their implications for the fashion and textile industry.

Keywords Lean manufacturing · Toyota production system · Lean supply chain · Sustainability · Fashion and textile

1 Introduction

Fashion and textile manufacturing has shifted from developed countries to developing countries (Nayak 2015a, b). This can be attributed to the increased labour charges, and utility costs that lead to higher cost of manufacturing in developed countries. For example, Australia, once famous for best quality woolen garment manufacturing, has no more manufacturing them. The manufacturing has been done

---

M. George · L. C. Nguyen · H. M. Nguyen
RMIT University Vietnam, Ho Chi Minh City, Vietnam

M. Akbari (✉)
James Cook University, Townsville, Australia
e-mail: reza.akbari@rmit.edu.vn

© The Author(s), under exclusive license to Springer Nature Singapore Pte Ltd. 2022
R. Nayak (ed.), *Lean Supply Chain Management in Fashion and Textile Industry*,
Textile Science and Clothing Technology,
https://doi.org/10.1007/978-981-19-2108-7_4

in the developing countries such as China, Vietnam, India and Bangladesh. The fashion and textile manufacturers in developing countries face very stiff competition due to globalization and cheap buying trend by many global fashion brands. It is very hard for many manufactures working on the traditional technologies to gain enough profit due to rising costs of labor, material and energy. As the manufacturing of fashion and textiles is labor and energy intensive, the manufacturers should focus on the best utilization of the labour force and reduce the energy wastage. Hence, the concept of lean manufacturing can meet these targets and increase the productivity and profitability.

Recently, there has been dynamic changes in the global fashion market place compared to the past. Large volume orders are shrinking, fashion and textile manufacturers are facing volatile and diverse demands in their supply chain (Nayak 2020). Fashion retailers like Zara and H&M have increasingly improved supply chain integration to enhance responsiveness to the markets and at the same time maintaining requirements of cost, quality and speed of delivery. A number of contextual factors such as growth of a middle income population and dominance of e-commerce service providers, have meant that fashion and textile manufacturers need to respond quickly to market changes and to produce high-quality but price competitive fashion items. Lean operations such as module development and standardization processes have been developed to handle flexible and disruptive demands of changing fast fashion trends.

This chapter aims to review the development of lean principles, and to discuss major techniques that can promote the fashion and textile industries. Section 1 provides an overview of lean operations and lean supply chains in fashion and textile industry. Section 2 describes a framework for analysis of several lean techniques, benefits and challenges in the fashion and textile industry. This section discusses the brief of manufacturing processes in the fashion and textile industries. How the lean concepts can be implemented in the fashion and textile supply chain has also been discussed. The case information from the most dominant fashion retailers including Benetton, Zara and H&M provide enhancement of the lean models. The final section offers conclusions and future directions.

1.1 An Overview of Lean Manufacturing

After the Second World War, Japan, in an effort to reduce the cost and increase the quality and quantity of production, adopted lean manufacturing techniques. Toyota's Taiichi Ohno and Shigeo Shingo, were part of finding a faster and cheaper way of producing automobiles, and pioneered the lean-approach. This not only enabled industries to maintain their competitiveness, and increase innovation by moving from traditional methods to a more flexible system with fewer resources and less wastage but also increased their profitability. The introduction of lean manufacturing methods by the Japanese car manufacturer Toyota Motors paved the way to a huge Japanese success in the industry and eventually led to change in the whole market through the strategy initiated in the car industry by Toyota Production Systems (TPS).

Observing the tremendous success of TPS, many other industries then started applying lean manufacturing to wider areas, ranging from electric and electronics, automobiles, missionary wood, machine tools, construction, ceramics and clothing. Depending on available funds, technology knowhow, demography, culture and skills, the adoption of lean manufacturing principles varied from company to company. Various factors have been noted as limiting some companies when introducing and comparing performances by using lean manufacturing across different industries.

Business organizations are increasingly looking for new and better ways of operating as they strive to remain competitive in a fast-changing global economy. This has meant that some companies are moving away from a traditional operation to more modern methods which have included lean operations, using fewer resources in a more efficient way to achieve elimination of waste. This innovative approach to reduce wastes has helped organizations to increase productivity using shorter cycle times, producing higher quality, for lower cost than non-lean systems.

The implementation of lean production has been achieved by professionally managing resources in such a way as to enhance quality, reduce lead times, reduce costs, and minimize waste while increasing productivity. In lean manufacturing, the word "input" is used to mean the physical quantity of resources used and their cost while the word "output" means the quality and quantity of the product sold along with corresponding customer services. From the main goal of minimizing waste, lean manufacturing helps a businesses maximise customer value, achieve the organization's goals through producing outstanding output while using less input. It is to be remembered that effective and efficient use of resources with a minimum of zero waste is one of the main pillars of lean manufacturing.

2 The Development of Lean Manufacturing/Production

Ongoing efforts to maximize productivity within the manufacturing system have been continuously advancing towards lean manufacturing. This has involved tremendous contributions by human personnel at different milestones in the development of the productivity-efficiency initiative. Therefore, the following chapter concentrates on highlighting significant cultural periods in lean manufacturing. Table 1 discusses the chronological developments of lean manufacturing starting from the eighteenth century.

2.1 A Conceptual Framework

One of the first attempts to establish a waste-reduction principle in lean manufacturing involved interchangeable parts (Čiarnienė and Vienažindienė 2012). An Interchangeable part is a concept of producing identical or nearly identical parts that can be inserted into an assembly without the extra effort of customization or craftwork.

Table 1 Chronological developments of lean manufacturing

Year	Invention	Inventor(s)
1799	Interchangeable parts	Eli Whitney
1850	The method was spread all over the industrialized world "American System of Manufacturing"	Eli Whitney
1881	Time Study and Standardized Work	Frederick Winslow Taylor
1911	Scientific Management	Frederick Winslow Taylor
1913	The assembly line technique of mass production	Henry Ford
1919	Motion-time study Process charting	Frank and Lillian Gilbreth
1920	Statistical process control (SPC)	Walter Shewhart
	The concept of implementing total quality management (TQM)	Edwards Deming, Joseph Juran, and Philip Crosby
1950s	Toyota production system (TPS)	Taiichi Ohno and Shigeo Shingo
1990	The popularisation of the lean manufacturing concept	James P. Womack, Daniel T. Jones, and Daniel Roos

The concept was first raised by Honore le Blanc in France in the mid-eighteenth century (Alder 1997). Notwithstanding, it was Eli Whitney in 1799 who popularized the concept and formed the first stage in Lean production and Just-in-Time (JIT) (Coulson 1944).

This was the remarkable outcome originating from a musket manufacturing contract that Whitney negotiated with the U.S. Army in 1798 (Woodbury 1960). Despite an unachievable goal required under the contract, the effort enabled compatible parts of muskets to be easily mass-produced and replaced, which reinforced a strategy involving a combination of interchangeable parts and mass production. Simultaneously, the strategy involving interchangeable parts standardized the manufacturing process and eliminated extraneous steps, allowing unskilled workers to create and repair weapons rapidly at a fraction of the usual cost.

Eventually, the concept of interchangeable parts spread to all areas of manufacturing and became well-known as the "American System of Manufacturing" (Sawyer 1954). This soon conceptualized the waste-elimination principle, since, before interchangeable parts, incompatibility of products had led to difficulties in assembly, causing expensive and unrecoverable costs for material and labour.

Despite the evolution and popularization of the interchangeable parts concept in America, the labour-intensive approach had prevented this initiative from performing to its optimal potential. Due to this limitation, Frederick Winslow Taylor had attempted to advance worker efficiency. His analyses were based on observing individual workers and their working methods and later published as Time Study and Standardized Work (Drucker 1999). All of the publications were produced as part of a new management class known as "Scientific Management" or "Taylorism", created to guide contemporary management boards with economic efficiency improvement, especially time reduction, in the area of productivity (Taylor 2004).

The Time Study was further developed by Gilbreths, Frank and Lillian, who monitored the behavioural and psychological aspects of workers when performing assigned tasks. While Frank Gilbreth aimed to minimize the worker's movements in production (ergonomy) through Motion Study, Lilian Gilbreth added the psychological perspective to the research from the study of worker motivation and its impacts on worker attitudes and overall working results (Gilbreth and Gilbreth 1916). These findings complemented Taylor's previous work and become known as the Motion-Time Study.

Another contribution to Taylor's Time Study, became known as the Process Mapping of Frank Gilbreth. This methodology concentrated on identifying and eliminating any non-value-added work throughout the work process, hence delivering benefits such as waste detection, depicting a process-flow, defining and standardization, and so on (Lee and Snyder 2007).

2.2 Henry Ford Production System

Yet it was not until the time when Henry Ford integrated an entire production process (known as mass production or flow production) that he was able to create a foundation for the concept of lean manufacturing. This was in 1913, when Ford created his revolutionary technique of production with consistently interchangeable parts with standard work and moving conveyances (Womack et al. 1990). In fact, from the view of both the public and the engineers, it was truly a breakthrough in the production industry with the advent of the moving assembly line (Mullin 2007).

Particularly, the fabrication process was sequenced with mechanization being used to generate a high quantity and detailed construction of the material stream, labour division, and careful quality control. The lined-up fabrication steps allowed the use of specialized machines and gauges to assemble the vehicle's components all within a few minutes, extracting perfectly fitting end-products thereupon. This was truly a revolution in the practice of the American manufacturing style that consisted of a finished product grouped through processes of the sophisticated subassembly and final assembly (Michelsen 2020). As a result, mass production proved to create many advantages, namely, accuracy, lower costs, fewer workers needed, faster producing and distributing processes, able to deliver higher levels of efficiency (Aristova and Chadeev 2018; Nayak and Padhye 2018).

Despite those benefits, the flow production had still not reached the optimal point of maximizing the productivity in manufacturing. Womack et al. (1990) stated that the weakness of the system derived from an inability to be flexible and restraints on the capability to produce variable and customized products. This prompted other automakers to update Henry Ford's methods in response to minimise those shortcomings.

Therefore, the manufacturing techniques of Henry Ford, along with the advent of a statistical process control (SPC) (the use of statistical techniques to control a process or production method) from Walter Shewhart and the progressive methodology on implementing total quality management (TQM) from Edwards Deming, Joseph Juran, and Philip Crosby in the U.S. and Kaoru Ishikawa in Japan from around 1920 onwards (Alghamdi 2016), can be considered as the foundation for further improvements in the manufacturing industry, as claimed by Dave (2020).

2.3 Toyota Production System (TPS)

It has been asserted that the lean manufacturing technique was only highlighted by the success of the Japanese giant, Toyota. Toyota's concept can be referred to as the Toyota Production System (TPS), which was suggested by Taiichi Ohno and Shigeo Shingo in the 1950s. The goal of TPS was to increase the efficiency of Toyota production to cut costs (Wahab et al. 2013). The TPS consisted of two main concepts: "Cost-cutting thorough removal of waste" and "full utilization of workers' capabilities" (Sugimori et al. 1997) (Fig. 1).

Two important pillars for the materialization of this system were JIT and Jidoka. The objective of JIT production was to produce a necessary item in a necessary time and quantity (Monden 1994). The principle of JIT required the "withdrawal by subsequent processes", the "one-piece production and conveyance", the "levelling

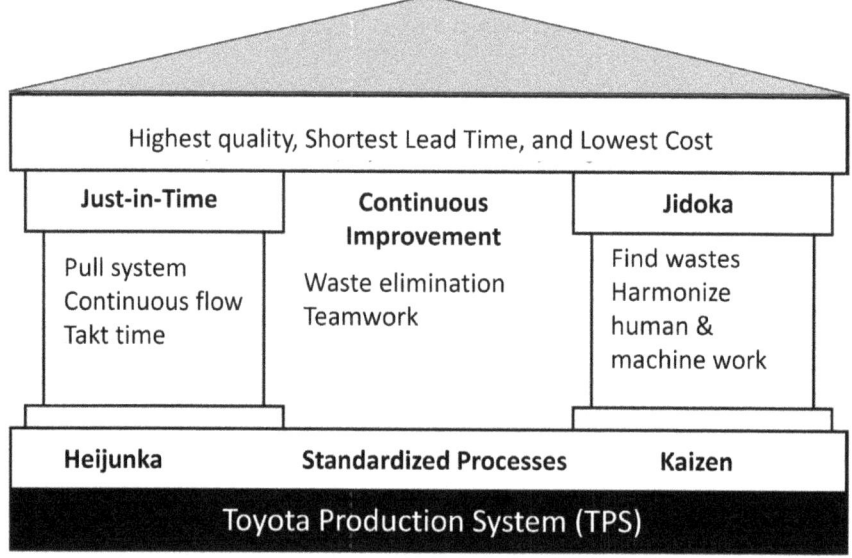

Fig. 1 Toyota production system (TPS)

of production", and the "elimination of waste from over-producing" (Sugimori et al. 1997). The second pillar of TPS, Jidoka, explained the situation when any possible unusual phenomenon or defective condition can halt the equipment or entire production line. The consequences of Jidoka were to refrain from redundant production and manage abnormal circumstances (Sugimori et al. 1997).

To achieve full utilization of workers' capabilities, it constructed a respect-for-human system that focused on the eradication of wasteful movements by workers, consideration of workers' safety, and inducements for workers to maximize their capabilities with preeminent responsibility and authority (Sugimori et al. 1997).

The catalyst that encouraged the invention of TPS was the need to reduce costs in the war-ravaged economy following War World 2 (Wahab et al. 2013). To be more specific, the consciousness and attitude of the Japanese workforce, the narrow domestic market, the pressure from foreign producers, and the lack of resources drove the changes in Toyota's traditional production (Womack et al. 1990).

The TPS is said to be another revolutionary production management system, along with the Taylor system (Scientific Management) and the Ford system (mass-assembly line), since it could still generate profit even during sluggish periods in the economy (Monden 1994). This concept was introduced to a broader audience after 1983 with the accentuation of mixed model production, preventive maintenance, and multifunction workers. In addition, the TPS has gone through tremendous changes in practices and improvements since the 1970s (Bhamu and Sangwan 2014).

3 The Development of Lean Supply Chains in Fashion and Textile

Since 1990, efficient managing of the entire supply chain has become the most important part of organizational success in the world (Akbari 2018; König and Spinler 2016). Supply Chain is defined as the planning and control of the end to end production and services, from raw materials all the way to the end customer (Akbari et al. 2017; Maleki Far et al. 2017). Lean Supply Chain (LSC) is a smart way of linking a set of organizations by managing the up and downstream flows of funds, information, services, and products satisfying the end-user by the reduction of cost and wastage (Vitasek et al. 2005). Lean manufacturing and Supply Chain Management can be classified into upstream activities such as procurement and supply, and downstream activities particularly distribution, including warehousing, transportation, retailing, etc., and the integration of Logistics and Supply Chain activities (Akbari and Ha 2020).

In general, the textile industry comprises fibre manufacturing, spinning, weaving, knitting and finishing while the fashion sector deals with clothing and accessories production (Maia et al. 2013). Bruce et al. (2004) define the fashion and textile industry as all activities "from chemical conglomerates producing dyes, detergents and artificial fibres, to healthcare companies producing heart valves, prosthetics, bandages, etc., to niche design-driven fashion companies". The industries of fashion and textile are characterized by diversity and heterogeneity with a volatile market and a wide range of products marked by short life cycles (Bruce et al. 2004; Nayak et al. 2020a, b). As the profit margin of fashion and textile is low, companies tend to produce these products quickly to fulfil the customers' orders and reduce the holding stock. Additionally, fashion and textile manufacturing units are among the most labour-intensive industries, which cost more in developed countries and thus are more likely to be moved to developing countries which can offer cheaper labour costs.

3.1 Lean and Garment Manufacturing Process

The typical garment manufacturing process is described in detail in Fig. 2. The input raw materials to a garment are fabrics and trims (such as buttons, zipper and thread). Firstly, the fabrics is being inspected for quality before going to a cutting section. Different types of fabrics are inspected by skilled people in a lighted board. The accuracy and efficiency of fabric inspection depends on the skill and quality of fabrics. During manual inspection, various fabrics faults (some shown in Fig. 3) are inspected and marked by chalk in the fabrics, which gives indication to the cutting section. Then fabrics are graded based on the points allocated depending on the severity of the fabric faults following various grading systems such as four-point,

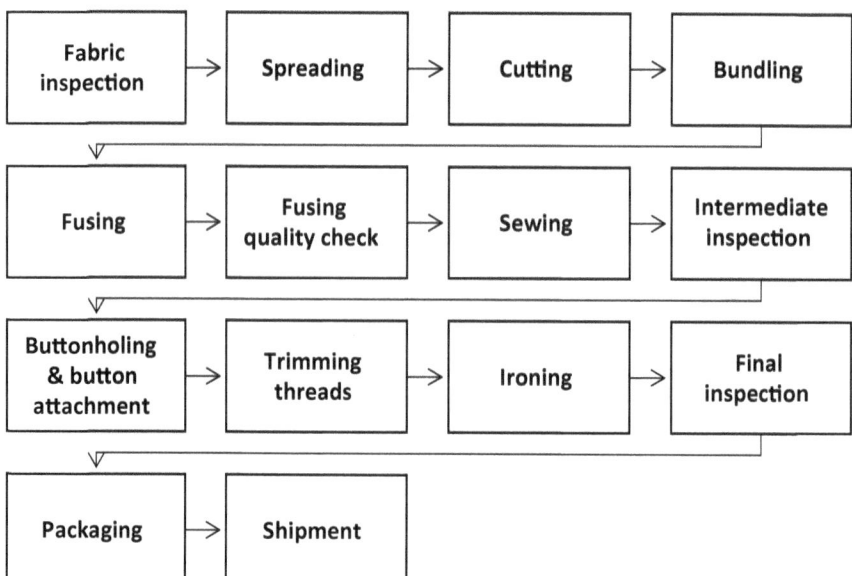

Fig. 2 Garment production flow chart

Fig. 3 Some fabrics faults found during fabric inspection

ten-point and Dallas system. If the total points exceed the specified value, the fabrics are rejected.

Fabric inspection plays an important role as it directly translated to the garment quality. Inorder to improve the efficiency and reduce the wastage, fabrics should be carefully inspected before garment manufacturing starts. However, some industries neglect the fabric inspection due to shortage of time and lack of facilities, hence, they encounter a lot of rejections or reworks. For these industries, the implementation lean manufacturing can help to reduce the amount of rejections. Although automatic fabric inspection systems are commercially available, which can provide consistent results at greater accuracy, several industries still rely on the manual inspection due to high cost of the automated machines. Digital cameras, and image analysis software are used for the automatic fabric inspection.

Automated fabric inspection systems can be classified as online and off-line systems. In the online monitoring systems, the inspection is performed in the weaving machine while it is being produced. In the off-line monitoring systems the fabric is inspected after being taken off from the loom, by mounting on an inspection frame where the quality of the fabric is analysed. Automated fabric inspection systems can inspect the fabrics in full width either at the batcher for greige fabrics or at the exit end of the finishing machine. Automated fabric inspection systems are designed to find and catalog defects in a wide variety of fabrics including greige, apparel, upholstery, furnishing, dyed, finished, denim and industrial fabrics. For getting better results in lean manufacturing by reducing waste, it is recommended to have automated fabric inspection systems.

Following the quality inspection, the next processes are spreading, cutting, and then bundling, which also influences the quality of garments. Mistakes in cutting or bunding of wrong sizes or different batches can lead to the rejection of garments. The next process is fusing, where some garment components are fused with interlining to provide stiffness or functionality to the fabrics. Then, they are passed to the sewing section where different parts of the garment are being assembled. The sewing section combines various cut components by different sewing machines such as lock-stitch and/or chain-stitch machines or button attachment, and buttonholing machines. The last section is finishing, which includes thread cleaning, washing and ironing. Broken needle parts are being checked by metal detection machines to ensure safety of consumers.

Finally, the quality of the garments are inspected following various quality standards before they go to the packaging section. Quality inspection plays an essential role in the manufacturing process as it assures the garment quality is being met by the manufactures. The inspection involves visual checking of the faults present in the garment or inspection of the measurements as per the technical specification. The visual inspection of garment (as shown in Fig. 4) is done to find any stitching or seaming defects, stains, and protruding threads present in the garment (Nayak 2015a, b). The inspection for measurements or different size specifications ensures the right fit of the garments. Experienced quality inspectors perform the garment analysis as agreed by the Terms and conditions between the manufacturer and the buyer.

In traditional manufacturing, the approach is to check the garments for defects once the garment manufacturing process has been completed. Hence, any defects in a process that could have eliminated by intermediate inspection goes to the next process. This increases the percentage defective garments and many cases the garments are being rejected, hence is a wastage of materials and labour. Reworking is a common problem in many garment manufacturing industries following the traditional methods, where the garment faults are rectified by opening the sewn components and reworking. On the otherhand, when lean manufacturing has been implemented the approach is to prevent any defective products entering to the next process. This is a preventive approach and confirms that the output from a previous process complies with the quality standards. Hence, the outputs of each process are inspected (intermediate inspection) to ensure the right quality product enters to the subsequent

Fig. 4 The visual inspection of suits in a garment industry

process. The implementation of lean manufacturing can ensure the production of right product in each step (Paneru 2011).

The next step after the quality inspection is the packaging of the garments that pass the quality inspection. During packaging, various tags are attached to the garment and the garments are wrapped in plastics. The garments are either folded (such as shirts, trousers) or kept in hangers (such as suits, coats etc. to avoid creases) depending on the type of the garment. Once, the garments are packaged, they are sent to the warehouse for stock verification and then shipment.

3.2 Lean and Textile Manufacturing Process

A brief discussion of garment manufacturing process and how lean manufacturing can be implemented to reduce the waste have been given in the above section. This section discusses the application of lean manufacturing in the spinning, weaving and textile chemical processes in brief. A brief discussion of the process sequence have been given in Figs. 5, 6 and 7 in the spinning, weaving and textile chemical processes, respectively. The principles of lean manufacturing can be implemented in the three different processes to eliminate wastes.

In spinning process, yan is manufactured from different types of fibres. The spinning process involves steps such as fibre mixing, blowroom, carding, combing (for combed yarn), drawing, roving and ring spinning for ring yarns. For rotor and airjet yarns, which gives high productivity directly the carded or drawn slivers can be used eliminating other process. However, there are certain limitations of the these processes in terms of yarn count, versatility and yarn properties. Various lean tools such as Kaizen 5S, Kanban can be implemented in the spinning processes to improve activities such as quick changeover, to improve cleanliness in the workplace, and promote cooperation among the workers in the workplace. Regular meetings, teamwork, effective communication can assist in implementing the lean manufacturing.

In weaving process fabric is manufactured from yarns by using different types of looms. The weaving process involves steps such as winding, warping, sizing, drawing-in and then weaving in a loom. An additional step of pirn winding is needed for shuttle looms. Various lean tools such as 5S, TPM (Total Productive Maintenance), Kaizen, Kanban, and Six Sigma can be used in weaving to improve the productivity and efficiency. The implementation of lean manufacturing in weaving industries can help to reduce the inventory, changeover time and waste in the loomshed. Skilled top management, team condition and effective communication

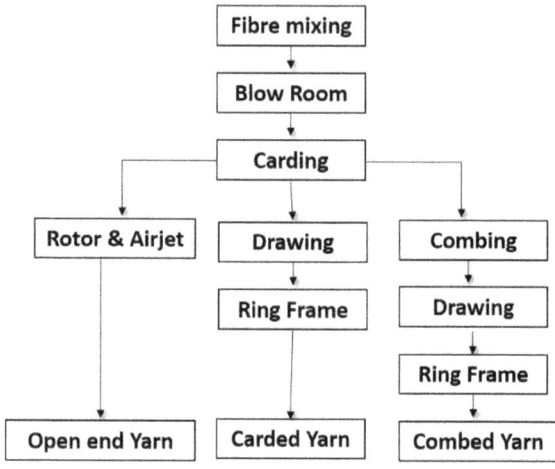

Fig. 5 The process sequence in yarn manufacturing

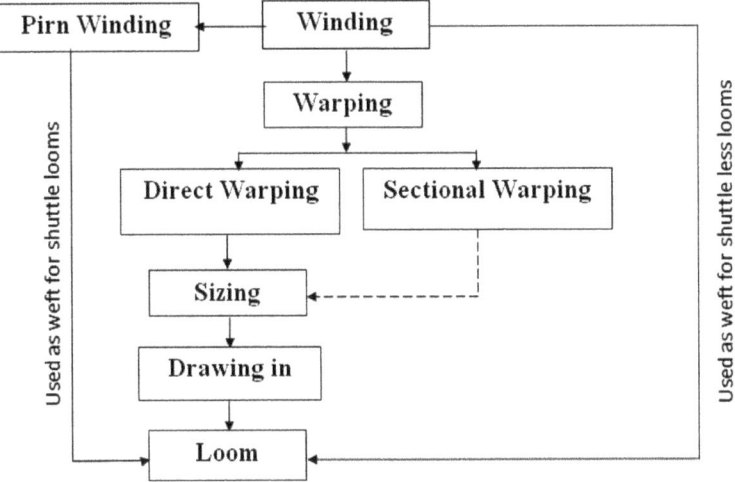

Fig. 6 The process sequence of the weaving process (--- line is optional process and solid lines are mandatory processes)

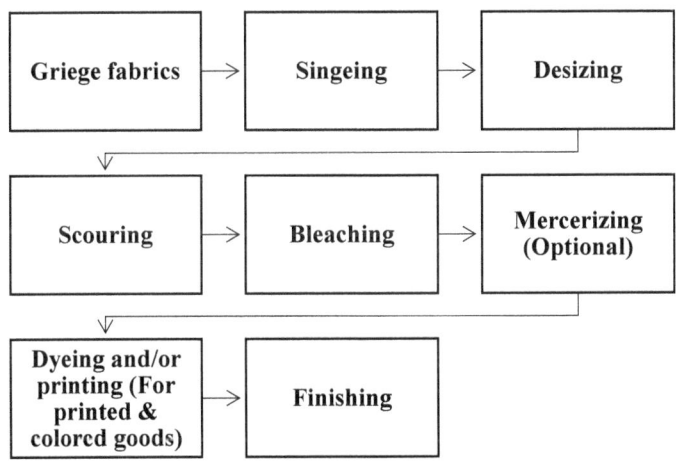

Fig. 7 Process sequence for textile chemical processing

is essential for successful lean manufacturing. Similarly, the other process, knitting, used for fabric manufacturing can be benefited by the implementation of lean manufacturing to improve productivity, reduce waste and improve fabric quality.

Textile chemical processing involves the processes to prepare the greige fabric ready for garment manufacturing. As such the fabric from loom state is not suitable to be used in garment manufacturing due to shrinkage issues and aesthetics. The chemical processing steps includes singeing, desizing, scouring, bleaching, mercerizing (optional for cotton fabrics), dyeing and/or printing and finishing. Lean manufacturing tools such as six-Sigma, 5S, SPC and Kaizen can be implemented to improve the quality of fabrics, reduce the hazards of chemical management, reducing inventory, waste reduction, and improve productivity.

3.3 Lean Implementation in Fashion and Textile Manufacturing-Research

As discussed above, lean manufacturing refers to the elimination of waste and non-value-added activities within the principle of production while still maintaining an appropriate customer satisfaction level. Such a concept has been applied in the fashion and textile industries to improve manufacturing productivity and efficiency in different contexts all around the world, including both developed countries and developing countries.

Examples of lean implementation in developed economies can be found in the US and Europe. Hodge et al. (2011) found that textile manufacturers benefited from implementing the lean concept by improving their production process and thus enhancing their competitiveness. The study summarized that the most common lean tools used by these manufacturers include 5S, Kaizen, Six-Sigma, Quick Changeover, Standardized Work, TPM, Value Stream Mapping (VSM) and Visual Management. Similarly, Manfredsson (2016) investigated cost reduction in manufacturing by applying lean thinking to small and medium textile enterprises in Europe.

Research has demonstrated four main improvements by applying the lean concept in such companies, i.e., information and communication, team collaboration, changes in stress levels and team performance. The author also suggests that such implementation should include cultural aspects rather than a tool perspective only. In a specific context, the lean concept was implemented in the Ethiopian textile industry by defining and eliminating the main waste sources of resource, production defect, human talent, excessive motion and overproduction (Mezgebe et al. 2013). Additionally, the implementation of lean manufacturing was proven to promote sustainable development in the Portuguese textile and clothing industry (Maia et al. 2019).

In contrast, Comm and Mathaisel (2005) demonstrated that lean implementation, which is suitable for the capital-intensive textile industry in the US and Europe, is also working well for labour-intensive textile companies in developing countries. The lean concept helped Chinese manufacturers in identifying the need to process bottlenecks and downtime, thus providing opportunities to reduce lead-time and increasing productivity and quality control. In this case, VSM was applied to observe how many touches (relate to material handling) were made for each product and qualify the added value of these touches. Detected excessive touches were eliminated or combined with others to reduce the total number of touches. As a result, the revised process provided a significant improvement in the manufacturing throughput.

Durand-Sotelo et al. (2020) also found that lean management could help Peruvian textile SMEs reduce the order fulfilment times. As a result of lean implementation, such companies experienced a reduction in problems, an increase in productivity and a decrease in manufacturing cycle time. They recommended that frequent training in lean manufacturing for workers was necessary, along with a need for continuous monitoring to create a long-term lean culture. Saleeshya et al. (2012) demonstrated the beneficial results of lean implementation in Indian textile manufacturing with the combination of Kanban, 5S, Kaizen, VSM and Poka-yoke.

The studies of Shah and Hussain (2016) and Bhutta et al. (2013) examined lean practice among Pakistan textile companies, and concluded that the most popular lean tools used in the sector included 5S, Kaizen, quality control circles and single minute exchanges of die (SMED). In the context of Sri Lanka, Kgdas et al. (2012) found that lean application in the knitting industry was more challenging than the garment sector because of the long-life product cycle time and the mass amount of production. However, success in lean implementation has helped these companies reduce lead time and achieve financial profits. The key factors for implementation in Sri Lanka included the method of introduction and the order of implementation (Silva et al. 2011).

Lean implementation in fashion and textile manufacturing follows the main six indicators: (1) elimination of waste; (2) continuous improvement; (3) multifunctional teams; (4) just-in-time production and delivery; (5) integration of suppliers; and (6) flexible information system (Sanchez and Pérez 2001; Karlsson and Åhlström 1995; Comm and Mathaisel 2005). These indicators will be discussed as follows.

3.4 Elimination of Waste in Fashion and Textile Manufacturing

The core process of lean manufacturing is defining different types of waste and non-value-added activities and removing them from the production process. Within the fashion and textile sectors, common wastes and recommendations for eliminating such wastes have been extensively discussed in the studies of Mezgebe et al. (2013) and Thakkar and Joseph (2015). Different types of wastes and their elimination has been discussed in the following section.

The first type of waste in this industry is utility waste. This type of waste is quite common in all industries and may come from several sources, including negligence by personnel, lack of resource utilization and another inefficiencies in the workplace. Examples of this type of waste included leaving lights on during daytime or after having left a room, PCs left running 24 h, machine run-off during jobs, ventilated rooms and running water tap. However, eliminating this type of waste can be achieved easily by posting reminder notices and monitoring the machine downtime when it is out of operation.

The second type of fashion and textile manufacturing waste is the overproduction of goods like fabrics and yarns. Such waste implies the fact of producing goods too soon or much more than necessary or that can be consumed. These wastes, need

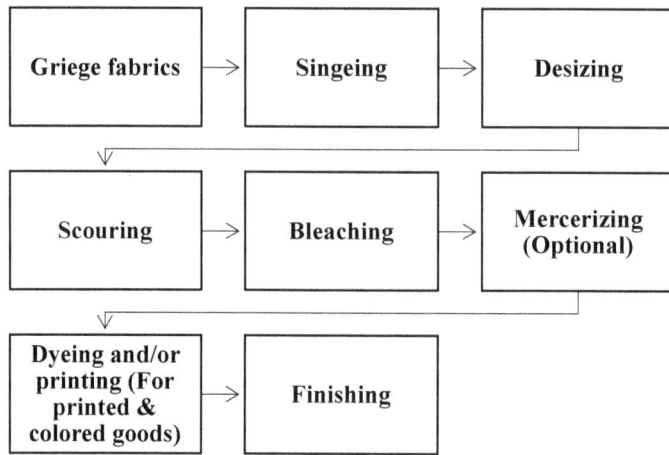

Fig. 8 Fabric inspection system in a garment industry

storage place, and labour, which increase the overall cost. To avoid this type of waste, Mezgebe et al. (2013) recommended that factories communicate better with their customers and focus more accurately on their dynamic demand.

The third type of waste is human waste, which refers to a failure to utilize the company's manpower. As fashion and textile manufacturing companies are labour-intensive, the gap between the number of employees required and the actual labour force has resulted in significant losses during the production process. In several instances, the laborforce has to wait for longtime due to improper work allocation and lack of synchronisation. Another form of human waste is loss of talent, or a waste of creativity when a company fails to recognize the ability of employees and that leads to a lack of suggestions for improvement generated within the company. Therefore, it is necessary to respect employees, and better connect with them through use of clear guidelines and compensation packages, motivating them to take ownership of their areas and processes, and encourage them to become involved in all improvement processes by offering package rewards.

The fourth type of waste is the waste of defective materials, consisting of both input and output defective materials. For example, defects of input materials for garment manufacturing might be caused by the defects present in the input fabrics, which is the major raw material used for garment manufacturing. Many of the garment manufactures outsource their fabrics from fabric suppliers. Due to busy schedule, the garment manufactures often skip the fabric inspection process, which lead to several defects in the sewn garment. The garment manufactures should have the fabric inspection system inhouse (as shown in Fig. 8), or they can get it done by the third party fabric inspecting agencies. Sometimes, the fabric inspection reports received from the fabric supplier can be referred as a guideline on the fabric quality. However, in some instances the fabric report may not be accurate as it is prepared favouring the fabric manufacturers.

The fifth type of wastes relates to the defects produced by technical reasons like damages in the loading/unloading process or lack of skilled employees. The output defects happen with the finished products possibly due to incapable processing, lack of control, incorrect orientation or incapable suppliers. Defect examples in textile manufacturing may be needle holes, stains and skiped stitches. To eliminate these wastes, garment companies must provide on-job training for employees and standardize operation processes. Islam (2021) recommended that Pareto analysis is useful to identify garment defects as 80% of defects often happen due to 20% of common causes.

The last type is a waste of motion and transportation, i.e., excessive travel by people and machines, caused by unorganized space and tools, constantly turning and moving products during assembly of heavy objects placed on high or low shelves. Inappropriate workflow management also causes excessive material movement in material warehouses, cutting, sewing and finished warehouses. This type of waste can be reduced by defining and removing non-value-added steps and standardizing working procedures.

3.5 Continuous Improvement

The next important indicator of the lean concept in the fashion and textile industry is how companies continuously improve their manufacturing processes. According to Sanchez and Pérez (2001), this cannot be done without the engagement of all employees and support from the management team. Through a case study in China, Shah and Hussain (2016) concluded that such improvement can be achieved if a company pays attention to training and develops improvement focus groups. This will lead to product quality enhancement and reduce the average processing times. Adopted from Sanchez and Pérez (2001), one could measure this continuous improvement process at fashion and textile manufacturing by recording the number of suggestions per employee each year, the proportion of suggestions to be implemented, the benefits of such implementation, the percentage of defective parts changed, the percentage of the non-working machine and the human resources used in quality control.

3.6 Multifunctional Teams

Another lean indicator in manufacturing is how employees can perform a variety of tasks when their teams are likely to rotate more than normal. This creates flexibility and adaptability by companies to accommodate changes in manufacturing levels. The implementation of multifunctional teams requires greater effort in training employees to accept more tasks. This implementation will also possibly reduce indirect costs as workers also take over the tasks of maintenance and handling materials (Karlsson and Åhlström 1995).

3.7 Just-in-Time Production

The JIT concept refers to the delivery of the right products with the necessary quantity only when they are needed to minimize the buffering level. JIT helps the factory to reduce inventory level, thus reducing the carrying cost and inventory management tasks. It increases product flexibility and reduces the working space and overheads. The implementation of JIT also reduces the set-up time and the lead time within the manufacturing (Gilmore and Smith 1996).

The JIT concept has a long history in the fashion and textile industry. Hokoma (2010) studied the awareness and implementation of the JIT concept in the Libyan textile industry. A survey was conducted and obtained 136 responses from 60 targeted companies in Libya. The result found that half of those companies did not implement JIT at all, 41.7% were not very sure about their JIT implementation and only 8.3% successfully applied JIT within their organizations. The main reasons for JIT absence was concluded to be that these companies were not familiar with JIT. Other significant reasons included a shortage of top management support and a fear of high expense for such implementation. Therefore, the study recommended the management body in the textile industry pay more attention to JIT, enhance the education and training programs and provide more support to achieve high manufacturing outcomes in textile manufacturing.

In another research, Yam (1994) discussed the JIT implementation at small Hong Kong garment factories in the 1990s as a solution to counter high land expenses, increasing labour costs and growing global competition from Central America, Indonesia and Eastern Europe. As explained in the study, the garment industry in Hong Kong at that time comprised 7737 factories, in which more than 80% were small with less than fifty employees. For these small companies, the demands were volatile in both quantity and style as they were more likely to manufacture products which were less economical than for large companies. Implementing JIT in Hong Kong was also challenging as the textile suppliers and their customers were located overseas. Yam (1994) concluded five main suggestions for successfully introducing JIT in the garment industry were as follows.

First, the JIT transition requires commitment and active involvement from senior management. Second, the company needs to strengthen the coordination with the supplier to achieve JIT delivery of raw material. The shipment of such raw material should be standardized for smooth handling with high-quality products for fewer inspection activities. The next key point for JIT is educating and training the employees, i.e., managers, technicians and operators to improve their flexibility with changes and non-productive skills like preventive maintenance. The case study demonstrated that education and training could help these small garment factories in Hong Kong overcome the issue of work culture from the majority of Chinese workers that was hindering teamwork. Two other suggestions included using group incentive plans to encourage teamwork in manufacturing and facilitate identifying the issues via teamwork.

3.8 Integration of Suppliers

Textile factories could also integrate with suppliers to achieve lean thinking in many disciplines, i.e., co-designing, integration of the information system for better sharing, exchanging technicians or developing long term contracts with suppliers. This implementation supports JIT practice, enhances productivity and contributes to the resilience of the fashion and textile supply chain (Nayak et al. 2019).

4 Flexible Information System

Last but not least, lean manufacturing requires a reduction of the hierarchy level within a company to facilitate the diffusion of information at all levels. A flexible information system should allow the timely delivery of information between top management and the production line, and between different company departments and groups of employees. Such flexibility of the information system could be measured by the frequency of the employees receiving information, the number of meetings between managers and employees, how many procedures are standardized and recorded and how many decisions workers could make without asking management. Excessive paperwork is a type of waste in manufacturing and developing a flexible information system could reduce the amount of time and efforts in processing documents by decentralizing responsibilities to field workers (Comm and Mathaisel 2005).

4.1 Lean Implementation to Achieve Sustainability in Fashion and Textile Industries

Implementing the lean concept in manufacturing not only has an economic impact on the fashion and textile industries but also contributes to sustainable practices by companies. Fashion and textile industries are known to consume a large number of natural or manmade resources, i.e., water, energy, natural fibres and dyes, and release chemical waste which highly contaminates the environment (Akbari et al. 2020; Maia et al. 2019; Nayak et al. 2020a, b). According to Moreira et al. (2010), appropriately implementing the lean concept would lead to eco-efficiency practices in this sector. Their study synthesized the literature and concluded that there existed a causal relationship between lean implementation and environmental performance enhancement. This shares similarity with the study of Maia et al. (2013), which demonstrated that by applying the lean concept, textile SME (small and medium-sized enterprise) companies could reduce the energy and water consumption, the use of raw material and thus reduce the environmental waste. They recommended the application of the toolkit by the USEPA (US environmental protection agency) (2018) in implementing lean strategies, i.e., waste elimination culture, Kaizen events and just-do-it, standard work, visual control, 5S and total productive maintenance. Ciccullo et al. (2018) also confirmed that the lean practices supported a sustainable paradigm in the garment sector. An example could be shortening the supply lead time would lead to the reduction in carbon dioxide emission and the creation of new jobs in the region. In another research, Maia et al.(2019) suggested such implementation should base on a holistic approach, which considers the four dimensions of people, ergonomics, sustainability and operation.

Sustainable business practise requires the following four enablers from the lean concept as listed by Caldera et al. (2019). First, green strategies and a lean concept should be integrated to achieve business sustainability. This shares similarity with Raj et al. (2017), who also provided guidance that lean production and environmental sustainability should be implemented hand in hand, given the case study of Indian apparel manufacturers. The second enabler is the continuous improvement in the manufacturing process by leveraging a lean concept and green practices. The approaches of 5S and VSM are recommended in this enabler. Next, the successful achievement of business sustainability requires stakeholder involvement, both internal and external. The former are the company's employees, managers, and owners, while the latter consist of suppliers, customers, society, central and local government. The last enabler requires the company to have streamlined processes in place, which help a company increase transparency and traceability. The lean concept supports streamlining by standardizing their processes and thus adhering to international methods. The study also synthesized six barriers to sustainable business practice, i.e., lack of education and awareness, lack of financial resources for implementing lean and sustainability strategies, high risk in the implementation, eco-economic decoupling and a weak regulatory environment.

5 Lean Operations: Benefits and Challenges in Fashion and Textile Industry

A Lean operation is a flexible system that uses considerably fewer resources (i.e., activities, human resources, materials, energy, and floor space) than a traditional system. Lean principles help fashion and textile firms become more competitive and sustainable. The apparel industry is characterized by a short product life cycle, relatively high variety, uncertain demand forecast, which often contributes to inefficiency in the supply chain (Taplin 2014). Literature indicates that, the value-added of garment in supply chain pipeline represented a small part of the total lead time (only 11 weeks out of the 66 weeks) that was required on the actual garment value-adding processes. The remaining idle times represent waste activities such as work-in-process and finished inventories at various stages in the system (Aftab et al. 2018). This inefficiency could lead to overstocking or backlog and stock-out for the fast-moving high demand items. Literature has estimated losses of $25 billion relating to the idle times (Aftab et al. 2017, 2018).

5.1 *Lean Goals and Supporting Techniques: Benefits and Challenges*

The goals for lean operations are to eliminate disruptions and to create flexible systems to meet diverse customer requirements. Due to rapid changes in customer requirements such as fast fashion, the aim of the lean systems is not only cost reduction but also to create a balanced, continuous and flexible flow of fabrics, trims, garment accessories and finished products. The key foundations of a lean system include (Stevenson 2005):

- Product design and process design
- Organizational and network
- Manufacturing control and planning.

Manufacturing control and planning consists of several tools such as level planning, resources allocation and push and pull planning. These tools have been well developed for a specific industry. This section discusses some key benefits and challenges of lean principles in the fashion and textile industry from product, process design and supply chain perspectives in lean operations (Fig. 9). Modular design and postponement are two major lean operations that can significantly impact on fashion and textile industry. The literature and real case information from the most prominent fashion retailers, Zara and H&M, will be used to distinguish the lean applications and their implications for the fashion and textile industry.

Due to rapidly changing fashion trends, garment manufacturers have faced volatile demands and diversity in customer requirements. Fashion retailers have increasingly pushed supply chain rationalization and improved channel integration to force garment and textile manufacturers to be more responsive to cost, quality and speed of delivery requirements. Besides, the growth of e-commerce platforms, fashion and garment manufacturers need to respond as fast as possible to models and quantity changes and to produce high-quality but at the same time cost competitive fashion items. Manufacturers need to consider a new production system that can handle flexible and relatively simple modular designs and use standardization concepts to enhance the efficiency of the production systems.

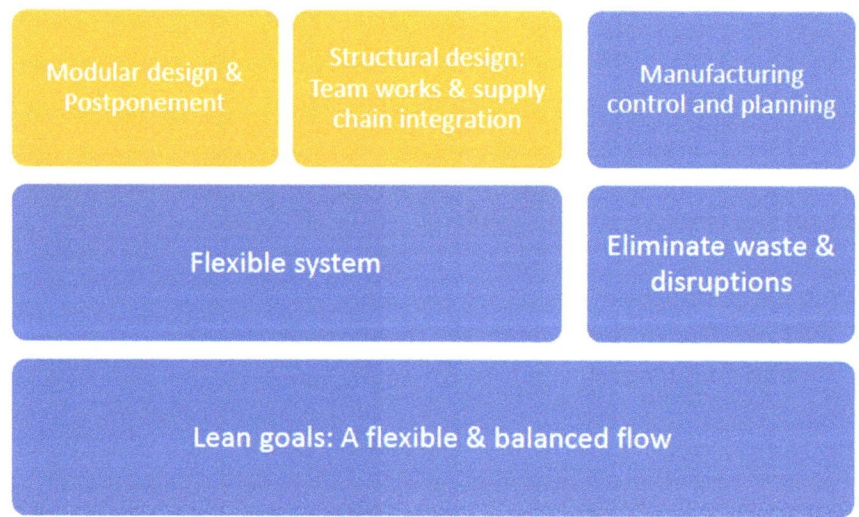

Fig. 9 Lean production:key foundations. *Source* Adapted from Stevenson (2005)

5.2 Lean Technique: Standard Parts and Processes

The use of standard parts and processes means that workers have fewer parts to deal with, lower training times and costs are reduced. These concepts of lean principle can be applied to purchasing, material handling, and checking quality, which are more routine and can be improved on an ongoing basis. The basis for this standardization allows garment manufacturers to follow the same protocol and the ability to use standard processing that will lead to cost reduction. The economy of scale and learning curve theories justify the efficiency gain in these processes.

Fast fashion companies like H&M and Zara all deploy to a different extent the standardization concepts in their design and manufacturing. Sweden's H&M and its business concept dominates the more affordable mass markets. Since its mission is to offer "fashion and quality at the best price", it relies upon large production volume and an efficient logistical system to keep costs low. When designing a product, firms tend to use a common standard such as virtual human body (VHB) standard from the International Organization for Standardization (ISO) specifically used in virtual garment systems in the apparel field, as suggested in ISO/TC 133/WG 2 (Kim et al. 2019). By using the same standard, the design process can produce a large amount of styles, demonstrating that production volume and an efficient logistical system can be competitive when sharing the same fixed costs such as initial investment and prototype development. Encouraging results after implementing these principles will lead the way to reach the critical mass that will allow companies to compete in these markets. Research indicates marvellous benefits of standard operating procedures, including a decrease in sample rejection level by 60%, reduction of work level for repairing works by 70%, eliminating bottleneck results in a reduction of excess motion and non-value added works by 50% (Sivakumar 2015).

Zara, the major fast fashion retailer and brand producer, utilizes process standardization to reduce new product development lead time and enhance the variety of fashion items. The company employs process standardization during fabric manufacturing, which often have a longer lead time than other processes. It happens during several stages including procurement, design and distribution. First, by accumulating undyed fabric procurement, buyer power has increased significantly due to the accumulation of inventory in the warehouse for undyed materials; secondly, the discount can be applied to major procurement processes, and being flexible to meet customer demand. Depending on the final marketing and order details, the company can adapt to different colours by dyeing in own manufacturing facilities (BusinessWire 2002).

Before the standardization process, there were several issues related to quality assurance on the pattern size, shape, and measurement, which eventually affect the quality of the garment. After standardization, firms minimize the quality assurance on the pattern size, shape, and measurement, which will be altered before mass production. Importantly, this process eliminates the non-value added activities (e.g., repeated measurement, duplication in tasks). Since fabric supply is an essential part of any fashion and textile supply chain, process standardization also helps significantly to reduce lead time in the fabric sourcing cycles (Aftab et al. 2017). Zara represents a successful model for fast fashion that offers a quick response to customers and meet a large variety of products due to its process standardization in the design phase of its products through developing standardized design modules (Aftab et al. 2017).

5.3 Lean Technique: Modular Design and Postponement

The American Apparel Manufacturing Association (AAMA) has defined modular manufacturing as "a contained manageable work unit of 5 to 17 people performing a measurable task (Wang et al. 1991). Modules are clusters of parts treated as a single unit. This greatly reduces the number of parts to deal with, simplifying assembly, purchasing, handling, and training. Standardization has the added benefit of reducing the number of different parts contained in the bill of materials (BOM) for various products, thereby simplifying the BOM lists. In a modular system, processes are grouped into a module instead of being divided into their smallest components. Modular design differs from the traditional method of production that relies on individual employees to perform tasks repetitively by using teams of workers to sew and assemble parts or all of a garment. As a rule, fewer numbers of multi-functional operators work on the machines which are arranged in a U-line (Aftab et al. 2017, 2018; Berg et al. 1996). This team-based approach brings several benefits such as reducing costs, enhancing workforce productivity, and the reaction to customer requirements (Taplin 2014).

In operations management, postponing/delaying final processing or manufacturing can be very effective when there are multiple demands (Stevenson 2018). The

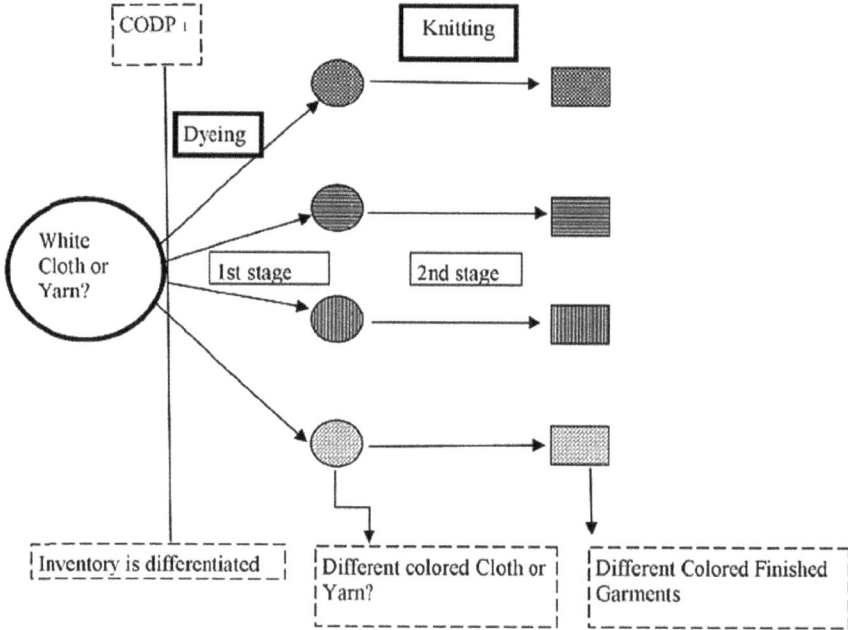

Fig. 10 Lean postponement in dyeing fabrics. *Source* Aftab et al. (2017) (CODP: customer order decoupling point)

lean technique can be used for alignment between operations and business differentiation strategy. When markets require a different variant of the final products, the critical mass of semi-finished inventories guarantees a lower level of production costs. Depending on the details of orders, a differentiated end product in terms of form, labelling, and identity could be finalized to meet each market. The example from famous Italian brand, Benetton, a major apparel manufacturer of sweaters is given here. Using two separate stages (1) dyeing; and (2) knitting, Benetton can hold on to raw materials of white yarns (original) and converted them into yarns of different colours later on. The dyeing operation can be separated into mass production, which has often been outsourced for lower operating costs. Depending on the details of final orders, a smaller amount of the dyed/coloured yarns were knitted into different finished garment products (different styles, colours and sizes). This is often called "make-to-order" which offers alignment to differentiation strategy. Figure 10 provides a schema of the two stages postponement lean techniques in sweater manufacturing.

This concept has been used in Zara and H&M production to enhance some degree of flexibility towards demand. H&M, on one hand, relies upon a similar differentiation of fashion-basic garments, but with low costs derived from largely sourcing production via sub-contractor networks in Asia, most recently in Bangladesh (Taplin 2014). Zara further advances the postponement into design function where modular concepts associated with a highly integrated "customer insight" intelligent unit. This unit or so-called vanilla box designs (Harlé et al. 2002) updates on trends and tastes that can trigger design functions to select the appropriate colour for final delivery. Vanilla box can be applied at the beginning of each selling season, the designers create a library of design modules that can be tested in the market. The concept design modules are delayed until confirmation of new trends and are stored only in electronic format, which postpones prototype and holding inventory costs (Venkatesh and Swaminathan 2004). Applying the standardization process and design can also guarantee cost benefits through economy of scale in designing and manufacturing delays.

6 Conclusions

This chapter highlighted the literature and case studies focusing on the development of lean operations in the fashion and textile industry. The key focus of the chapter is to examine several approaches to enhance lean benefits such as delivery time, reduce various wastes, increase productivity and improve responsiveness to customer requirements to ensure smooth production and distribution of fashion and textile products. Several techniques have been used to reduce waste, minimize disruption and create a balanced flow of products in garment manufacturing, spinning, weaving and fabric chemical processing. A detailed description of implementation lean manufacturing in garment industries has indicated with illustrations from fashion industries. Similarly, the scope of lean manufacturing in various textile manufacturing operations has been discussed in brief. Modular design and postponement have been at the forefront in supporting lean manufacturing. Information from Zara, and H&M confirm the importance of lean thinking in the fashion and textiles industry.

References

Aftab M, Yuanjian Q, Kabir N (2017) Postponement application in the fast fashion supply chain: a review. Int J Bus Manage 12(7). https://doi.org/10.5539/ijbm.v12n7p115

Aftab MA, Yuanjian Q, Kabir N, Barua Z (2018) Super responsive supply chain: the case of Spanish fast fashion retailer Inditex-Zara. Int J Bus Manag 13(5). https://doi.org/10.5539/ijbm.v13n5p212

Akbari M (2018) Logistics outsourcing: a structured literature review. Benchmark Int J 25(5):1548–1580. https://doi.org/10.1108/BIJ-04-2017-0066

Akbari M, Ha N (2020) Impact of additive manufacturing on the Vietnamese transportation industry: an exploratory study. Asian J Ship Logist 36(2):78–88. https://doi.org/10.1016/j.ajsl.2019.11.001

Akbari M, Clarke S, Maleki Far S (2017) Outsourcing best practice-the case of large construction firms in Iran. In: Proceedings of the informing science and information technology education conference, Ho Chi Minh City, Vietnam. http://proceedings.informingscience.org/InSITE2017/InSITE17p039-050Akbari3237.pdf

Akbari M, Ha N, Majo G (2020) Chapter 4: role of logistics service providers in the supply chain. In: Nayak R (ed) Supply chain management and logistics in the global fashion sector: the sustainability challenge, Routledge, London

Alder K (1997) Innovation and amnesia: engineering rationality and the fate of interchangeable parts manufacturing in France. Technol Culture 38(2):273–311. https://www.jstor.org/stable/3107124

Alghamdi H (2016) Toward better understanding of total quality management (TQM). J Bus Econ Policy 3(4):29–37. http://jbepnet.com/journals/Vol_3_No_4_December_2016/5.pdf

Aristova NI, Chadeev VM (2018) A methodology for estimating the benefits of mass production automation. Autom Remote Control 79(2):366–371. https://doi.org/10.1134/S0005117918020157

Berg P, Appelbaum E, Bailey T, Kalleberg AL (1996) The performance effects of modular production in the apparel industry. Ind Relat J Econ Soc 35(3):356–373

Bhamu J, Sangwan KJ (2014) Lean manufacturing: literature review and research issues. Int J Oper Prod Manag 34(7):876–940. https://doi.org/10.1108/IJOPM-08-2012-0315

Bhutta MKS, Rosado-Feger AL, Huq F, Muzaffar A (2013) Exploratory study of adoption of lean management practices in Pakistani textile firms. Int J Ser Oper Manag 15:338–357. https://doi.org/10.1504/IJSOM.2013.054446

Bruce M, Daly L, Towers N (2004) Lean or agile: a solution for supply chain management in the textiles and clothing industry? Int J Oper Prod Manag 24(2):151–170. https://doi.org/10.1108/01443570410514867

BusinessWire (2002) Applied textiles selects V3 systems enterprise supply chain execution. Business Wire

Caldera H, Desha C, Dawes L (2019) Evaluating the enablers and barriers for successful implementation of sustainable business practice in 'lean' SMEs. J Clean Prod 218:575–590. https://doi.org/10.1016/j.jclepro.2019.01.239

Čiarnienė R, Vienažindienė M (2012) Lean manufacturing: theory and practice. Econ Manag 17(2):726–732. https://doi.org/10.5755/j01.em.17.2.2205

Ciccullo F, Pero M, Caridi M, Gosling J, Purvis L (2018) Integrating the environmental and social sustainability pillars into the lean and agile supply chain management paradigms: a literature review and future research directions. J Clean Prod 172:2336–2350. https://doi.org/10.1016/j.jclepro.2017.11.176

Comm CL, Mathaisel DF (2005) An exploratory analysis in applying lean manufacturing to a labor-intensive industry in China. Asia Pac J Mark Logist 17(4):63–80. https://doi.org/10.1108/13555850510672430

Coulson T (1944) The origin of interchangeable parts. J Franklin Inst 238(5):335–344. https://doi.org/10.1016/S0016-0032(44)90044-X

Dave PY (2020) The history of lean manufacturing by the view of Toyota-Ford. Int J Sci Eng Res 11(8):1598–1602. https://www.ijser.org/researchpaper/The-History-of-Lean-Manufacturing-by-the-view-of-Toyota-Ford.pdf

Drucker PF (1999) Knowledge-worker productivity: the biggest challenge. Calif Manage Rev 41(2):79–94

Durand-Sotelo L, Monzon-Moreno M, Chavez-Soriano P, Raymundo-Ibañez C, Dominguez F (2020) Lean production management model under the change management approach to reduce order fulfillment times for Peruvian textile SMEs. In: IOP conference series: materials science and engineering. IOP Publishing, p 012023

Gilbreth FB, Gilbreth LM (1916) The effect of motion study upon the workers. Ann Am Acad Pol Soc Sci 65(1):272–276

Gilmore M, Smith D (1996) Set-up reduction in pharmaceutical manufacturing: an action research study. Int J Oper Prod Manag 16(3):4–17. https://doi.org/10.1108/01443579610110459

Harlé N, Pich M, Van der Heyden L (2002) Marks and Spencer and Zara: process competition in the textile apparel industry. Insead

Hodge GL, Goforth Ross K, Joines JA, Thoney K (2011) Adapting lean manufacturing principles to the textile industry. Prod Plan Control 22:237–247. https://doi.org/10.1080/09537287.2010.498577

Hokoma RA (2010) The current awareness of Just-In-Time techniques within the Libyan textile private industry: a case study. In: International conference on computer, electrical, and systems sciences, and engineering, Amsterdam, Holland

Islam A (2021) Implementation of lean manufacturing in garments factory. https://ordnur.com/apparel/implementation-lean-manufacturing-garments-factory/

Karlsson C, Åhlström P (1995) Change processes towards lean production: the role of the remuneration system. Int J Oper Prod Manag 15(11):80–99. https://doi.org/10.1108/01443579510102918

Kgdas G, Piyanka W, Jayathilake L, Perera H, Gamage J (2012) Challenges and potential impact of applying lean manufacturing techniques to textile knitting industry: a case study of a knitting factory in Sri Lanka. In: Proceedings of 3rd international conference: engineering, project and production management

Kim HS, Choi HE, Park CK, Nam YJ (2019) Standardization of the size and shape of virtual human body for apparel products. Fashion Textiles 6(1):33. https://doi.org/10.1186/s40691-019-0187-z

König A, Spinler S (2016) The effect of logistics outsourcing on the supply chain vulnerability of shippers. Int J Logist Manag 27(1):122–141. https://doi.org/10.1108/IJLM-03-2014-0043

Lee Q, Snyder B (2007) The strategos guide to value stream & process mapping. Enna Products Corporation

Maleki Far S, Akbari M, Clarke S (2017) The effect of IT integration on supply chain agility towards market performance (a proposed study). In: Proceedings of the informing science and information technology education conference. Ho Chi Minh City, Vietnam. https://doi.org/10.28945/3749

Maia LC, Alves AC, Leão CP (2013) Sustainable work environment with lean production in textile and clothing industry. Int J Ind Eng Manag 4:183–190

Maia LC, Alves AC, Leão CP (2019) Implementing lean production to promote textile and clothing industry sustainability. Lean engineering for global development. Springer

Manfredsson P (2016) Textile management enabled by lean thinking: a case study of textile SMEs. Prod Plan Control 27:541–549. https://doi.org/10.1080/09537287.2016.1165299

Mezgebe TT, Asgedom HB, Desta A (2013) Economic analysis of lean wastes: case studies of textile and garment industries in Ethiopia. Int J Acad Res Bus Soc Sci 3:101. https://doi.org/10.6007/IJARBSS/v3-i8/123

Michelsen KE (2020) Industry 4.0 in retrospect and in context. In: Collan M, Michelsen K-E (eds) Technical, economic and societal effects of manufacturing 4.0. Palgrave Macmillan, Cham, Switzerland, pp 1–14. https://doi.org/10.1007/978-3-030-46103-4_1

Monden Y (1994) Toyota production system: an integrated approach to just-in-time, 2nd edn. CRC Press

Moreira F, Alves AC, Sousa RM (2010) Towards eco-efficient lean production systems. In: International conference on information technology for balanced automation systems. Springer, pp 100–108

Mullin JB (2007) Henry Ford and field and factory: an analysis of the Ford sponsored village industries experiment in Michigan, 1918–1941. J Am Plann Assoc 48(4):419–431. https://doi.org/10.1080/01944368208976814

Nayak R (ed) (2020) Supply chain management and logistics in the global fashion sector: the sustainability challenge. Routledge.

Nayak R, Padhye R (eds) (2015a) Garment manufacturing technology. Woodhead Publishing

Nayak R, Padhye R (2015b) Introduction: the apparel industry. In: Garment manufacturing technology. Woodhead Publishing, pp 1–17

Nayak R, Padhye R (2018) Introduction to automation in garment manufacturing. In: Automation in garment manufacturing. Woodhead Publishing, pp 1–27

Nayak R, Akbari M, Maleki Far S (2019) Recent sustainable trends in Vietnam's fashion supply chain. J Clean Prod 225:291–303. https://doi.org/10.1016/j.jclepro.2019.03.239

Nayak R, Panwar T, Nguyen LVT (2020a) Sustainability in fashion and textiles: a survey from developing country. Sustainable technologies for fashion and textiles, pp 3–30

Nayak R, Nguyen LT, Panwar T, Ulhaq I, George M (2020b) Standards, organizations and lean concept in managing sustainable fashion supply chains. Supply chain management and logistics in the global fashion sector, pp 183–215

Paneru N (2011) Implementation of lean manufacturing tools in garment manufacturing process focusing sewing section of Men's Shirt. https://www.theseus.fi/bitstream/handle/10024/34405/Paneru_Naresh.pdf?sequence=1&isAllowed=y

Raj D, Ma YJ, Gam HJ, Banning J (2017) Implementation of lean production and environmental sustainability in the Indian apparel manufacturing industry: a way to reach the triple bottom line. Int J Fashion Design Technol Edu 10:254–264. https://doi.org/10.1080/17543266.2017.1280091

Saleeshya P, Raghuram P, Vamsi N (2012) Lean manufacturing practices in textile industries–a case study. Int J Collaborative Enterp 3:18–37. https://doi.org/10.1504/IJCENT.2012.052367

Sanchez AM, Pérez MP (2001) Lean indicators and manufacturing strategies. Int J Oper Prod Manag 21(11):1433–1452. https://doi.org/10.1108/01443570110407436

Sawyer JE (1954) The social basis of the American system of manufacturing. J Econ History 14(4):361–379. https://www.jstor.org/stable/2114247

Shah ZA, Hussain H (2016) An investigation of lean manufacturing implementation in textile sector of Pakistan. In: Proceedings of the 2016 international conference on industrial engineering and operations management, pp 8–10.

Silva N, Perera H, Samarasinghe D (2011) Factors affecting successful implementation of lean manufacturing tools and techniques in the apparel industry in Sri Lanka. https://doi.org/10.2139/ssrn.1824419

Sivakumar P (2015) Standardization of apparel manufacturing industry focusing on "cutting section". Textile and finance, standardization of apparel manufacturing industry 1–11

Stevenson WJ (2005) Operations management. McGraw-hill

Stevenson WJ (2018) Operations management, 13th edn. McGraw-Hill Education, New York, NY, 605. https://www.worldcat.org/title/operations-management/oclc/958356980

Sugimori Y, Kusunoki K, Cho F, Uchikawa S (1997) Toyota production system and Kanban system Materialization of just-in-time and respect-for-human system. Int J Prod Res 15(6):553–564. https://doi.org/10.1080/00207547708943149

Taplin IM (2014) Global commodity chains and fast fashion: how the apparel industry continues to re-invent itself. Compet Chang 18(3):246–264

Taylor FW (2004) Scientific management. Routledge

Thakkar A, Joseph S (2015) Lean manufacturing in apparel industry. https://www.academia.edu/15688560/Lean_Management_in_Apparel_Manufacturing

United States Environmental Protection Agency (2018) Create space independent publishing platform

Venkatesh S, Swaminathan JM (2004) Managing product variety through postponement: concept and applications. In: The practice of supply chain management: where theory and application converge. Springer, pp 139–155

Vitasek K, Manrodt K, Abbott J (2005) What makes a lean supply chain. Supply Chain Manag Rev 9 (7):39–45. https://trid.trb.org/view/774087

Wahab ANA, Mukhtar M, Sulaiman R (2013) A conceptual model of lean manufacturing dimensions. Procedia Technol 11:1292–1298. https://doi.org/10.1016/j.protcy.2013.12.327

Wang J, Schroer BJ, Ziemke MC (1991) Understanding modular manufacturing in the apparel industry using simulation

Womack JP, Roos D, Jones DT (1990) The machine that changed the world: the story of lean production—Toyota's secret weapon in the global car wars that is now revolutionizing world industry. Simon and Schuster

Woodbury RS (1960) The legend of Eli Whitney and interchangeable parts. Technol Culture 1(3):235–253. https://www.jstor.org/stable/3101392

Yam M (1994) The Hong Kong garment industry. In: The 90's: survival of small factories through the 'just-in-time' route. Textile Asia XXV(June):110–118

Standardized Work in Fashion Industry

Ashish Bhardwaj

Abstract There are standard operating procedures to perform many jobs in fashion and textile industries. The industrial engineering department is responsible to set the standard methods and standard time for manufacturing garments. Time and motion study are used to establish standard time and standard methods, respectively. The establishment of standard time and standard methods of doing jobs reduces wastage in manufacturing processes. This chapters discusses the details of time and motion study, which are used in garment industries to establish standard time and standard methods. The fundamental principles relating to time and motion study in garment manufacturing has been covered in this chapter. This chapter also discusses about the terminologies such as productivity, efficiency, cycle time, changeover time, work balancing, time measurement, economy of motion and time management. Various causes of idle time and bottlenecking are also covered in this cheaper.

Keywords Time study · Motion study · Idle time · Productivity and efficiency

1 Introduction

In apparel manufacturing process, it is essential to have standard methods and standard time for completing any work. Time study and motion study are done by the industrial engineering section in a garment industry to establish these standards (Bedi 2016; Bhattacharya 2010; Chary 2019). For meeting the exact delivery date, time and motion study are very essential tool. As the production of specific style is performed by different operators, it is essential to know the operating efficiency of individual workers (Norman 2010; Paneerselvam 2012). Hence, it is essential to have standard methods of doing the jobs. Similarly, in order to calculate the productivity (i.e., number of pieces of garments produces in a unit time), it is essential to know the standard time for each garment. In order to set the standard time (SAM-standard allowed minutes), time and motion study are mandatory (Barnes 1980).

A. Bhardwaj (✉)
Fashion & Apparel Engineering, TIT & S, Bhiwani, India
e-mail: ashishbha21@gmail.com

Time study and motion study can be used to estimate the lead time, which is essential to calculate the exact delivery date of products (Mundel 1978; Niebel 1976). Time study and motion study can be used to reduce the events of idle time and maximize the productivity. Without any standard time and standard method, the manufacturing process will be encountering excessive waste. Hence, time study and motion study can be used to eliminate process wastes, which is the fundamentals of lean manufacturing. This chapter discusses the methods of doing the time and motion study in apparel manufacturing.

Similarly, productivity has become a buzzword in recent years (Bedi 2016). It is critical for the success of industrial firms as well as the country's economic development. High productivity refers to completing tasks in the quickest time possible with the least number of inputs, without losing quality, and with the least amount of waste.

Work-study is the foundation for creating a work system (Curie 1977). The goal of work design is to figure out how to get the job done in the most efficient way possible. This work-study aims to improve existing and planned work methods as well as develop work performance standards. Method study and work measurement are two methodologies that fall under the umbrella of work-study.

> Method study is the systematic recording and critical assessment of existing and planned work techniques in order to develop and use easier and more effective methods while lowering costs.
>
> Work measurement is the application or techniques designed to establish the time for a qualified worker to carry out a specified job at a defined level or performance.

2 Productivity

Productivity is the quantitative relationship between what we produce and the resources we utilise to produce it, i.e., the arithmetic ratio of output to resources (input). Productivity can be expressed in the following way (Bhattacharya 2010):

$$\text{Productivity} = \frac{\text{Output}}{\text{Input}}$$

The efficiency of a manufacturing system is known as productivity. It is the principle that governs the production system's management. It's a measure of how well the production factors (land, capital, labour, and energy) are used.

2.1 Factors Influencing Productivity

2.1.1 Controllable (or Internal) and Uncontrollable (or External) Factors Influencing Productivity (Chary 2019)

(A) **Controllable (or Internal) Factors**

1. **Product factor**: In terms of productivity, the utility of a product is used to determine how well it meets production needs. A product's cost benefit factor can be improved by either increasing the benefit at the same cost or lowering the cost for the same benefit.
2. **Plant and equipment**: Plant and equipment play a significant role in increasing productivity. Productivity rises as the plant's availability rises as a result of adequate maintenance and a reduction in idle time. Productivity may be improved by paying attention to factors such as use, age, modernization, cost, and investment (Fig. 1).
3. **Technology**: Productivity is increased to a higher level when using innovative and cutting-edge technology. Material handling, storage, communication systems, and quality control all benefit from automation and information technology. The following are some of the technology factors to think about:
   (i) Plant size and capacity,
   (ii) On-time input supply and quality,
   (iii) Production planning and control,
   (iv) Repairs and maintenance,
   (v) Waste reduction, and
   (vi) Efficient material handling system.

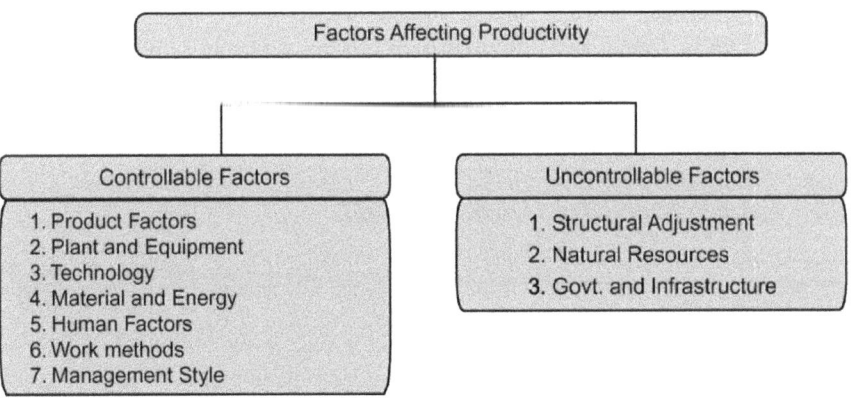

Fig. 1 Factors influencing productivity

4. **Material and energy**: Efforts to reduce material and energy usage result in significant productivity gains.
   (a) Selection of quality material and right material.
   (b) Control of wastage and scrap.
   (c) Effective stock control.
   (d) Development of sources of supply.
   (e) Optimum energy utilisation and energy savings.
5. **Human factors**: Human competence and skill are the foundations of productivity. Employees' ability to work efficiently is influenced by a variety of elements such as their education, training, experience, aptitude, and so on. Employees' motivation has an impact on productivity.
6. **Work methods**: Improving work methods increases productivity. Work study, industrial engineering approaches, and training are examples of areas where work methods can be improved, resulting in increased production.
7. **Management style**: This has an impact on the organization's design, communication, policies, and processes. A more adaptable and dynamic management style is a better way to boost production.

(B) **Un-controllable (or External) Factors**

1. **Structural adjustments**: Economic and social changes are both part of structural adjustments. The following are significant economic changes:
   (a) Shift in employment from agriculture to manufacturing industry,
   (b) Import of technology, and
   (c) Industrial competitiveness.

   Women's engagement in the labour force, education, cultural values, and attitudes are some of the elements that have a substantial impact on productivity increase.
2. **Natural resources**: Manpower, land, and raw material resources are all important factors in increasing productivity.
3. **Government and infrastructure**: Government policies and programmes have a large impact on government productivity practises, transportation and communication power, and fiscal policies (e.g., interest rates, taxes).

2.2 Total Productivity Measure (TPM)

It's based on the tangible output and input data. The model can be used by any manufacturing or service enterprise (Bedi 2016; Norman 2010).

$$\text{Total productivity} = \frac{\text{Total tangible output}}{\text{Total tangible input}}$$

$$\text{Total tangible output} = \text{Value of finished goods produced}$$
$$+ \text{Value of partial units produced}$$
$$+ \text{Dividends from securities}$$
$$+ \text{Interest} + \text{Other income}$$

$$\text{Total tangible input} = \text{Value of (human} + \text{material} + \text{capital} + \text{energy}$$
$$+ \text{other inputs) used.}$$

Both the firm's output and input must be expressed in the same measuring unit. The best method to express them is in terms of volume or monetary value.

2.3 Partial Productivity Measures (PPM)

Partial productivity measures are expressed as follows, depending on the particular input (Bhattacharya 2010):

$$\text{Partial productivity} = \frac{\text{Total output}}{\text{Individual input}}$$

1. Labour productivity $= \frac{\text{Total output}}{\text{Labour input}}$
   Labour input is measured in terms of man-hours.
2. Capital productivity $= \frac{\text{Total output}}{\text{Capital input}}$
3. Material productivity $= \frac{\text{Total output}}{\text{Material input}}$
4. Energy productivity $= \frac{\text{Total output}}{\text{Energy input}}$

One of the fundamental drawbacks of partial productivity metrics is that they place too much focus on one input item, causing other inputs to be undervalued or disregarded.

2.4 Productivity Improvement Techniques

(A) **Technology Based**
   1. **Computer Aided Design (CAD), Computer Aided Manufacturing (CAM), and Computer Integrated Manufacturing Systems (CIMS):** CAD stands for computer-aided design of products, processes, or systems. The influence of CAD on human productivity is important because of the following benefits:
      (a) The speed with which various designs are evaluated,
      (b) Reduction of the danger of malfunctioning, and
      (c) Error reduction (Fig. 2).

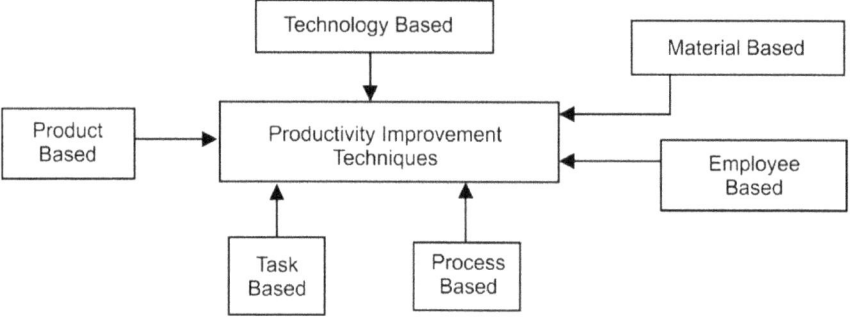

Fig. 2 Technology based productivity improvement techniques

>CAM is extremely useful for industrial design and control. Line balancing aids in improving the efficiency of the production system.
>(a) Production Planning and Control
>(b) Capacity Requirements Planning (CRP), Manufacturing Resources Planning II (MRP II) and Materials Requirement Planning (MRP)
>(c) Automated Inspection.
>
>CIMS involves automatic line balancing, machine loading (scheduling and sequencing), automatic inventory control, and inspection, in addition to the techniques such as:
>(a) Robotics
>(b) Laser technology
>(c) Modern maintenance techniques
>(d) Energy technology
>(e) Flexible Manufacturing System (FMS).

(B) Employee Based

1. Individual and group financial and non-financial incentives
2. Employee advancement.
3. Job creation, job expansion, job enrichment, and job rotation
4. Worker participation in decision-making
5. Quality Circles (QC), Small Group Activities (SGA)
6. Personal development.

(C) Material Based

1. Material control and planning
2. Logistics and purchasing
3. Storage and retrieval of materials
4. Quality material purchase and source selection
5. Elimination of waste.

(D) Process Based

1. Methods engineering and work simplification
2. Job design, job evaluation, and job safety
3. Human factors engineering.

(E) Product Based

1. Value analysis and value engineering
2. Product diversification
3. Standardisation and simplification
4. Reliability engineering
5. Product mix and promotion.

(F) Task Based

1. Management style
2. Communication in the organisation
3. Work culture
4. Motivation
5. Promotion group activity.

2.5 Work Study

"Work study" is a broad phrase that encompasses all methodologies, method studies, and work measuring methods used to examine human work in all of its forms (Jhamb 1988; David 1980). Work study leads to a methodical analysis of all elements affecting the efficiency and economy of the situation under consideration in order to improve (Fig. 3). Work study can help to achieve the objectives of lean manufacturing as it is based on the elimination of unnecessary motion.

Work study is a technique for increasing a company's production efficiency (productivity) by eliminating waste and superfluous procedures. It's a method for identifying non-value-adding actions by looking into all the variables that affect the job. It is the only method for establishing time standards that is both accurate and systematic. It will add to profit because the savings will begin immediately and continue throughout the product's lifespan.

Work study includes method study and work measurement. Motion study is a component of 'Method study', and work measurement is often known as 'Time study'.

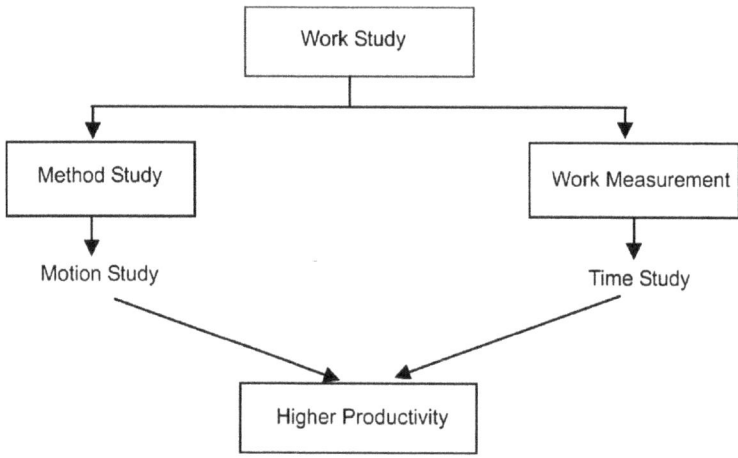

Fig. 3 Framework of work study (Jhamb 1988)

2.6 Advantages of Work Study

The following are some of the benefits of work study:

1. It aids in maintaining a steady production flow with minimal disruptions.
2. It aids in the reduction of product costs by removing waste and superfluous activities.
3. Improved management-worker interactions.
4. Meets the delivery deadline.
5. Reduced rejects and scrap, as well as better utilisation of the organization's resources.
6. Assists in the improvement of working conditions.
7. A more functional workplace layout.
8. Contributes to the standardisation and simplification of existing processes or methods.
9. Assists in determining the typical time for an operation or job, which is useful in workforce and production planning.

3 Method Study

The industrial engineer can use method study to analyse each operation in detail. The fundamental goal of method study is to minimise needless operations and find the optimal way to complete the task (Adam and Ebert 1978; Riggs 1987; Geneva Indian Adaptation International Labour Office 2015).

Method study is also known as job design or method engineering. The term "method engineering" refers to a collection and analysis of approaches aimed at increasing the efficiency of individuals and machines.

"Method study is the systematic recording and critical assessment of existing and proposed ways of performing work as a means of developing and applying easier and more effective techniques to decrease cost," according to the British Standards Institution (BS 3138). The separation of an operation or procedure into its component pieces and their systematic investigation is what method study is all about. It is critical to have the correct mental attitude when conducting the method study. The following points or ability should be considered while selecting operators for method study.

1. The ambition and tenacity of the operator to get things done.
2. The ability of the operator to produce outcomes.
3. Operator's awareness of the variables at play.

The goal of the method study is to improve work methods by analysing processes and operations, such as:

1. Manufacturing operations and their sequence.
2. Workmen.
3. Supplies, tools, and gauges.
4. Physical facility layout and workstation design.
5. Man-to-man movement and material handling.
6. The working atmosphere.

3.1 Objectives of Method Study

The major objective of method study is to final alternative ways to do a job. The following are the detailed objectives of method study techniques:

1. Present and analyze accurate information about the situation.
2. To seriously evaluate the facts.
3. To come up with the best feasible solution under the circumstances based on a careful evaluation of the facts.

Finding better methods to accomplish a job is at the heart of method study. By eliminating superfluous activities, avoidable delays, and other forms of waste, it creates value and improves efficiency (Jhamb 1988). The increase in efficiency is achieved through the following methods:

1. Improved workplace layout and design.
2. More efficient and improved work practises.
3. Efficient use of people, machines, and materials.
4. A better final product design or specification.

3.2 Scope of Method Study

The scope of method study isn't limited to the manufacturing sector. Method study techniques can also be used effectively in the service industry. It can be used in places like offices, hospitals, banks, and other service businesses. Method study can be used successfully in manufacturing sectors in the following areas:

1. To improve work processes and procedures.
2. To figure out the appropriate order in which to do tasks.
3. Improve layout and smooth material flow with minimal backtracking.
4. To improve working conditions and, as a result, labour productivity.
5. To break up the monotony of the job.
6. To increase the efficiency of plant and material utilisation.
7. Waste and inefficient activities are eliminated.
8. To lower production costs by shortening operation cycle times.

3.3 Steps Involved in Method Study

The eight steps that make up the fundamental technique for method study are as follows. Figure 4 shows the entire procedure for conducting the method study (Niebel 1976).

1. DEFINE the boundaries of the work to be studied.
2. DIRECTLY OBSERVE the key facts regarding the job and collect any extra data that may be required from acceptable sources.
3. EXAMINE how the work is done and question its goal, sequence of events, and technique of execution.
4. DEVELOP the most feasible, cost-efficient, and effective strategy by incorporating input from all parties involved.
5. EVALUATE several options for developing a new enhanced method by comparing the cost-effectiveness of the selected new method to the current method of performance.
6. DEFINE the new approach in a clear manner and present it to all stakeholders, including management, supervisors, and workers.
7. IMPLEMENT the new procedure as a normal practise and train the people who will be using it.
8. MAINTAIN the new approach and introduce control procedures to avoid returning to the old approach.

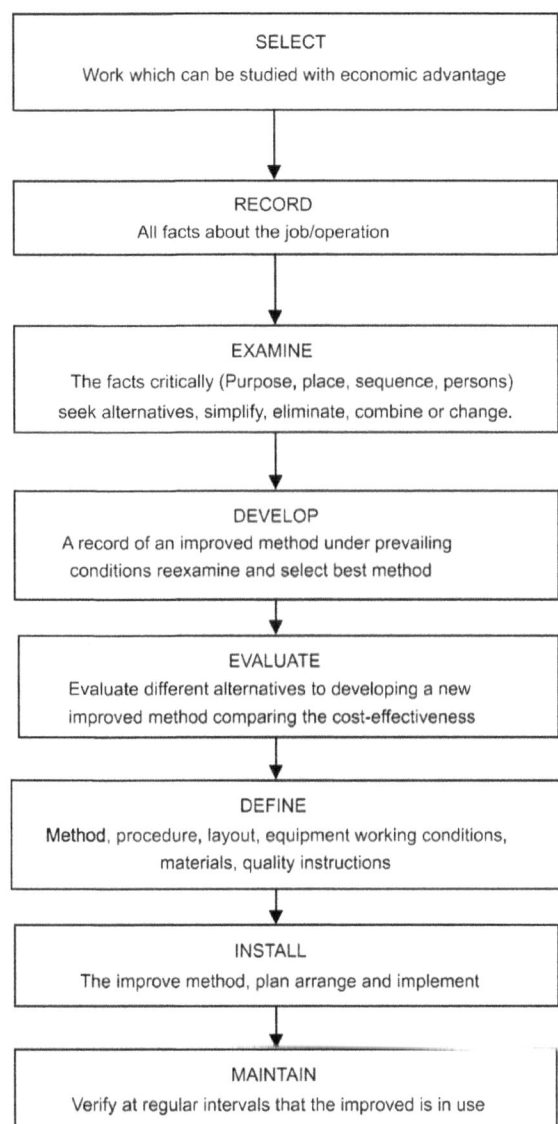

Fig. 4 Method study procedure (Jhamb 1988)

3.4 Selection of the Job for Method Study

For method study, cost is the most important criterion for selecting a job, process, and department. A job is chosen for the method study so that the suggested technique accomplishes one or more of the following outcomes:

(a) Improvement in quality with lesser scrap.
(b) Increased production through better utilisation of resources.

(c) Elimination of unnecessary operations and movements.
(d) Improved layout leading to smooth flow of material and a balanced production line.
(e) Improved working conditions.

3.5 Considerations for Selection of Method Study

The job for the method study should be chosen based on the following factors (Barnes 1980; Mundel 1978; Niebel 1976):

A. Economic aspects,
B. Technical aspects, and
C. Human aspects.

A. **Economic Aspects**

Cost and time are the important factors in the economic aspects of method study. If adequate results are not obtained, the entire operation will be a waste of time. As a result, the money spent should be justified by the money saved. The guidelines below can be utilised to help in selecting a job:

(a) Bottleneck operations which are holding up other production operations.
(b) Operations involving excessive labour.
(c) Operations producing lot of scrap or defectives.
(d) Operations having poor utilisation of resources.
(e) Backtracking of materials and excessive movement of materials.

B. **Technical Aspects**

The method study man should be careful enough to select a job in which he has the necessary technical knowledge and expertise. A person selecting a job in his area of expertise is going to do full justice. Other factors which favour selection in technical aspect are:

1. Job having inconsistent quality.
2. Operations generating lot of scraps.
3. Frequent complaints from workers regarding the job.

C. **Human Aspects**

Method study is a change because it will influence the way the job is currently done and is not totally supported by the workers and the union. In method study, human factors are crucial. Here are some examples of circumstances when the human aspects should be prioritised:

1. Workers complaining about unnecessary and tiring work.
2. More frequency of accidents.
3. Inconsistent earning.

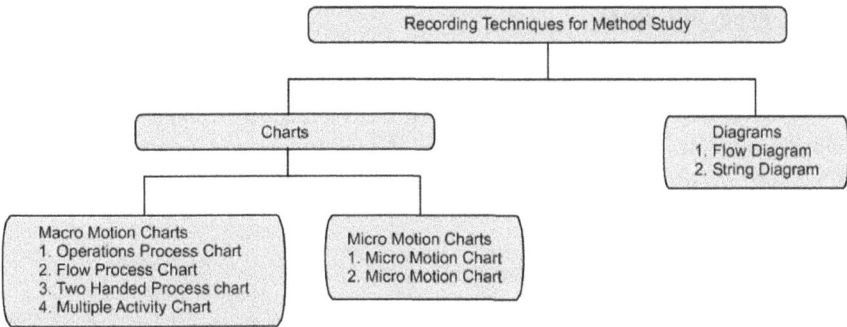

Fig. 5 Recording techniques for method study

3.6 Recording Techniques for Method Study

Following the selection of the task to be analysed, the next step in the fundamental technique is to capture all details pertaining to the existing method (Jhamb 1988). It is crucial to have some way of recording all of the necessary details about the existing approach in order to visualise the actions chosen for inquiry in their totality and to improve them through subsequent critical analysis. Records are extremely useful for comparing the effectiveness of the proposed better procedure before and after. The recording techniques are intended to make the process of recording easier and more consistent. Charts and diagrams are utilised to do this (Fig. 5), which are discussed in the following section.

3.7 Charts Used in Method Study

Use of charts is the most common way to keep track of information. Method study charts are used to record the activities that make up the jobs (Curie 1977). The charts must be prepared with considerable care so that they are accurate. The data it displays should be simple to comprehend and recognise. These charts are used to track the movement of an operator or a piece of work (i.e., in motion study). The chart should have the following information.

(a) Adequate description of the activities.
(b) Whether the chart is for present or proposed method.
(c) Specific reference to when the activities will begin and end.
(d) Time and distance scales used wherever necessary.
(e) The date of chart preparation and the name of the person who does the chart.

3.7.1 Types of Charts

It is separated into two types: (A) macro motion charts and (B) micro motion charts.

For macro motion analysis, macro motion charts are utilised, and for micro motion analysis, micro motion charts are employed. A macro motion study may be measured with a stop watch, however, a micro motion study cannot be done with a stop watch.

(A) Macro Motion Charts

Following four charts are used under this type:

1. Operation Process Chart

It's also known as an outline process diagram (Niebel 1976; Curie 1977). By capturing only the major operations and inspections involved in the process, an operation process chart provides a bird's eye view of the entire process. Only two symbols are used in the operation process chart: operation and inspection. The operation process chart is used for:

(a) Visualise the complete sequence of the operations and inspections in the process.
(b) Know where the operation selected for detailed study fits into the entire process.
(c) In operation process chart, the graphic representation of the points at which materials are introduced into the process and what operations and inspections are carried on them are shown.

2. Flow Process Chart

A flow process chart depicts the flow of work for a product or a portion of it via a work centre or department, with events represented by appropriate symbols. It is an enlargement of the operation process chart, which depicts operations such as inspection, storage, delay, and transportation. Process charts, on the other hand, are divided into three categories:

- Material type: Which shows the events that occur to the materials.
- Man type: Activities performed by the man.
- Equipment type: How an equipment is used.

The flow process chart is useful:

(a) To reduce the distance travelled by men (or materials).
(b) To avoid waiting time and unnecessary delays.
(c) To reduce the cycle time by combining or eliminating operations.
(d) To fix up the sequence of operations.
(e) To relocate the inspection stages.

Flow process charts, like operation process charts, are made up of symbols that are stacked one on top of the other according to the order in which the activities occur and are connected by a vertical line. On the right hand side of the activity symbol, a brief description of the activity is printed, and on the left hand side, time or distance is supplied.

3. **Two Handed Process Chart**

The most complex sort of flow chart is a two-handed (operator process chart), in which the activities of the employees' hands are recorded in connection to one another. The two-handed process chart is often limited to tasks done at a single location. This also provides a graphical picture of the worker's manual tasks in a synchronised manner. The following are examples of how these charts might be used:

- To visualise the complete sequence of activities in a repetitive task.
- To study the work station layout.

4. **Multiple Activity Chart**

It's a graph in which the activities of multiple subjects (worker or equipment) are recorded on a similar time scale to demonstrate their interdependence. A multi-activity chart is created:

- To study idle time of the men and machines,
- To determine number of machines handled by one operator, and
- To determine number of operators required in teamwork to perform the given job.

(B) **Micro Motion Charts**

Micro motion chats are used for short duration and repetitive works. The motions involved in the operations require very small time and it is rather difficult to measure by the stop watch. However, the time needed by these micro motions cannot be neglected due to repetitive nature of the job. As the micro motion charts are not used widely in time study, they are not discussed in detail here.

3.8 Diagrams Used in Method Study

The flow process chart depicts the sequence and nature of movement but not the path of those movements. There are frequently undesired qualities in movement paths, such as congestion, backtracking, and needless long movements. A representation of the working area in the form of flow diagrams or string diagrams can be produced to record these needless features:

1. To study the different layout plans and thereby; select the most optimal layout.
2. To study traffic and frequency over different routes of the plant.
3. Identification of back tracking and obstacles during movements.

Diagrams are of two types: 1. Flow diagram and 2. String diagram.

1. Flow Diagram

A flow diagram is a depiction of a working area that shows the location of numerous activities that are defined by numbered symbols and are related with a specific flow process chart that is either man-made or machine-made.

The routes taken in transportation are depicted by connecting the symbols in a sequence with a line that symbolises the subject's journey or movement as closely as possible. Following are the procedures to make the flow diagram:

1. The layout of the workplace is drawn to scale.
2. Relative positions of the machine tools, work benches, storage, and inspection benches are marked on the scale.
3. Path followed by the subject under study is tracked by drawing lines.
4. Each movement is serially numbered and indicated by arrow for direction.
5. Different colours are used to denote different types of movements.

2. String Diagram

A string diagram is a scale layout drawing in which the length of a string is used to record the extent and pattern of movement of a worker operating inside a constrained space over a period of time (Jhamb 1988). A string diagram's main purpose is to provide a record of an existing set of conditions so that determining what is actually happening is as simple as feasible.

One of the most useful characteristics of the string diagram is the ability to compute the real distance travelled during the study session by matching the length of the thread used to the drawing scale. As a result, it makes easier to compare different layouts or methods of conducting work in terms of the amount of travel required.

The fundamental advantage of a string diagram over a flow diagram is that it may easily indicate respective motions between work stations that are harder to trace on a flow diagram. Following are the procedures to draw string diagram:

1. A layout of the work place of factory is drawn to scale on the soft board.
2. Pins are fixed into boards to mark the locations of work stations, pins are also driven at the turning points of the routes.
3. A measured length of the thread is taken to trace the movements (path).
4. The distance covered by the object is obtained by measuring the remaining part of the thread and subtracting it from original length.

3.9 Symbols Used in Method Study

Gilberth invented the graphical style of recording in order to portray data in a clear and unambiguous manner so that they could be grasped quickly and easily. Symbols can be used instead of written descriptions. The following symbols are used in method study.

O OPERATION
☐ INSPECTION
→ TRANSPORTATION
D DELAY
∇ STORAGE

Operation O

When one or more of an object's attributes are altered purposefully, it is called an operation (physical or chemical). This denotes the most important steps in a technique, operation, or process.

An operation advances the work one step closer to completion. Turning, drilling, and milling are some examples of operations. Some other examples include:

- The result of a chemical reaction.
- Welding, brazing, and riveting are operations that can be used in method study.
- Lifting, loading, and unloading are also considered as operations.
- Obtaining supervisory directives.
- Taking dictation is a task that requires a lot of concentration.

Inspection □

An inspection occurs when an object is examined and compared with standard for quality and quantity. The inspection examples are:

- Visual observations for finish.
- Counting quantity of incoming materials.
- Checking the dimensions.

Transportation →

- The transfer of workers, materials, or equipment from one location to another is referred to as transportation.
- Material movement from one workstation to another.
- Worker movement from one to the other place to collect tools and materials.

Delay D: (Temporary Storage)

- When the next planned operation does not happen right away, there is a delay in the process.

Storage ▽

- When an object is maintained in an authorised custody and is safeguarded from unauthorised removal, it is said to be in storage.

4 Motion Study

Motion study is part of method study where analysis of the motion of an operator or work will be studied by following the prescribed methods.

4.1 Principles of Motion Study

A number of principles relating to the economy of movements have been developed as a consequence of experience and serve as the foundation for the creation of enhanced workplace procedures (Mundel 1978; Niebel 1976). These were first utilised by Frank Gilbreth, the father of motion analysis, and were later modified and amplified by Barnes, Maynard, and others. The principles are categorised into three categories such as:

(a) Uses of human body.
(b) Arrangement of the workplace.
(c) Design of tools and equipment.

(A) **Uses of Human Body**

1. Both hands should start and finish actions at the same time.
2. Except during intervals of relaxation, the two hands should not be inactive at the same time.
3. Arm motions should be made at the same time.
4. Hand and body actions should be performed at the lowest rating that allows the work to be completed satisfactorily.
5. Momentum should be used to aid the worker, but it should be kept to a minimum if muscular effort is required to resist it.
6. Continuous curving movements are preferred than straight line motions with abrupt direction changes.
7. 'Ballistic' (or free-swinging) movements are quicker, easier, and more precise than confined or regulated movements.
8. Rhythm is necessary for the seamless and automatic execution of a repetitive task. Wherever possible, the task should be organised to allow for an easy and natural rhythm.
9. Work should be arranged so that eye movements are confined to a comfortable area, without the need for frequent changes of focus.

(B) **Arrangement of the Workplace**

1. To encourage habit formation, all tools and supplies should be assigned to specific and fixed locations.
2. To save time in searching, tools and materials should be pre-positioned.
3. Materials should be delivered as close to the site of usage as feasible using gravity fed bins and containers.
4. Tools, supplies, and controls should be kept as close to the worker as practicable within a maximum working space.
5. Materials and tools should be placed to allow for the most efficient motion sequence.
6. Whenever possible, drop deliveries or ejectors should be employed to eliminate the need for the worker to dispose of finished parts with hands.
7. Adequate lighting should be given, as well as a chair with adjustable height that allows for healthy posture. Workplace and seat heights should be adjusted to allow for alternate standing and sitting.

(C) **Design of Tools and Equipment**
1. The colour of the workplace should be different from the colour of the job to reduce eye strain.
2. Wherever possible a jig, fixture, or foot-operated device may be used to 'hold' the work piece, the hands should be relieved of all effort of 'holding' the work piece.
3. Wherever practical, two or more tools should be merged.
4. When each finger performs a certain movement, such as typing, the burden should be divided according to the fingers' natural capacities.
5. Handles on screwdrivers and cranks should be constructed to allow the maximum amount of surface area of the hand to come into contact with the handle.
6. The position of the levers, crossbars, and wheel bars should be such that the operator can handle them with the least amount of body movement and the greatest mechanical advantage.

5 Work Measurement

'Time study' is another word for work measurement. Measurement of work is critical for both planning and controlling of operations. We can't estimate delivery schedules or costs without the work measurement data, and we can't calculate the capacity of facilities. We are unable to estimate the rate of output, as well as labour utilisation and efficiency. It's highly likely that incentive schemes and standard costs for budget control will be rather difficult to implement.

5.1 Objectives of Work Measurement

Only a small portion of work measurement's entire application is used as a basis for incentives. Work measurement's goals are to establish a solid foundation for:

1. Weighing the pros and cons of various approaches.
2. Determining the proper personnel skill requirements (manpower requirement planning).
3. Planning and control.
4. Costing that is realistic.
5. Incentive compensation plans.
6. The date on which the products will be delivered.
7. Cost-cutting and cost-control measures.
8. Identifying workers who aren't up to par.
9. Educating new employees.

5.2 Techniques of Work Measurement

For the purpose of work measurement, work can be regarded as one of the two categories as mentioned below (Jhamb 1988):

1. **Repetitive work**: The sort of work in which the principal operation or combination of processes repeats itself over and over again during the duration of the job. These rules apply to work cycles that are only a few minutes long.
2. **Non-repetitive work**: It covers several types of maintenance and construction work, in which the work cycle is rarely repeated in exactly the same way.

Various techniques of work measurement are:

1. Time study (stop watch technique),
2. Synthetic data,
3. Work sampling,
4. Predetermined motion and time study (PMTS), and
5. Analytical estimating.

Out of the five techniques, time study and work sampling involve direct observation and the remaining are data based and analytical in nature.

1. **Time study**: A work measurement approach for documenting the times and rates of work for the elements of a certain task performed under specific conditions and analysing the data to calculate the time required to complete the job at the defined level of performance. In other words, time study refers to the practise of measuring time with a stopwatch (discussed in Sect. 6).
2. **Synthetic data**: A work measurement technique that totals element times gathered previously from time studies on other tasks containing the elements in question or from synthetic data to build up the time for a job or parts of the job at a predetermined level of performance.
3. **Work sampling**: A technique in which a large number of observations of one or more machines, processes, or workers are made over a period of time. Each observation captures what is happening at that precise moment, and the percentage of observations recorded for a specific activity, or delay, is a measure of the percentage of time that those activities are delayed.
4. **Predetermined motion time study (PMTS)**: A time-based work measurement technique in which times for basic human motions (classified according to the nature of the motion and the conditions under which it is performed) are used to build up the time for a job at a predetermined level of performance. Methods Time Measurement (MTM) is the most often utilised PMTS.
5. **Analytical estimating**: A work measurement technique in which the time required to complete pieces of a job at a set level of performance is estimated using a combination of knowledge and practical experience with the elements in question, as well as synthetic data.

The work measurement techniques and their applications are shown in Table 1.

Standardized Work in Fashion Industry

Table 1 Work measurement techniques and their applications (Barnes 1980; Mundel 1978)

Techniques	Applications	Units of measurement
1. Time study	Short cycle repetitive jobs. Widely used for direct work	Centi minute (0.01 min)
2. Synthetic data	Short cycle repetitive jobs	Centi minutes
3. Work sampling	Long cycle jobs/heterogeneous operations	Minutes
4. PMTS	Manual operations confined to one work centre	TMU (1 TMU = 0.006 min)
5. Analytical estimating	Short cycle non-repetitive job	Minutes

5.3 Time Study

Work measurement is another name for time study. It is required for both operational planning and control. Time study is defined as "the application of methodologies aimed to establish the time for a competent worker to carry out a specific job at a set level of performance," according to the British Standard Institution.

5.3.1 Steps involved in Time Study

The simplest technique for determining precise time standards is stop watch time. They are cost-effective for repetitive tasks. The following are the steps in conducting a time study:

1. Select the work to be studied.
2. Obtain and record all the information available about the job, the operator and the working conditions likely to affect the time study.
3. Breakdown the operation into elements. An element is a instinct part of a specified activity composed of one or more fundamental motions selected for convenience of observation and timing.
4. Measure the time (by means of a stop watch) taken by the operator to perform each element of the operation. Either continuous method or snap back method of timing could be used.
5. At the same time, assess the operator's effective speed of work relative to the observer's concept of 'normal' speed. This is called performance rating.
6. Adjust the observed time by rating factor to obtain normal time for each element

$$\text{Normal time} = \frac{\text{Observed time} \times \text{Rating}(\%)}{100}$$

7. Add the suitable allowances to compensate for fatigue, personal needs, contingencies. etc. to give standard time for each element.

8. Compute allowed time for the entire job by adding elemental standard times considering frequency of occurrence of each element.
9. Make a detailed job description describing the method for which the standard time is established.
10. Test and review standards wherever necessary. The basic steps in time study are represented by a block diagram in Fig. 6.

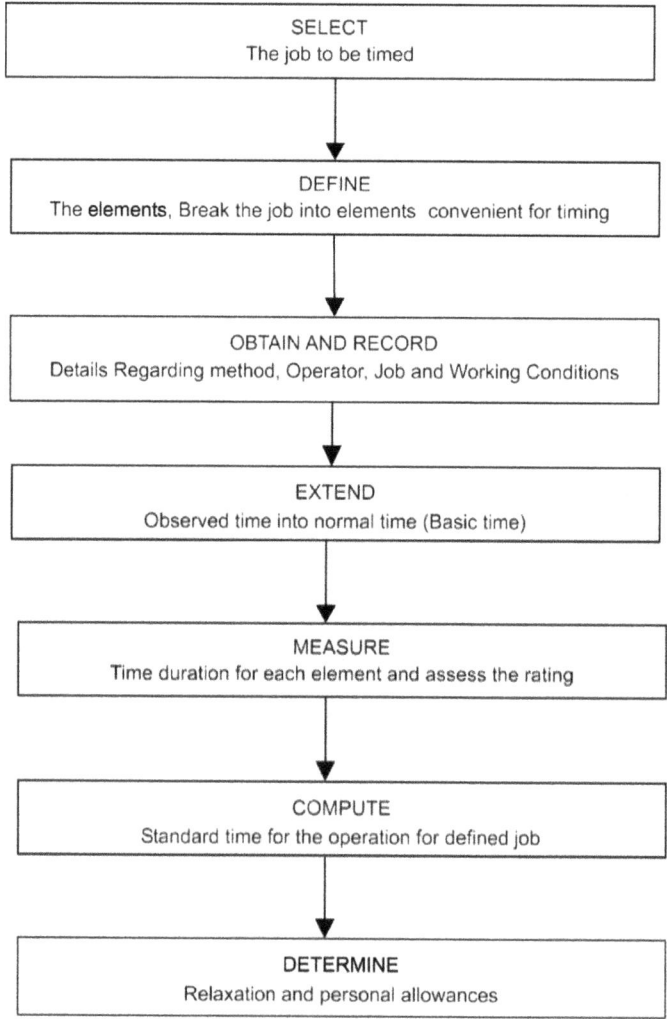

Fig. 6 Steps in time study

Standardized Work in Fashion Industry

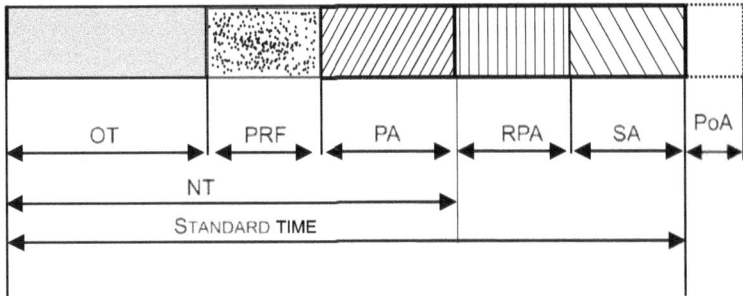

Fig. 7 Components of standard time. OT—Observed Time, PRF—Performance Rating Factor, NT—Normal Time, PA—Process Allowances, RPA—Rest and Personal Allowances, SA—Special Allowances, PoA—Policy Allowances

5.3.2 Computation of Standard Time

The time allotted to an operator to complete a task under specified conditions and at a set level of performance is known as standard time. As indicated in Fig. 7, various allowances are applied to the normal time as needed to obtain the standard time. The length of time required to do a unit of work:

(a) under current working conditions,
(b) using the stated method and machinery,
(c) by an operator capable of performing the work properly, and
(d) at a standard pace.

Thus, basic constituents of standard time are:

1. Elemental (observed time).
2. Performance rating to compensate for difference in pace of working.
3. Process allowance and special allowance.
4. Rest and personal allowance.
5. Policy allowance.

6 Allowances

There are no concessions for the worker in the regular time for an operation (Mundel 1978; Niebel 1976). Even though the most practical and effective strategy has been established, it is impossible to work throughout the day. Even with the best working style, the job will still require human effort, thus some time must be set out for fatigue recovery and rest. Allowances must also be granted to allow the employee to take care of operator's personal needs. The allowances can be further divided into five categories:

(1) Relaxation allowance,
(2) Variable allowance,
(3) Interference allowance,
(4) Contingency allowance, and
(5) Policy allowance.

6.1 Relaxation Allowance

Allowances for relaxation are calculated to allow the worker to recover from exhaustion. Relaxation time is the time added to the normal workday to allow the worker to recuperate from the physiological and psychological impacts of performing specific tasks under specific conditions, as well as to attend to personal needs. The amount of the allowance will be determined by the job's nature. Relaxation allowances are of two types: fixed allowances and variable allowances. **Fixed allowances** constitutes:

5.1.1 Personal needs allowance: This allowance is intended to allow the operator for the time spent away from the workplace to attend personal needs such as drinking water, smoking, and hand washing. Women, on the other hand, demand a larger personal allowance than men. A reasonable personal allowance for men is 5%, and for women it is 7%.

5.1.2 Basic fatigue allowance: This allowance is given to compensate for the energy expended while working. Generally, 4% of the basic time is used as the fatigue allowance.

6.2 Variable Allowance

A variable allowance is given to an operator who is working in unfavourable working conditions that cannot be rectified, causing additional stress and strain on the job. An operator who is engaged in medium or heavy work and operating under unusual conditions receives a variable fatigue allowance in addition to the fixed allowance. The amount of variable fatigue allowance varies depending on the company.

6.3 Interference Allowance

It is a time allowance included into the job's work content to compensate the operator for the unavoidable loss of production caused by the stoppage of two or more machines operating at the same time. This allowance is applicable to jobs that are controlled by a machine or a process. The operator's interference allowance is proportional to the number of machines assigned. The machine's interference increases the amount of work to be done.

6.4 Contingency Allowance

A contingency provision is a tiny amount of time that can be added to a regular schedule to accommodate reasonable and anticipated work or delays. Because of its sporadic or irregular frequency, exact measurement is impractical. This provision accounts for slight unavoidable delays as well as modest extra effort on occasion. Some of the examples for contingency allowance are:

- Tool breakage that entails removing the tool from the holder as well as any other procedures required to replace the tool in the holder.
- Power failures of small duration.
- Getting the required tools and gauges from the central tool store. The contingency allowance should not be more than 5%.

6.5 Policy Allowance

Policy allowances are not a true component of the time study and should be used with extreme caution and only under specific circumstances. The policy allowance is usually made to align normal times with the obligations of a pay agreement between companies and trade unions.

The policy allowance is a non-bonus increment applied to a standard time (or a component part of it, such as work content) in order to offer a suitable level of remuneration for a set level of performance under unusual conditions. As a result of the inefficient operation of a division or part of a plant, policy allowance is necessary in the time study.

7 Line Balancing

Line balancing is a production approach that involves matching the production rate to the Takt time by balancing human and machine time. The rate at which components or goods must be manufactured in order to meet consumer demand is known as the Takt time.

The rate at which components or goods must be manufactured in order to meet consumer demand is known as take time.

If production time is exactly equal to Takt time for a certain production line, the line is properly balanced. Otherwise, bottlenecks and surplus capacity should be removed by reallocating or rearranging resources. To put it another way, the number of personnel and machines assigned to each task in the line should be rebalanced to achieve the highest possible production rate.

7.1 Benefits of Line Balancing

1. **Reduce waiting waste**

One of the eight categories of waste in Lean manufacturing is waiting waste (David 1980; Adam and Ebert 1978). Any idle time that occurs when operations are not perfectly synchronised is referred to as the waiting waste. Waiting waste, for example, happens while operators are waiting for resources or someone else to finish their duty. Another form of waiting waste is equipment downtime, which is defined as time when the equipment is not in use.

Line balancing guarantees that all operators and machinery are working in harmony. No machine or operator should be overworked or idle. Line balancing decreases waiting waste by decreasing downtime, which is one of the eight wastes in lean manufacturing.

2. **Reduce inventory waste**

Another type of waste is inventory waste, which refers to a surplus of raw materials, unfinished components, or finished goods. Inventory waste shows inefficient capital allocation in an organization.

Line balancing standardises production, making it considerably easier to avoid inventory build-ups. Line balancing guarantees that there is minimal work in progress by reducing idle time. Finally, it ensures on-time delivery by bringing production time closer to the Takt time (Riggs 1987; Geneva Indian Adaptation International Labour Office 2015).

3. **Absorb internal and external irregularities**

Line balancing is a technique for reducing variances in a production line. A well-balanced manufacturing line is both steady and adaptable to changes. For example, if consumer demand shifts, and thus the Takt time shifts–operations can be promptly adjusted via line balancing. The effects of making modifications to a well-balanced production line are predictable. As a result, modifying the line to adapt the output rate is significantly easier.

4. **Reduce production costs and increase profit**

Worker and machine performance are totally synchronized as a result of perfect line balancing. Standing idle is not compensated in any way. Every machine is put to its greatest capability. In other words, the capacity of both labour and machines is maximised. Process efficiency translates to lower costs and higher earnings.

7.2 What is the best way to achieve line balancing?

a. **Calculate Takt time**

Knowing your Takt time is critical because the purpose of line balancing is to match production rate to Takt time. Tulip, the front-line operations platform that connects people and machines, comes with a Takt Time App and a Takt Time Dashboard that make calculating and tracking Takt time a much simpler process. Takt time is calculated by dividing available shift working time by the rate of consumer demand every shift.

b. **Perform time studies**

A comprehensive guide on how to perform time studies has been discussed in earlier section. Time studies are used to determine how long it takes to complete each task on a manufacturing line. To put it another way, you want to know how much time employees and machines spend on each step of a process.

Keep in mind that while using a stopwatch and a clipboard to conduct time studies, there are now far better solutions. Data collection and storage have been changed by IoT (Internet of things) connectivity and cloud computing. Manufacturers may now undertake autonomous and continuous time studies using sensors and production software, avoiding human bias and the sample size effect.

c. **Identify bottlenecks and excess capacity**

When it's time to examine the data from time studies, keep track of which steps take longer than they should. Late deliveries, hefty transportation expenses, and disgruntled customers are all consequences of exceeding the Takt time. Also, keep track of which areas are taking less time than others. In some locations, there is a surplus of capacity.

d. **Reallocate resources**

To begin, think about task precedence, or the order in which tasks must be completed. If a step requires a specific part, the operator must ensure that item is complete before proceeding to the next step. A Precedence Diagram can be very useful.

Then restructure tasks to eliminate bottlenecks and extra capacity. For example, reallocate resources–workers and equipment–from portions of the line that have surplus capacity to bottlenecks. In other words, seek to reduce the workload where there are blockages, and redistribute it to regions where surplus capacity may be filled by absorbing more work. This will lower the amount of time people waste waiting in situations where there was an overabundance of capacity. It will also aid in the improvement of production flow when bottlenecks exist.

Organize basic jobs into groups to reduce operator idle time and maximise machine and equipment use. Distribute the workload among the operators in the most rational manner possible, taking into account the data obtained on operator performance. To achieve synchronisation, each group of tasks should be finished in the same amount of time.

Consider whether you have an excessive number of or insufficient number of workstations. Line balancing may increase process efficiency to the point where the line has extra capacity. Removing workstations or combining operations could be useful.

The organization should attempt to decrease the imbalance between workers and workloads anywhere with numerous operators doing consecutive activities and operating as a unit. In production lines, proper work layout and allocation help maximise output at the appropriate time.

5. **Make other improvements**

Analyzing the quantitative data on production lines will almost certainly uncover more areas to improve the balance of lines.

Three parameters can be changed to optimise the process: operator time, machine time, and setup time. The organization may provide additional training to employees who take longer to finish jobs or make transitions easier to reduce changeover times. They can also upgrade machines or ensure that operators follow standard operating procedures (SOPs) for machine setup and maintenance.

Many Lean approaches can also aid in the reduction of variation in the production lines. For example, 5S and visual management help to establish a unified workspace, which cuts down on time spent hunting for tools and boosts process efficiency. Poka-yoke, often known as error-proofing, is a method for detecting flaws early and improving output consistency.

8 Conclusions

Line balancing is a critical optimization challenge in the manufacturing industry. Organizations may minimise Lean manufacturing wastes and generate more value by enhancing the efficiency of their lines by using time and motion study. In the last two decades, digitisation and technological advancements has led to easier approaches of completing time study and motion study in apparel manufacturing. However, the fundamental principles are still remaining the same. A large number of industries are moving to establish their standard methods and standard time by time study and motion study. Both these tools are essential to eliminate wastes from manufacturing processes in apparel industry. Time and motion study are the essential parts of lean manufacturing to eliminating waste, increase productivity and efficiency. This chapter has discussed the fundamentals of time and motion study in apparel industries to meet the objectives of lean manufacturing.

References

Adam EE Jr, Ebert RJ (1978) Production and operations management. Prentice Hall, Englewood Cliff, NJ
Barnes RL (1980) Motion and time study design and measurement of work, 7th edn. Wiley, New York
Bedi K (ed) (2016) Production and operations management. Oxford University Press, New Delhi
Bhattacharya DK (2010) Industrial management. Vikas Publications, New Delhi
Chary SN (2019) Production and operations management. Mc Graw Hill, New Delhi
Curie RM (1977) Work study, 4th edn. ELBS and PITMAN Publishing
David JS (1980) Productivity engineering and management. Tata McGraw Hill, New Delhi
Geneva Indian Adaptation International Labour Office (2015) Introduction to work-study, 3rd revised edition. Geneva. ISBN-13: 978-8120406025
Jhamb LC (1988) Work study and ergonomics. Everest Publishing House, Pune
Mundel ME (1978) Motion and time study, 5th edn. Prentice Hall, Englewood Cliffs, NJ
Niebel BW (1976) Motion and time study, 6th edn. McGraw-Hill Education
Norman G (2010) Production and operations management. Pierson Publications, Mumbai
Paneerselvam R (2012) Production and operations management. Prentice, New Delhi
Riggs JL (1987) Production system, planning analysis and control. Wiley

5S and Its Implications in Fashion and Textile Industry

Irfan Ulhaq, Majo George, and Rajkishore Nayak

Abstract 5S is a systematic and visual approach to improve the working culture at the workplace. The implementation of the 5S has been seen central to increase the cleanliness in operational as well as office environment. Furthermore, this approach is a steppingstone to direct staff members to improve their capabilities and standardise the operational processes by using 5 stages of the methodology, from where the name has been derived. The 5S methodology is a very appropriate way to initiate and achieve the process of continual improvement. This chapter introduces the basic foundations of the 5S systems and its role in the workplace standardisation. By reviewing the previous published work, the book chapter has stated importance of the 5S approaches using a hypothetical case. The chapter further highlights the key areas the fashion and textile industries should focus on adopting 5S system in their organisations. In the last section of chapter managerial practices and appropriate mechanisms are discussed for successful implementation.

Keywords 5S · Lean management · Cleanliness · Standardise · Fashion and textile

1 Introduction

The concept of lean management is no longer a fad, as it has received considerable attention from researchers and practitioners. Globalization and innovation have scaled competitive environment to supply chain. The purpose of supply chain management in today's competitive environment is not limited to cost minimization. Expanded value creation relies on the concept of lean management and waste reduction through adoption of strategies and tools which could help firms to overcome the supply chain waste.

I. Ulhaq · M. George
School of Business and Management, RMIT Vietnam, Ho Chi Minh City, Vietnam

R. Nayak (✉)
School of Communication & Design, RMIT Vietnam, Ho Chi Minh City, Vietnam
e-mail: rajkishore.nayak@rmit.edu.vn

Thus, new agenda in the supply chains literature is pushing firms to focus on value-addition by eliminating waste in customer-centric markets. In several emerging supply chain themes, lean philosophy offers an impetus to integrate this perspective as part of organizational strategy. Exploiting the tools in supply chain management and logistics are the only option to answer the competitive environment. It is important for the supply chain managers to understand these innovative methods to overcome the issue after quality management, waste minimalization, cost efficiency, an organizational performance improvement. Lack of knowledge on such practices could bring enormous challenges to the organizations participating in the supply chains as the business can easily lose the customers and market share in the competitive environment.

Role of 5S is paramount for manufacturing industries such as fashion and textiles for waste minimisation in the improvement of work (Warwood and Knowles 2004). Fashion and textile industries are one of the most innovative but a fast-paced industry due to changing consumer behaviours and fashion trends (Nayak et al. 2020a, b). Such changes in the industry are like living organisms change in a flexible relationship with these customers centric industry. Meanwhile, the pressure to improve the quality and reduce the cost of manufacturing operations in the textile industry mounts in a competitive marketplace. As stated by the International Labor Organization (ILO), the major threat faced by the global fashion and textile industry is competitive pressure due to low-cost and presence of foreign manufacturers in the marketplace. Consequently, the fashion and textile manufacturers have sought to improve their manufacturing processes and reduce their operating costs, as they are able to easily compete with the foreign manufacturers. As a result of these challenges, the fashion and textile industry must go beyond traditional supply chains to adopt a streamlined and modernized version of supply chains (Nayak et al. 2020a, b).

Studies have shown the positive effects of Lean methods for a long time in several industries (Secchi and Camuffo 2016) Lean 5S is used combined with other methods to reduce non-compliance in orders delivered. In addition, 5S is also applied as part of a lean strategic sourcing, supplier approval to manage down times of manufacturing services (Kattman et al. 2012). The financial benefits of 5S are also discussed in the fashion and textile industry through the study of labour indicators, whose results indicate a 59% increase in productivity. However, enterprises should develop an organizational culture of change to meet waste reduction objectives.

As part of the lean supply chain and waste minimization drive, companies have focused on different methods to overcome such issues. The 5S is a Japanese technique to improve the operational efficiency, and to minimize waste. This chapter focuses on the lean methodology, and particularly discusses 5S system in detail. The chapter will introduce the basic concepts, benefits of the 5S system in the context of fashion and textile supply chain, and 5S methodology as part of the supply chain operations. Furthermore, the chapter focuses on the 5S, and its related aspects in the fashion and textile supply chain.

2 Historical Background of 5S System

Along with several quality management approaches; the 5S is one such approach used by many organizations to achieve the purpose of total quality environment. 5S is *"a system of steps and procedures that can be used by individuals and teams to arrange work areas in the best manner to optimize performance, comfort, safety, and cleanliness"* (Peterson and Smith 1998). Hirano (1995) describes the 5S system quality by optimising supply chain waste, focus on improving productivity using visual cues for more consistent operational results. The 5S system is commonly implemented to contribute through standardization of processes, with improved configuration. The 5S philosophy applies to any job suitable for visual mastery and lean production.

Since the introduction of the Total Quality Management (TQM) in the 1980s, there has been a considerable degree of implementations of 5S system. Japanese firms have widely practiced the 5S technique and believe that it can help in every aspect of a business. Adopting the 5S methodology is found to be very useful as part of lean and quality management practices (Warwood and Knowles 2004). The 5S methodology forms the basis in establishing a quality work environment which focuses on both physical and mental aspects as well as forms the basis of first impressions of clients.

The 5S system is derived from the letter "S" used to create an appropriate workplace for visual inspection and lean manufacturing (Warwood and Knowles 2004). In simple terms, the 5S methodology helps a workplace to remove items that are no longer required, organisation of the items to optimize efficiency and flow, clean the area to identify problems more easily (Shamsi 2014). Furthermore, the objective of 5S methodology is to implement colour coding and labels to remain consistent with other domains (standardization) (Ho 2006). Currently, 5S methodology has enabled companies to achieve international standards, namely TQM, ISO 9000, ISO 14001 and OHSAS 18001 (Occupational Health and Safety Assessment Series).

Application of 5S in the daily life of people and in the workplace is commonly unnoticed. People practice sorting, ordering and cleaning as part of daily routine such as wastebaskets, towels, and tissues in convenient and familiar places (Hirano 1995). When our home environment becomes crowded and disorganized, we tend to function less efficiently. Similarly, 5S has been found reliable approach in Lean manufacturing. The 5S philosophy was accepted as a workable approach to achieve operational stability in supply chains in order to successfully support continuous improvement efforts (Kattman et al. 2012). Furthermore, it also enhances the ecofriendly practices by contributing into sustainable manufacturing towards meeting green initiatives in organisations (Vimal and Vinodh 2013). 5S implementation along with other lean tools not only improves the cleanliness and safe working environment but also increases the productivity by removing non-value adding actions. In addition, employees using this method achieve increased efficiency by cutting down searching effort and meeting demands on the foundation of the 'right product' at the 'right time' (Punnakitikashem and Buavaraporn 2012). The subsection will provide the brief overview of five (5) pillars of 5S System.

3 The 5S Methodology

3.1 The First Pillar: Seiri (Sort)

In 5S system the first pillar of is known as sort. In Japanese seiri (sort) relates to material management. The arrangement used to keep each material in the company where it belongs is called Sort (Peterson and Smith 1998). Defective or infrequently used equipment and disorganized equipment cause demolition of workplace control and reduced work efficiency (Hirano 1995). Therefore, the required and unnecessary materials available at the workplace should be sorted and graded. To improve the availability of work machinery and equipment; certain workstations such as machines, tools and materials should be kept available in an order for effortless access (Hirano 1995; Peterson and Smith 1998). The five pillars of 5S system are shown in Table 1.

The first pillar of 5S if applied properly improves the workflow, lower-level customer complaint and effective communication among staff members. Moreover, this pillar in the 5S can bring cost saving by effective use of space, and improved working environment (Singh et al. 2014).

3.2 The Second Pillar: Seiton (Set in Order)

This pillar of 5S supports workers and employees to design space so parts could be neatly arranged, and identification is made easier for daily use in operations (Patel and Thakkar 2014). This pillar helps the workers to develop set of arrangement performed at the workstations (machines, tools, hand tools, materials to be used, etc.), which must be kept in a readily accessible location as required (Kocaalan 1999). Celebi (1997) suggested important key points in the context of lean management and in the context of set-in order. Firstly, operations when carried out, the material transition paths, and the storage method must be considered in this step. Secondly, solutions such as a shelf order in proportion to the height of the classified material and drawers instead of big sized cupboards and boxes can gain efficiency in terms of stocking.

Table 1 Explanation of 5S in Japanese and English

Japanese S	Translation	English S
Seiri	Organization	Sorting
Seiton	Neatness	Simplifying access or Set in order
Seiso	Cleaning	Shine or Sweeping
Seiketsu	Standardization	Standardization
Shitsuke	Discipline	Sustain or Self-discipline

Recreated from Peterson and Smith (1998)

Thirdly, stock areas, shelves, and drawers as well as materials should be labelled. In case the dimension and type of the product change, then special vehicles may be used in machine adjustments. Further, if the size and type of the product change, special vehicles may be used for the adjustment of the machinery.

3.3 The Third Pillar: Seiso (Shine)

Seiso emphasizes on the removal of dirt, trash, medical wastes, and other contaminants from the workplace (Hirano 1995). As such, Shine means that we keep everything wiped, swept, and clean always. Indeed, dust, dirt and waste are a source of disorder, indiscipline, inefficiency, defective production, and occupational accidents (Patel and Thakkar 2014). To perform efficient tasks, it is essential to create a clean and regular working and living environment. Celebi (1997) describes a two-step detailed cleaning approach, cleanliness of workplace as well as cleanliness of machines, hardware, and tools. Dirt and dust cause corrosion and early demolishment of machine, and its components get rusty impacting on the operations (Vimal and Vinodh 2013).

A major purpose of Shine is to turn the workplace into a clean, bright, safe, and sanitary place where everyone will enjoy working. Another key purpose is to keep everything in top condition so that when someone needs to use something, it is ready to be used. Many organizations following Lean principles have already abandoned the inadequate tradition of annual "year-end" or "spring" cleanings due to regular practice of Shine. (Celebi 1997; Kanamori et al. 2016).

In case of detailed cleaning, some advantages can be obtained. These can be summarized as follows (Shah and Hussain 2016):

(i) Dirt and dust causes bad operation, therefore, dirt and dust sources are removed.
(ii) As a result of making the workplace more proper to the working conditions, the morale of the personnel improves.
(iii) The abnormal cases such as lubricant leakage, spillage and shop floor wastes are recognized immediately.
(iv) As a result of psychological impact, the reactions and performances of the personnel get better.
(v) Through providing a safer working environment, the danger contained in the works decrease.

To realize shining through an effective system, the names of the personnel who are responsible from the cleaning of each zone, each department and each point of the factory should be clearly determined and written at proper places.

The shining time should be very short to obtain effective utilization. The best times for cleaning are the beginning of shift, end of shift or after meal. All personnel should be well trained about cleaning and participate in cleaning.

Cleaning should become a daily activity at the workplace. The dust, dirt and wastes are the source of untidiness, indiscipline, inefficiency, faulty production, and workplace accidents. Therefore, workplace should be cleaned at regular intervals. Every tool and equipment should be restored at their own places after being used (Patel and Thakkar 2014).

3.4 The Fourth Pillar: Seiketsu (Standardize)

Standardization supports staff in managing and detecting anomalies in the processes, helping to develop a quick solution (Jaca et al. 2014). Established and implemented standards in the form of procedures and instructions should be available to maintain order in the workplace. They should be highly communicative, clear and readily comprehensible by the personnel. Signals and marks should be highly visual and clear enough to guide operators and manage machines in an easy way. Standard procedures should be adhered by all staff members and its essential that all employees involved in the process should be given included in training and preparation of these standards (Peterson and Smith 1998). Training and guidance in these procedures to operational and shop floor workers offers a possibility to consider each aspect of the operation. In the aim of assuring all the easy access, obligatory standards should be found in constant and visible places (Dudek-Burlikowska, 2007). The premise of 5S lays the foundation so standards should not be limited to production related activities but must be extended to administrative functions such as human resources management, financial, and procurement (Peterson and Smith 1998).

3.5 The Fifth Pillar: Shitsuke (Sustain)

The implementation of 5S can improve processes and operations if standards used and applied are sustained in the workplace and operations. To sustain the results of applied standards, organisations will require employees to adopt the changes as part of processes and should comply with the regularity rules in cleaning and sorting (Gapp et al. 2008). Change in the norms, and habits brings the discipline and success to the use of lean methods. Therefore, creating a culture of learning and encouraging continuous training and disseminating information is essential (Jaca et al. 2014; Ulhaq et al. 2017).

In this regard, firms can adopt training centre close to the workplace to share and enhance knowledge with new employees are train the skills related to their workstation. Moreover, psychomotor exercises, ergonomic movements, networking, and informal knowledge sharing activates could be initiated (Ulhaq et al. 2020). Such practices help the firms to reduce the number of non-conforming products' improvements in the internal communication, and through this to improve in the trustful relations (Gapp et al. 2008). Moreover, firms need to establish and execute routine inspections of these implemented changes in the practices adopted on monthly basis. Summary of five pillars of 5S is given in Table 2.

Table 2 Summary of five pillars of 5S

Sorting	Simplifying	Sweeping	Standardizing	Self-discipline
• Determining the frequency of usage for every item in the workplace. • Marking the items that are not used. • Disposing of the nonessential items. This may include recycling, donation, or auction. • Eliminating sources of clutter and unwanted items.	• Arranging items in the work area and establishing guidelines. • In the sorting step, items are classified by frequency of usage. In the simplifying step, items are placed by frequency of usage.	• Visually and physically sweeping the work area to ensure that everything is in its proper location. • Take action for missing or misplaced items. • Visually and physically sweeping to identify and correct repeated violations, and repeated housekeeping problems. • Tools out of place, manuals out of sequence, inventory in incorrect area.	• Labels on a shelf are formatted the same way. • Retrieving and returning items and information are uniform, it is easier for everyone in the work group to locate them quickly.	• Routine practice of all the steps that precede it. • Self-discipline is having all associates doing their part to carry out the 5S actions they have agreed upon.

4 Implementation of 5S in Fashion and Textile Industry

4.1 Tools for Implementing the 5S Lean Method

The 5S audit is a standard instrument for assessing and verifying the 5S approach on a regular or ongoing basis. In fashion and textile industries, 5S can be implemented successfully to improve the workplace operations by changing the existing systems and standardising the new processes. Several operations in these industries are done by practices that has a lot of potential for improvement. It is the role of lean management, not the auditor, to develop, implement, and continuously monitor the 5S implementation. Following a 5S audit, it is possible to see if the situation has improved or deteriorated in comparison to earlier auditing (Filip 2010). A standard sheet prepared to be used in identifying the production zones is another tool used to apply the 5S lean technique in fashion and textile industries (Table 3).

Implementing the 5S technique entails assigning some tasks and responsibilities to the team members involved in the process activities. Following the establishment of defined duties and responsibilities for implementing the 5S, it was necessary to create a standardization form to check the process in all workplaces (automated and manual), as a tool for implementing the 5S.

5 Implementing the 5S Method: Proposed Model

The proposed implementation of 5S in fashion and textile industry has four phases which is presented in Fig. 1.

Table 3 Standard of marking the production areas (Reproduced from Filip and Marascu 2015)

Role	Delineates the walkways	Delineates the storage areas of the processing parts	Delineates the storage areas of finished parts	Delineates the storage areas of rework parts	Delineates the storage areas of rejected parts	Delineates the storage areas of trolleys, carriages, boxes, devices, molds	Delineates the storage areas of waste products	Delineates the dangerous areas	Delineates the fire extinguishers areas
Marking colors	yellow	yellow	green	orange	red	gray/blue	black	yellow/black	red/white

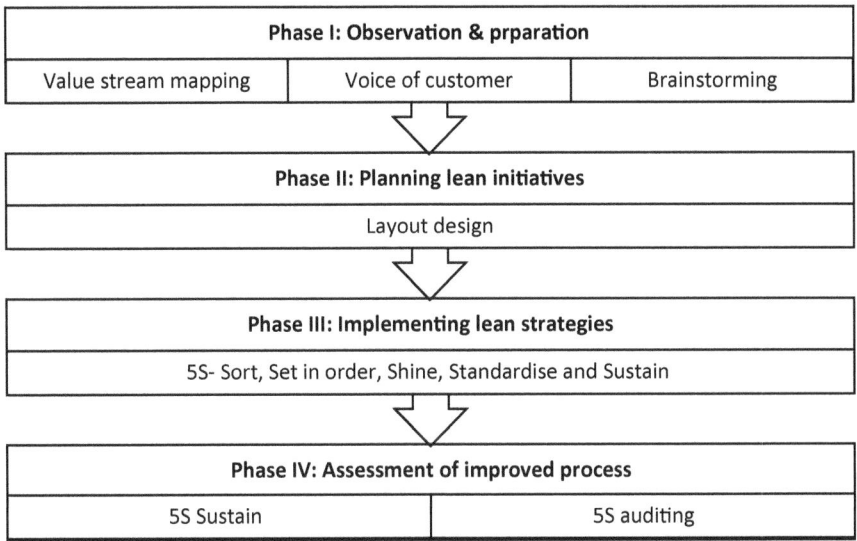

Fig. 1 Model of proposed 5S implementation

5.1 Phase I: Observation and Preparation

This step entails observing the process to detect issues such as inefficient procedures, area delays, and product loss. From the client's perspective, a value stream map should be created to identify the areas that are impacted (Dotoli et al. 2015). Its benefit is that it represents the flow of materials, allowing it the most excellent tactic for spotting improvement opportunities and differentiating between operations that add value and those that do not.

5.2 Phase II: Planning Lean Initiatives

The ideas generated in Phase I are selected in this phase, and an execution plan is developed to implement the improvements. The required approvals are acquired, and the management are notified of the plan to be implemented and the expected outcomes.

5.3 Phase III: Implementing Lean Strategies

This phase is net implementation, including the execution of the first four Ss, sort, set to-order, shine, and standardize. The 5S implementation manual described by (Punnakitikashem and Buavaraporn 2018) should be followed.

5.4 Phase IV: Assessment of Improved Process

The efficacy, efficiency, relevance, and impact of the enhancements are evaluated in this phase. It also includes the last S, which stands for sustain. This is the most crucial and difficult phase since it is necessary to maintain the progress accomplished over time, which necessitates the participation of all levels of workers involved in the process.

6 Hypothetical 5S Case Study in Textile Industry

6.1 Phase I

In this chapter, we have examined company A which performs spinning, slashing, warping, and weaving for producing a wide range of items from draperies to denim, with capabilities to offer full package. Regarding the current context, in one of the company's plants, the operation was producing 3 times more waste than the initial goal thereby bearing the inefficiencies including operational costs. This was due to a lack of procedures in the warehouse. The implementation of 5S was done to improve the performance.

6.2 Phase II

The department's production was halted for three days due to the original event. To demonstrate their dedication to the effort, the corporation elected to cease production in this area. They wanted everyone to take the occasion seriously. Everyone on the 5S crew was paid his/her regular rate to come in and perform on the event, but it was a completely voluntary event. The gathering drew a total of roughly 30 people, including everyone who worked in the area.

6.3 Phase III

6.3.1 First S—Sort

The first stage was to clear the clutter from tables, workspaces, and equipment, removing any objects that were not required for the operation and any unnecessary equipment. To isolate these things from the usual production area, they adopted the red tag strategy.

Red tags are a visual tool that is used during sorting phase of a 5S initiative. In such a phase, items are either gotten rid of, kept, or placed in a red tag auction area. The red tag is used for items that are either gotten rid of or placed in auction (Fig. 2).

6.3.2 Second S—Set to Order

The next stage was to use indicators to designate places and limitations for equipment and item storage. Walkways and other storage places were also marked with indicators such as lines and identifying markers. Empty beam storage places, for example, were marked off with lines on the floor, which not only indicated to the worker where to store these goods, but also set a limit on the number of beams that could be stored, because beams were not to be put past the floor line (refer to Table 1). Tools and equipment that are used frequently throughout the workday were located and stored close to their point of use, making it easier for workers to collect them when needed.

Fig. 2 A 5S red tag used for lean manufacturing

6.3.3 Third S—Shine

When top management decided to participate in this 5S event, one of their main goals was to clean up and organize the environment. To eliminate garbage and grease, the flooring and machines were thoroughly cleaned.

6.3.4 Fourth S—Standardize

The next step was to guarantee that standard working procedures were in place, and that everyone in the area had received 5S training and understood the organization's aims. The important procedures were documented and made available to any employee who had a query regarding their involvement in the process.

6.4 Phase IV

6.4.1 Fifth S—Sustain

The company implemented a 5S audit system to maintain the changes made to the production area where the 5S event had taken place. This audit method is used to ensure that 5S is followed consistently throughout the area, as well as that the procedures and activity boards are kept up to date.

Another important action taken to sustain 5S was the weekly meeting established after the 3-days event. These weekly meetings brought about suggestions for improving the process and work environment for the people.

7 Facilitating Factors: 5S Adoption in Fashion and Textile Industry

Mass customization trend has been accelerating in the fashion and textile industries since customers' preferences and personalities are becoming more demanding besides basic needs such as convenience or cleanliness (Rai et al. 2017; Fralix 2001). Although studies in the developed countries and 5S implementation has shown benefits in the manufacturing companies, fashion and textile industries need to follow these protocols to stay responsive and profitable (Jayaram et al. 2014; Raji et al. 2021). Specifically, increased availability of options and substitutes has affected customer demand, pushing firms to standardise processes (Merritt and Zhao 2020). According to authors such as Tapia-Cayetano et al. (2020), Bayhan et al. (2019), Bhutta et al. (2013) implementing lean management is the most effective way to address the issues in apparel supply chain management and the industry's production environment in order to increase productivity including:

- Waste reduction in the fashion and textile industry through the rise in material and inventory utilization.
- Warehousing cost reduction through the increase in labour efficiency and the decrease in material loss.
- Working environment improvement.
- Accelerating the flow of products on the manufacturing line.
- The barriers to implement lean manufacturing.

In Vietnam, there are various successful lean implemented case studies in the textile industry. Some of the industries include Hung Yen Garment Corporation, Garment Corporation No.10, Nha Be Garment Corporation, Viet Tien, or Hoa Tho Garment Joint Stock Company. At that time, the cutting and sewing departments of Hung Yen Garment corporation were overloaded but the output yield was poor. According to Mr. Nguyen Xuan Duong, General Director of Hung Yen Garment Corporation, the company values workers' roles so that they are aware of the benefits of this approach. Therefore, after three weeks of deployment, their productivity has increased 21% compared to previous operation performance. It shortens working hours, raises workers' income, and fosters a contemporary and professional work condition. The utilization of Lean philosophy drastically reduces waste and helps organizations save money. Furthermore, because this is a "dynamic" management style, it is used differently at each organization and should not be administered in a rigid and clichéd manner. It aids in increasing the market competitiveness of businesses.

There is no doubt that the utilization of lean management delivers multiple improvements in the textile production line. Change in the processes requires several improvements at organizational levels to meet the needs of lean practices such as 5S. The section below highlights key areas for textile firms to focus and reap the benefits of 5S implementations.

7.1 Human Factors

Lean manufacturing demands a significant shift in the attitudes and behaviours of both management and employees. Any new implementations that require a close and integrated collaboration between managers and workers will ensure a smooth ride in the whole process (Lodgaard et al. 2016). The management team has to spend time enhancing their skills and knowledge so as not to obstruct any significant success circumstances; understanding about lean management is necessitated through formal and informal training.

Without solid guidance and sufficient commitment from leaders, the effectiveness and efficiency of development programs will be heavily influenced, and the enthusiasm of the whole system will sharply enfeeble (Silva et al. 2011). From the employee side, people must be appropriately notified since any new alterations could create FOMO (Fear of missing out) effects in the workforce or even resistance to change (Sakthi et al. 2019); Jadhav et al. 2014). Besides, training for workers is a big question for businesses when it comes to 5S implementation. To be more specific, from technical and financial perspectives, adapting a new working flow takes time, effort, and money for the labor force to acquaint with changing conditions. As a result, there will be an unwilling attitude existing in some part of employees threatening the stability of the whole system (Panwar et al. 2015). Such mentioned negative elements could struggle the business and implementation outcomes of 5S or other concepts.

7.2 Customer-First Strategy

This is an enabling element that involves exceeding customer expectations through the products and services provided. The research of Ojha (2021) underlines the importance of customer focus in the company strategy. It is essential to absorb and exploit consumers' insights via the identification of all criteria that individual values including quality, cost, delivery, flexibility, lead time, innovation in product and process, and striving to accomplish them is vital in order to effectively implement lean management (Islam et al. 2013). As such, the business is able to provide the maximum merit to consumers resulting in distinguishing its products and services from rivals, maintaining client loyalty, and running a flexible production system (Dong et al. 2012).

7.3 Management and Leadership

When manufacturing companies shift toward the customer focus goal, it is the responsibility of top management to make tactical changes in their internal operations (Ilyas et al. 2020). Excellent governance is one of the critical factors that facilitate the success of lean implementation (Adikorley et al. 2017). From the managerial perspective, their dedication is crucial since they have the right to decide the path, adjust the goal and allocate the appropriate financial resource and investment (Hodge et al. 2011). Top management must be the pioneer in change by considering their employees as cooperative and willing to collaborate since this will improve knowledge development and effective skills within the workforce (Jadhav et al. 2014). Hence, the information flow within the internal company should be enveloped to all stakeholders through effective communication before, during the implementation, and further continuous checkups and optimization post-implementation (Latif and Vang 2021). Managers and seniors could guide their employees through all techniques of lean philosophy to increase the knowledge regarding the advantages of the practice resulting in the spurt of adoption (Martínez-Jurado and Moyano-Fuentes

2014). In light of that, great leadership for managerial positions could enhance relationships with employees within an organization leading to a strong business culture.

7.4 Organizations Culture

Another element that endangers the success of lean implementation is organizational culture (Nguyen et al. 2017). Organizations are fundamentally a social system of competing interests that work together to achieve their strategic goals (Snyder et al. 2016). Having a culture is about driving overall performance across all levels of the organization (Alkhoraif and McLaughlin 2018). This means that to build a successful lean enterprise, a company must create an organizational culture based on trust through employee empowerment, because employees are viewed as the pillar that keeps the organization running; thus, they are critical to the success of lean implementation (Araújo et al. 2019; Sigler and Pearson 2000). Such lean change has an impact on the entire system's business structures which should be cautious in the deployment process appear to be no simple task. To be more specific, lean management involves removing waste activities and functions in the operation so non-value-added employees will be transferred or fired (Chan and Tay 2018). Many companies that want to embrace lean spend a large number of resources on lean efforts but fail due to the fact that individuals in the manufacturing company do not accept lean progressively compared to their old routines. As a result, the value gained by adopting lean is reversed over time. Subsequently, a lean corporation must adopt a methodology that promotes long-term thinking and individual respect.

7.5 Staff Compliance

Lean approaches require continual monitoring and assessment to reap its benefits. It is essential for fashion and textile industry to establish non-negotiable guidelines for staff members. Although 5S is a Japanese approach, firms in the other countries establish the mandatory rules for staff to adhere the 5S system. Clear guidance and policies should be narrated for staff to manage their own workplace and tools. The 5S is an approach to standardise workplace and processes, and as part of this lean drive, managers must approach their staff by guiding its importance (Jaca et al. 2014). In this regard a more informal and social communication strategy could be used to share and disseminate during the first phase of implementation to overcome staff rigidity issues.

7.6 Collaborative Tools

Implementation of lean practices is not an one time agenda and it requires continuous changes due to turbulent supply chain environments. Therefore, offering a safe climate and encouraging worker participation is foundation for 5S success. As highlighted in manufacturing industries use the Kaizen boards to trigger creation of ideas as part of standards improvements (Jaca et al. 2014). Staff feedback is found to be a valuable tool to enhance the business processes. To get the seriousness in such creative exercises, management can introduce incentives. Staff in the textile companies can share visual metaphors, diagrams, and pictures as part of their idea sharing. Since Kaizen is often seen as collaborative activities it will also complement the communication among employees. Moreover, supervisors and stakeholders can also mentor to strengthen the micro improvements towards their success.

8 Barriers to Implementing Lean Manufacturing in the Fashion and Textile Industry

There is no doubt that the utilization of lean manufacturing delivers multiple improvements in the textile production line. However, there are some obstacles that influence lean implement success since "one size can't fit all". The work of Ramdass (2015) affirmed that human factors and organizational culture account for the breakdown of lean management. Additionally, implementing new procedures into the current stable manufacturing operation of textile companies is not a simple job resulting in the failure of lean employment schemes at the beginning and the turnaround of numerous "successfully implemented companies" (Fernando and Ratnayake 2021).

8.1 Human Factors

Lean manufacturing demands a significant shift in the attitudes and behaviors of both management and employees. Any new implementations that require a close and integrated collaboration between managers and workers will ensure a smooth ride in the whole process (Lodgaard et al. 2016). The management team must spend time enhancing their skills and knowledge so as not to obstruct any significant success circumstances; understanding about lean management is necessitated through formal and informal training. Without solid guidance and sufficient commitment from leaders, the effectiveness and efficiency of development programs will be heavily influenced, and the enthusiasm of the whole system will sharply enfeeble (Silva et al. 2011).

From the employee side, people must be appropriately notified since any new alterations could create FOMO effects in the workforce or even resistance to change

(Sakthi et al. 2019); Jadhav et al. 2014). Besides, training for workers is a big question for businesses when it comes to 5S implementation. To be more specific, from technical and financial perspectives, adapting a new working flow takes time, effort, and money for the labor force to acquaint with the new normal which could push to the doubt from top management about the feasibility of the project. As a result, there will be an unwilling attitude existing in some part of employees threatening the stability of the whole system (Panwar et al. 2015). Such mentioned negative elements could struggle the business and implementation outcomes.

8.2 Organization Culture

Another element that endangers the success of lean implementation is organizational culture (Nguyen et al. 2017). Although it is an enabler in adopting management, it may appear to be no simple task and far from it. Such lean change has an impact on the entire system's business structures which should be cautious in the deployment process. To be more specific, lean management involves removing waste activities and functions in the operation, so, non-value-added employees will be transferred or fired. Many companies that want to embrace lean spend many resources on lean efforts but fail due to the fact that individuals in the manufacturing company do not accept lean progressively compared to their old routines. As a result, the value gained by adopting lean is reversed over time. Subsequently, a lean corporation must adopt a methodology that promotes long-term thinking and individual respect.

To be brief, although lean manufacturing practices bring various benefits, fashion and textile businesses must consider all the enablers and barriers before deciding to implement processes with or without consulting firms (Table 4).

Table 4 Summary of enablers and barriers to adopt lean manufacturing in textile industry

Enablers	Barriers
Customer focus scheme through mass customization	Loss in commitment from managers
Top management collaborations, planning and support	Unfeasibility and inappropriate planning and employment process
Embracing employee sanction	Resistance to change from employees
Proper training to adapt with new changes	Risk to lean implementing failure due to unsuitable organization culture

9 Conclusion

In summary, this book chapter focuses on understanding the role of lean management especially 5S in the fashion and textile industry. Regarding 5S implementation, the case study of company A has shown the process of four phases with the purpose of optimizing the procedures in the warehouse to eliminate the current operation wastes. Regarding lean manufacturing, we acknowledge various benefits for apparel businesses that they gain when employing this technique including the elimination of waste, increase productivity and improve working conditions inside the manufacturing facilities. Henceforth, the consumers' demand for mass customization is growing quickly in the fashion and textile industry, requiring manufacturers to stay competitive in the market with a lean strategy. Customer focus, management perspective, and organizational culture are elements motivating the implementation of lean practices. However, this is not easy work as human factors and organizational culture are obstacles that businesses should take into consideration to avoid any needless failures.

References

Adikorley RD, Rothenberg L, Guillory A (2017) Lean Six Sigma applications in the textile industry: a case study. Int J Lean Six Sigma 8(2):210–224

Alkhoraif A, McLaughlin P (2018) Lean implementation within manufacturing SMEs in Saudi Arabia: organizational culture aspects. J King Saud Univ Eng Sci 30(3):232–242

Araújo R, Santos G, da Costa JB, Sá JC (2019) The quality management system as a driver of organizational culture: an empirical study in the Portuguese textile industry. Qual Innov Prosper 23(1):1–24

Bhutta MKS, Rosado-Feger AL, Huq F, Muzaffar A (2013) Exploratory study of adoption of lean management practices in Pakistani textile firms. Int J Serv Oper Manag 15(3):338–357

Bayhan HG, Demirkesen S, Jayamanne E, (2019) Enablers and barriers of lean implementation in construction projects. In IOP Conference Series: Materials Science and Engineering (Vol. 471, No. 2, p. 022002). IOP Publishing.

Celebi H (1997) 5S and total productive maintenance with total quality perspective. M.Sc. thesis, Istanbul University, Institute of Science, Istanbul, Turkey

Chan CO, Tay HL (2018) Combining lean tools application in kaizen: a field study on the printing industry. Int J Product Perform Manag 67(1):45–65

Dong B, Jia H, Li Z, Dong K (2012) Implementing mass customization in garment industry. Syst Eng Procedia 3:372–380

Dotoli M, Epicoco N, Falagario M, Costantino N, Turchiano B (2015) An integrated approach for warehouse analysis and optimization: a case study. Comput Ind 70:56–69

Dudek-Burlikowska M (2007) Quality estimation of sale process with usage of quality methods in chosen company. J Achiev Mater Manuf Eng 531–534

Fernando TD, Ratnayake V (2021) Barriers for lean implementation in apparel industry. IEEE, pp 620–625

Filip F-C (2010) Theoretical researches regarding management methods and techniques for conducting the internal audit. Ann DAAAM Proc

Filip F, Marascu-Klein V (2015) The 5S lean method as a tool of industrial management performances. IOP Publishing, p 012127

Fralix MT (2001) From mass production to mass customization. J Text Appar Technol Manag 1(2):1–7

Gapp R, Fisher R, Kobayashi K (2008) Implementing 5S within a Japanese context: an integrated management system. Manag Decis 46(4):565–579

Hirano H (1995) 5 pillars of the visual workplace. CRC Press

Ho SK (2006) Management art and science: from 5-S to 6-σ. Int J Manag Sci Eng Manag 1:63–70

Hodge GL, Goforth Ross K, Joines JA, Thoney K (2011) Adapting lean manufacturing principles to the textile industry. Prod Plan Control 22(3):237–247

Islam MM, Khan AM, Islam MM (2013) Application of lean manufacturing to higher productivity in the apparel industry in Bangladesh. Int J Sci Eng Res 4(2):1-10

Ilyas S, Hu Z, Wiwattanakornwong K (2020) Unleashing the role of top management and government support in green supply chain management and sustainable development goals. Environ Sci Pollut Res 27(8):8210–8223

Jaca C, Viles E, Paipa-Galeano L, Santos J, Mateo R (2014) Learning 5S principles from Japanese best practitioners: case studies of five manufacturing companies. Int J Prod Res 52(15):4574–4586

Jadhav R, Mantha S, Rane S (2014) Exploring barriers in lean implementation. Int J Lean Six Sigma 5

Jayaram J, Choon Tan K, Laosirihongthong T (2014) The contingency role of business strategy on the relationship between operations practices and performance. Benchmarking Int J 21(5):690–712

Kanamori S, Shibanuma A, Jimba M (2016) Applicability of the 5S management method for quality improvement in health-care facilities: a review. Trop Med Health 44:1–8

Kattman B, Corbin TP, Moore LE, Walsh L (2012) Visual workplace practices positively impact business processes. Benchmarking Int J 19(3):412–430

Kocaalan M (1999) Improving and increasing machine performance loy using total productive maintenance (TPM) approach. M.Sc. thesis, Gazi University, Institute of Science and Technology, Ankara

Latif MA, Vang J (2021) Top management commitment and lean team members' prosocial voice behaviour. Int J Lean Six Sigma

Lodgaard E, Ingvaldsen JA, Gamme I, Aschehoug S (2016) Barriers to lean implementation: perceptions of top managers, middle managers and workers. Procedia CIRP 57:595–600

Martínez-Jurado PJ, Moyano-Fuentes J (2014) Key determinants of lean production adoption: evidence from the aerospace sector. Prod Plan Control 25(4):332–345

Merritt K, Zhao S (2020) An investigation of what factors determine the way in which customer satisfaction is increased through omni-channel marketing in retail. Adm Sci 10(4):85

Nayak R, Akbari M, Far SM (2019) Recent sustainable trends in Vietnam's fashion supply chain. J Clean Prod 225:291–303

Nayak R, Nguyen, LTV, Panwar T, George M, Ulhaq I (2020a) Sustainable supply chain management. Supply chain management and logistics in the global fashion sector: the sustainability challenge, pp 1–32

Nayak R, Nguyen LT, Panwar T, Ulhaq I, George M (2020b) Standards, organizations and lean concept in managing sustainable fashion supply chains. Supply chain management and logistics in the global fashion sector: The Sustainability Challenge, pp 183–215. Routledge

Nguyen NTD, Dang DT, Lan HTP (2017) Achieving the successful lean implementation at manufacturing companies in Vietnam: awareness of critical barriers. Econ Bus Adm 7(1):89–106

Ojha R (2021) Enablers of lean for manufacturing excellence: an interpretive structural modelling and analysis. Vision

Panwar A, Nepal BP, Jain R, Rathore APS (2015) On the adoption of lean manufacturing principles in process industries. Prod Plan Control 26(7):564–587

Patel VC, Thakkar H (2014) A case study: 5S implementation in ceramics manufacturing company. Bonfring Int J Ind Eng Manag Sci 4(3):132–139

Peterson J, Smith R (1998) The 5S pocket guide. Productivity Press

Punnakitikashem P, Buavaraporn N (2018) The important enablers of lean implementation towards organizational performance in financial services. Bus Manag Rev 9(4):241–248

Rai K, Lodha R, Dalpati A (2017) Mass customization is new trend for textile industries. Int J Dev Res 7(2):11600–11603

Raji IO, Shevtshenko E, Rossi T, Strozzi F (2021) Industry 4.0 technologies as enablers of lean and agile supply chain strategies: an exploratory investigation. Int J Logist Manag 32(4):1150–1189

Ramdass K (2015) Integrating 5S principles with process improvement: a case study. IEEE, pp 1908–1917.

Sakthi NT, Jeyapaul R, Vimal KEK, Mathiyazhagan K (2019) Integration of human factors and ergonomics into lean implementation: ergonomic-value stream map approach in the textile industry. Prod Plan Control 30(15):1265–1282

Secchi R, Camuffo A (2016) Rolling out lean production systems: a knowledge-based perspective. Int J Oper Prod Manag 36(1):61–85

Shah ZA, Hussain H. An investigation of lean manufacturing implementation in textile sector of Pakistan, pp 8–10

Shah ZA, Hussain H (2016) An investigation of lean manufacturing implementation in textile industries of Pakistan. In Proceedings of the International Conference on Industrial Engineering and Operations

Shamsi HS (2014) 5S conditions and improvement methodology in apparel industry in Pakistan. IOSR J Polym Text Eng 1(2):15–21

Sigler TH, Pearson CM (2000) Creating an empowering culture: examining the relationship between organizational culture and perceptions of empowerment. J Qual Manag 5(1):27–52

Silva N, Perera C, Samarasinghe D (2011) Factors affecting successful implementation of lean manufacturing tools and techniques in the apparel industry in Sri Lanka. SSRN Electron J

Singh J, Rastogi V, Sharma R (2014) Implementation of 5S practices: a review. Uncertain Supply Chain Manag 2(3):155–162

Snyder K, Ingelsson P, Bäckström I (2016) Enhancing the study of lean transformation through organizational culture analysis. Int J Qual Serv Sci 8(3):395–411

Tapia-Cayetano L, Barrientos-Ramos N, Maradiegue-Tuesta F, Raymundo C (2020) Lean manufacturing model of waste reduction using standardized work to reduce the defect rate in textile MSEs

Ulhaq I, Maqsood T, Khalfan M, Le TL (2017) Development of a conceptual framework for knowledge management within construction project supply chain. Int J Knowl Manag Stud 8(3–4):191–209

Ulhaq I, Maqsood T, Khalfan M, Le T (2020) Exploring use of mobile messaging applications in the Vietnamese construction industry. In: academic conferences international limited, pp 1021–XXIII

Vimal K, Vinodh S (2013) Development of checklist for evaluating sustainability characteristics of manufacturing processes. Int J Process Manag Benchmarking 3(2):213–232

Warwood SJ, Knowles G (2004) An investigation into Japanese 5-S practice in UK industry. TQM Mag 16(5):347–353

Kaizen Applications in Fashion and Textile Industries

Majo George, Vuong Nguyen Dang Tung, Le Phan Thanh Truc, Nguyen Minh Ngoc, and Le Khac Yen Nhi

Abstract This chapter offers comprehensive insights into the *Kaizen* method, which is also called continuous improvement, and its application in real-life situations, especially the urgent concerns in the fashion and textile industry. From a starting point amid World War II (WWII), *Kaizen* has evolved over time and reached the ultimate goal of waste reduction by concentrating on the problem as a highly prioritized target. This chapter has been prepared from the information collected from various resources such as the latest journal papers, books, book chapters, and information available on various company websites. Nevertheless, these book chapters and journal papers depict similar levels of continuous improvement, consisting of improvements in individual works, group works, and the complete system. Similar to many models consisting of Jidoka, Kanban and 5S (applied to achieve the lean principles), Kaizen methodology has also been recognized as one of the tools for achieving the lean manufacturing principles. Going through three phases of Kaizen implementation, from planning to execution and control stages, businesses can benefit from *waste reduction, increment in output quantity and quality, and optimization in operation and production*, which are discussed in this chapter. Indeed, the application of Kaizen methodology has been an effective solution for the fashion and textile industry with actual evidence from Bangladesh, Peru, and India. Specifically, Kaizen has resolved problems of fabric faults, maintained ergonomic standards and workers' comfort with the invention of shoulder pad gadget in a sewing line, and advanced productivity, thereby highlighting its advantageous functions in the fashion and textile industry, which are also discussed in this chapter.

M. George (✉) · V. N. D. Tung · L. P. T. Truc · N. M. Ngoc · L. K. Y. Nhi
School of Business and Management, RMIT Vietnam, Ho Chi Minh City, Vietnam
e-mail: majo.george@rmit.edu.vn

1 Introduction

The Kaizen method has always been an ambivalent topic and received numerous perspectives from a variety of researchers. On the subject of Kaizen principles, despite being influenced by three different perspectives, there is a consistency in arguments to point out six principles of Kaizen, in which the three top core principles, including (1) teamwork, (2) focus on process and standards, and (3) commitment from top management, are important to shape the success of Kaizen application. On the aspect of Kaizen events, the dissimilar viewpoints of researchers lead to heterogeneous classifications.

Regarding the fashion and textile industry, despite acquiring various chances to widen and develop, it is still facing certain challenges stemming mostly from the manufacturing process and the requirement of health and safety guarantees for its labour force and an environmentally friendly production process (Nayak and Padhye 2015; Nayak et al. 2021). Some of the common problems related to labour-intensive processes are low productivity, idle time, reworking, inability to meet the sustainability standards, health and safety concerns (Nayak et al. 2019, 2020). It is asserted that these problems have been solved by the application of Kaizen, a worldwide philosophy regarding continuous improvement allowing both individuals and businesses to thrive. Kaizen implementation will be discussed further, especially in the context of the fashion and textile industry with underlying current problems and certain case studies will be provided to clearly demonstrate the benefits.

1.1 *Definition of Kaizen*

In Japanese, the word Kaizen (改善) can be broken down into two smaller portions, in which Kai means "change" and Zen refers to "better" or "good". Thus, this can be inferred literally from the word's meaning that Kaizen implies for "change for the better" (Macpherson 2008). Hence, as Lillrank (1995) asserted, Kaizen could be either understood as continuous improvement or as its principle. From that basis, the term has been recognized beyond this meaning, eventually becoming a worldwide philosophy allowing both individuals and businesses to thrive. Seeing from a cooperative angle, Kaizen works as a systematic strategy for a company to achieve competitive advantages with the focus being centred on cumulative improvements. According to García-Alcaraz et al. (2017), businesses can continually and simultaneously achieve higher levels of advancement and renovation in long-term operations, attain better quality and productivity and reduce cost and other wastage. Womack et al. (1990) defined Lean Manufacturing in their book "The Machine that Changed the World" as an "innovative production system" that is a perfect combination of craft and mass production, which costs less than the first and is more flexible than the latter. Excessive utilization of both labour and machinery allows more scope for detection of errors, adjustment, and resulting improvements. In that sense, it appears indisputable that Kaizen has also formed one of the major foundations of Lean Manufacturing.

The concept of Kaizen, through its evolution from the time it began to be applied by Toyota (discussed later in the section on the "History of Kaizen"), has been the subject of analysis from many perspectives. One of the pioneers in discovering the term was Masaaki Imai. In his first book about Kaizen (1986), Imai defines it as "continuous improvement that involves every individual within an organization". Since then, the concept has been expanded with further discussion by other authors. This began with the advent of the first definition of Kaizen which was based on the close correlation between a company and its staff. From that, Elgar and Smith (1994) emphasized the importance of fostering staff involvement. This viewpoint was supported by Bessant and Caffyn (1997). Specifically, they believe that employees' active participation would significantly contribute to the overall development of a company.

Another point of view concentrated on the characteristics of the continuous process of identifying and assuring improvements for corporate goals, as demonstrated by Brunet and New (2003). From this perspective, Kaizen was considered as a "philosophy of life" (Wittenberg 1994) and an "individual spirit" (Brunet 2000). In this way, Kaizen was viewed as a spiritual force for the motivation of workers and allowed improvement to be measured at a micro-level. It asserted the essential need for individuals to reconcile their own values with their surroundings. In other words, the first perspective illustrated the active position of a corporation to consider staff involvement and foster this as a management approach (Malloch 1997), the second perspective focused more on the individual and asserted the harmony between oneself and nature (the workplace they work for) was fundamental in creating a positive impact and improvement within a corporation (Brunet 2000).

On the other hand, authors such as Barnwell or Webley, and Cartright viewed Kaizen from a more holistic perspective. They believed that Kaizen related more to a wide range of practices that simultaneously resolved a set of objectives including quality control, cooperative labour-management, workers' cooperation, quality circles implementation, and minimization of waste and defects (Webley and Cartwright 1996; Barnwell 2007). In this sense, Kaizen has contributed an essential role for the development of Total Quality Management (TQM) tools (Macpherson 2008). Some of the tools include Kanban, Poka-yoke, Quality Circles, or 5S in respect of Japanese Methods, and including Reliability Centred Maintenance, Lean Manufacturing, Just-In-Time (JIT) inventory management, and Six Sigma for Western methods.

With differing perspectives, it would be much easier to construe Kaizen from an integrated notion. In this respect, Kaizen, as it has continuously evolved, has stimulated different perceptions depending on when, where, and how it has been used (Tozawa and Bodek 2012). Imai (2007) for example, has developed his own interpretation of Kaizen, thinking of it as a continuous improvement in every single moment and every single aspect within the company that involves every personnel and covers whatever happens to a company. Therefore, it varies from an incremental to a substantial innovation. This description was shared by the Western authors (Suárez-Barraza et al. 2011). Noticeably, Hamel (2010) expanded that view by concluding that Kaizen existed in many forms, from an event, a philosophy, a mindset, a performance to a means for achieving an optimal point in carrying on a business.

1.2 History of Kaizen

The history of Kaizen can be traced back to Japanese quality management, beginning at one of the most momentous milestones in Japanese history—World War II (WWII) (Table 1).

Before and during WWII, Japanese manufacturers had prioritized product quantity over quality and reliability as part of an effort to prepare for wartime mobilization. The adroit balance between quality and quantity was possibly a combination of several management methods that had been applied and adapted from Western know-how.

In 1931, Japanese engineers studied American statistical analysis, being able to develop a scroll system of control charts for production management. Turning to 1934, the quality management and statistical analysis for the production process were introduced into Japan by management expert Kanzo Kiribuchi. From 1942, more and more interest from both academic statisticians and military managers were put into quality control techniques as a result of the onslaught of WWII, and up to 1945, it

Table 1 The evolution of *Kaizen*

Important milestone	Historical event
During World War II, 1943	Development of American Training Within Industry program and Job Methods as the progenitor of Kaizen
After World War II	In 1954, Joseph Juran suggested Japanese expand and upgrade the concept of quality control Toyota first implemented the Kaizen philosophy by having groups of workers performing similar works, meeting regularly to solve work-related problems
1955–1965	Japan refashioned the national industry, creating a 100% Japanese Total Quality Control (TQC) Japanese government amended the labour contract with benefit distribution's guidelines and lifetime employment's opportunity—background for national Kaizen application
1962	Toyota introduced the Toyota Production System with JIT and Kaizen approach
1986	The term became widely known via Masaaki Imai's book about its potential for business competitiveness's enhancement through improving efficiency and productivity
Today	Kaizen has gradually become a crucial part in the manufacturing system not only in Japan but also in other countries around the world

was officially applied in the military workshops. However, it was not until 1947 when the American Edwards Deming arrived in the country to conduct education classes and programs that the Japanese (mainly members of the Union of Japanese Scientists and Engineers) became conversant with the American quality theory and adopted it in an industrial application. As well as an industry philosophy and an introduction to statistical analysis to industrial production, Denning also brought the Japanese a belief that product quality could be developed even in the middle of the destruction of WWII (Macpherson 2008).

After WWII, Joseph Juran (an American quality management consultant) emphasized the need for Japan to expand the concept of quality control in a more comprehensive way that could cover every aspect in operation and should be associated with every member of an organization, not just statisticians (Tsutsui 1996). From 1955 to 1965, Japan inaugurated a renaissance to refashion its national industry. The result was stunning with the new appearance of 100% Japanese Total Quality Control (TQC), forming a solid basis for subsequent quality revolutions in Japan.

It was during this period of vigorous change, that the concept of Kaizen was used and spread widely, especially after WWII, when Toyota, for the first time, conducted with groups of workers doing a set of similar works and meeting regularly to detect, analyze and resolve any problems in the production process. Along with the emergence of the industrial movement in the 1950s, the concept became very popular.

According to Berhe (2021), at this time, when governmental authorities recognized the country's contemporary problems of an unproductive management system and labour shortage, they looked for a solution by connecting the workforce with the companies' manufacturing processes. From that time, the first step was to create a basis under the labour contract that included a promise of lifetime employment and specific standards of benefit distribution in regard to the development of the company. This then became the solid cornerstone for later Kaizen activities that could better ensure both the terms of the contract and increase the workforce's confidence (Brunet 2000). It was first analyzed and presented by Imai (1986) with the potential to enhance Toyota's competitiveness by boosting its efficiency and productivity amidst increased competition in the context of rising globalization. Ever since then, Kaizen has eventually grown into an integral part of not only the manufacturing system in Japan but also in other countries as well, substantially contributing to its manufacturing success (Ashmore 2001).

Huntzinger (2002) demonstrates evidence of the evolution of Kaizen suggesting that in 1943, the development of the American Training Within Industry program, and Job Methods during WWII were the progenitors of Kaizen and lean manufacturing. In particular, fundamental techniques from these programs, namely learning by doing, supervising the development, coaching, and improving productivity through methods rather than technologies, contributed to the ultimate goal of waste reduction in the workplace and were significant factors in the Kaizen philosophy taking off.

2 Principles of Kaizen

Indeed, Japanese manufacturers had been acquainted for decades with the superiority in developing and applying Kaizen principles, making it as business strategy. Some outstanding techniques could be mentioned are the total productive maintenance (TPM), total quality management (TQM), and Toyota Production System (TPS). In implementing these principles, Japanese manufacturing workers had elaborated a number of variant methods and principles for their application (Murata and Katayama 2010). Kaizen was seen as the "umbrella" that covered most of the Japanese systematic practices (Kiran 2017). It was **identified as "management philosophy"** (first perspective) with a set of principles and tools for cooperative management (Suárez-Barraza et al. 2011). Yet, that was only one perspective of Kaizen. As noted above, interpretations of the Kaizen concept appear in various forms. It would be remiss not to consider alternative perspectives of Kaizen when discussing the principles.

The second perspective views Kaizen as a TQM's constituent. As stated by Deming (1986), continuous improvement (CI) is a part of TQM, and since Kaizen has a strong correlation with CI, it could be inferred that Kaizen is also an integral part of TQM. In saying Kaizen works as an element of TQM, Deming illustrates a process of constant examination of administrative, technical and operational processes so as to find better methods of working. In addition, other authors such as Ho and Fung (1994) or Hellsten and Klefsjö (2000) have also supported the idea of Kaizen being a valuable element in TQM when it is used as a management system.

The last perspective conceives Kaizen merely as a theoretical principle for improvement techniques. In the literature reviews of Suárez-Barraza et al. (2011), they found that the first perspective of Kaizen was apprehended at a micro level within an organization, as a so-called "management philosophy" with a focus on every single aspect of an organization, including work operations, manufacturing processes and worker involvement. Nevertheless, with this new perspective, its emphasis is placed on a macro level standing from an organisational standpoint. Being viewed as an ancillary principle to support the organization's improvement process means there is more of a focus on the organization as a whole (Lareau 2003).

By putting all the three perspectives (as above) together, it can be concluded that there are six integrated principles of Kaizen, namely:

1. **Teamwork**
2. **Focus on process and standards**
3. **Commitment from top management**
4. **Elimination of waste in workplace management**
5. **Education and training**
6. **Proposing and applying improvements.**

Among these, it would be reasonable to assume the first three principles are the core ones, while the others function as subsidiary techniques and reinforce the top core principles, thereby gaining an overall success of Kaizen application. Therefore, the first three principles are discussed in the following section. This order of precedence also aligns with Leseure's notion (2010).

- **Teamwork (Involvement)**

When looking into the organisational milieu, it would be meaningless to adopt only one employee when implementing the philosophy. Accordingly, it is crucial for all personnel and departments to embrace and perform the Kaizen method. In the journal publication by Vieira et al. (2012), the authors emphasized the essential role of personal improvement. That means when it comes to work, everyone should be involved in the improvement process. In that way, the improvement will be conducted more comprehensively throughout every single aspect of the organization with more active participation from the employees to the leadership board. That will help to improve the capability to boost productivity and meet improvement expectations, producing products of higher quality and shorter delivery time.

- **Focus on process and standards (Small and incremental development)**

As mentioned above, the Kaizen principle is not in favour of radical transformation. In fact, its core method is to follow small yet frequent attempts so as to better manage and improve the quality of practices. As stated by Saleem et al. (2012), the Kaizen process focuses on a bottom-up approach, which requires small, incremental changes over an operating system. This approach will help to address all possible problems in a more absolute manner. In fact, this is seen as a distinguishing feature of Kaizen (and differs from TQM). Focusing on small improvements is actually a result of Kaizen's ongoing effort to achieve operational excellence because continuous improvements will accumulate within the overall performance of an organization. From that, it encourages individual responsibility, enhances learning while working with everyone, and strengthens discipline in cooperation as well (Singh and Singh 2015).

- **Commitment from top management (Commitment and persistence)**

It would be challenging in attaining improvement without incentives that are delivered in a clear, strong, and sustained manner to pursue and execute improvement. A similar approach is adopted just right in the workplace's philosophy. Vento et al. (2016) assert the crucial role of commitment from the management board as well as from the human resource development in achieving maximum benefits from implementing Kaizen. As noted above, Kaizen is a long-term strategy for a business to reach an optimal state where improvement is constant, by all the workers, and in every facet within an organization. Hence, without a strong commitment and persistence from all workers, especially the management board, the continual organizational strategy would remain ineffective, counterproductive, and even pointless.

After assuring the key principles, an organization should look towards achieving the remaining fourth, fifth, and sixth principles, in order to meet the most satisfactory expectations of Kaizen. In the effort to continue detecting flaws and improving operational processes, the need to eliminate *Muda* (waste) in **Gemba** (office) is expected as part of the improvement and innovation process (Wittenberg 1994). Proper education, training, and diffusion of the Kaizen philosophy and methods among the organization are also necessary before and during the implementation to make staff sufficiently well prepared to be part of significant change. Finally, proposing and applying improvements with suitable techniques and practices for specific circumstances within each organization is the last step in fulfilling the series of Kaizen principles.

3 Concepts of Kaizen

Along with the evolution of Kaizen, research has suggested different approaches to illustrate "continuous improvement" with various methods of classification. Yet, they all agree on the scope of Kaizen, that it comprises improvements in individual works, the group works, and the factory as a whole. From the viewpoint of **Masaaki Imai** (1986), continuous improvement can be implemented using three kinds of kaizen: management-oriented, group-oriented, and individual-oriented kaizen. First, **management-oriented kaizen** implies improvements in the factory layout by assessing company strategies and all aspects of business operations to advance the manufacturing system. Second, **group-oriented kaizen** is described by quality circles, in which workers are formed in groups with the aim of detecting problems, analysing the causes, and later establishing new standards and procedures to restrain the recurrence of problematic works within the same process. Third, **individual-oriented kaizen** focuses on providing qualified skillsets for workers to assist them in designing proper solutions for hidden problems in day-to-day works, thereby enhancing productivity in related working areas.

The interpretation by Capers Jones (2017) is similar. From his perspective, Kaizen is divided into two concepts, system kaizen and process kaizen, based on the process focus and the target responsibilities (Fig. 1). This separates the duties between

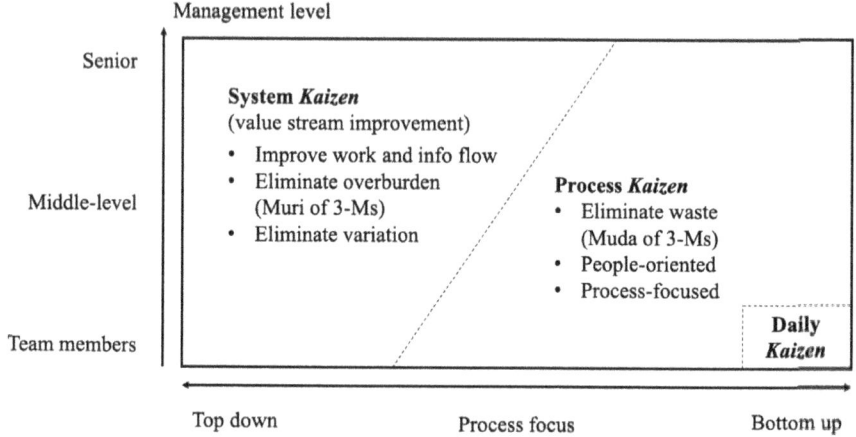

Fig. 1 The Kaizen concepts of Jones (2017)

management levels and applies appropriate approaches to eliminate a certain component in the 3-M model [Muda (waste), Muri (overburden) and Mura (unevenness)]. Specifically, **system kaizen**, otherwise known as flow kaizen, is executed with a top-down approach under the supervision of mid-to-senior management. This enables improvements for the overall value stream and eliminates "Muri" (overburden) in the 3-M model. By contrast, **process kaizen** exercises the bottom-up approach by concentrating on quality circles and improving individual or group works from repetitive production mistakes. This allows producers to achieve the elimination of "Muda" (waste). In addition, under certain circumstances, a rare case of process type, **daily kaizen**, is required to tackle out-of-the-ordinary events "Mura" (unevenness) that demand immediate attention but are minor in scope.

The research of Burton and Boeder (2003) suggests a different classification of Kaizen events. However, it still aligns with the aforementioned concepts. Kaizen events are classified as project kaizen and process kaizen based on the duration of implementation and the target outcome. **Project kaizen** focuses on value stream improvement by advancing the value of multiple production functions within the same stream. This type of Kaizen requires weeks of implementation and a top-down approach to fully develop an overall value stream. On the other hand, **process kaizen** is introduced as a short-duration, employee project with the goal of rapid targeted process improvement. These initiatives involve a small group or individual workers working intensely within a limited amount of time to detect and resolve problems at an equal level of difficulty. Driven by time constraints, process kaizen is further split into two types, **kaizen blitz**, which takes several days to complete, and **kaizen super blitz**, which demands immediate solutions within one to eight hours (Fig. 2).

Yet another perspective from Taiichi Ohno (1988), the inventor of the Toyota Production System, has formed a sequence for Kaizen implementation, which consists of Sagyo kaizen, Setsubi kaizen, and Kotei kaizen. Starting with **Sagyo**

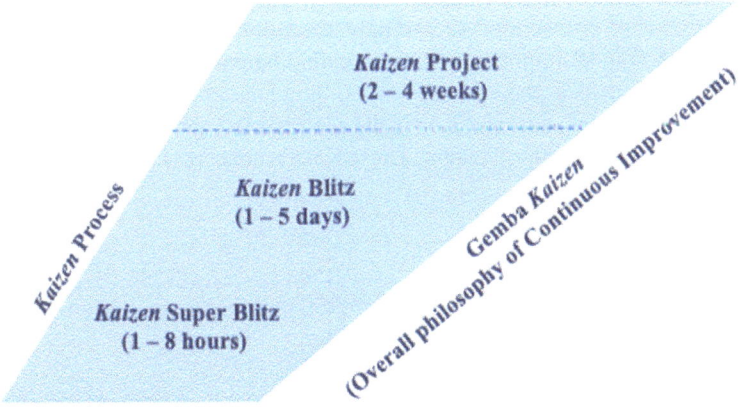

Fig. 2 The Kaizen concepts of Burton and Boeder (2003)

kaizen, also known as operations or manual work improvement, this is a people-oriented concept that concentrates on training programs for lower-level employees to utilize existing resources and enhance their working efficiency. After this prior step is completed, **Setsubi kaizen** (equipment improvement) is executed by encouraging workers to make use of the available machinery in new and creative ways. Finally, when the operations of the previous phases are running smoothly, **Kotei kaizen** (process improvement) becomes attainable with a strong linkage between the efficiency of human and capital resources.

The work of Shigeo Shingo (1988) offers similar improvements based on the priority of targets. Starting with "**easier**", which implies a requirement of simplification in the production line for working employees, higher productivity becomes achievable when time consumption in complex manufacturing systems can be mitigated. The second target, "**better**", will then become obtainable with improvement in the quality of the operation and/or product. Next, further improvement activities, or a "**faster**" initiative, can be implemented which will boost production efficiency. Finally, in the context of fulfilling the previous targets, a "**cheaper**" process becomes attainable, with cheaper or lower costs of goods sold per unit.

The Kaizen models of Hiroyuki Hirano (2009) further visualize the concepts and connect the previous theories mentioned through an explanation of vertical and horizontal improvements. The first concept, **point kaizen** is similar to operations improvement carried out by observing the movements of each worker to identify and eliminate common causes of workplace accidents. This is a vertical improvement and facilitates a simpler, more efficient, and less wasteful production. The second, **line kaizen** is an improvement that accumulates all the related point improvements and connects and paves the way for an improvement inflow. But even so, it is crucial to avoid applying rigid rules to force a connection, or to neglect intrinsic relationships and skip important stages. This model line also shows a vertical development being applied within a specific type of production line segment or product. The third, **plane kaizen** is achieved by utilizing the effectiveness of a line model and applying it to other parts of production. Accordingly, the whole factory floor is improved by this follow-the-model improvement, or so-called lateral or horizontal development of model lines across a plant. Finally, when the plane improvement has developed across multiple departments and even in the enterprise's supply chain, **cube kaizen** is obtained as a systemic improvement. These Kaizen concepts are discussed in Fig. 3.

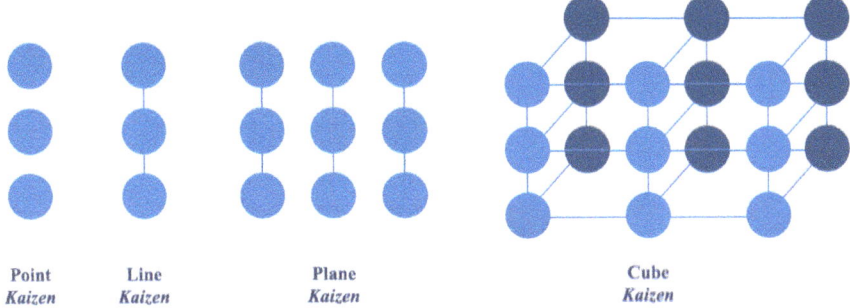

Fig. 3 The Kaizen concepts of Hirano (2009)

4 Techniques and Practices of Kaizen

According to Imai (2007), there are no standard techniques or practices that are vital for the implementation of Kaizen. There have been numerous overlapping and corresponding techniques or practices that apply the methodology of Kaizen. These techniques can be listed as below and some of these concepts have been discussed in other chapters. However, a brief of the techniques has been discussed here.

- 5S method (workplace organization)
- Jidoka (autonomation)
- 5 Whys technique
- Elimination of seven wastes (Muda)
- Poka-yoke (error proofing)
- PDCA cycle (Plan-do-check-act)
- Total productive maintenance (TPM)
- JIT system
- Total quality management (TQM)

5S method (workplace organization)

The 5S method has been prominent in the textile industries of Brazil and India. It has been applied successfully to operational processes, informational control, and especially within an organization as a whole (Carvalho et al. 2017; Juneja et al. 2012). But before introducing the 5S method, some advanced practices that are used to apply the Kaizen methodology must be briefly described.

Jidoka (autonomation)

The term Jidoka can be described as "Intelligent autonomation" or "autonomation with a human touch" (Lina and Ullah 2019). When using Jidoka, mistakes that occur in operational processes can be immediately identified, addressed, and corrected. A defect in a processed component is automatically identified by a machine. The

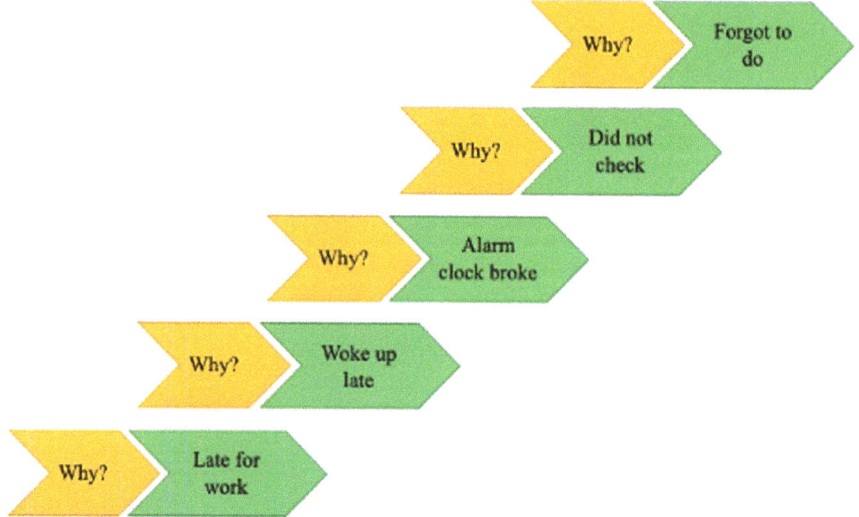

Fig. 4 The 5 Whys technique

production line then promptly pauses the working mode in order to allow time to assure the required quality of the product.

5 Whys technique

This was invented by the founder of Toyota and became popular in the 1970s as part of the Toyota Production System (TPS). The 5 Whys technique promotes a systematic process to identify the reason for a problem, thereby leading to an accurate solution. By asking "why" for each question, the user can identify and distinguish the symptoms, and in this way identify the root cause of a problem. Appropriate steps can then be taken to prevent it from happening again (Lina and Ullah 2019). The 5 Whys technique is illustrated below in Fig. 4.

Elimination of 7 waste (Muda)

The term "**Muda**" can be translated as "waste". This technique will extract any non-added value action or processed component that is not necessary for customer satisfaction or to enhance the quality of the final product. The seven classes of "**Muda**" (Lina and Ullah 2019), are:

(1) Overproduction
(2) Inventory
(3) Waiting
(4) Motion
(5) Transportation
(6) Rework
(7) Over-processing.

Fig. 5 The PDCA cycle

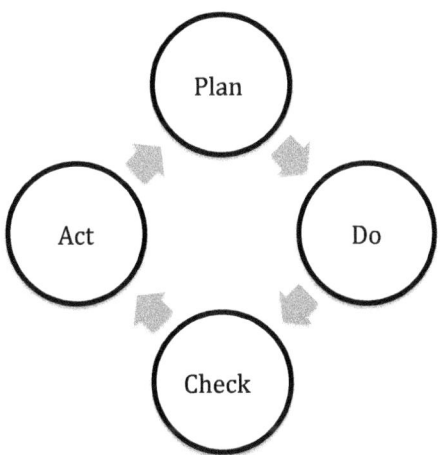

Poka-yoke

Poka-Yoke is a Japanese term that means avoiding mistakes (mistake-proofing). It is a basic mechanism that was developed as part of the Toyota Production System (TPS) to help avoid (Yoke) mistakes (Poka). This method is designed to prevent a product from being approved when it is defective or faulty, due to an error in a prior phase and helps maintain a stable process.

PDCA cycle

PDCA is a repetitive management method of continuously improving the production process. The cycle consists of four phases, Plan-Do-Check-Act (Fig. 5). At the Plan stage, the manufacturer must identify the needs of customers, the problems that need solving, and the causes, set goals, establish the scope of a plan, and collect appropriate data. At the Do stage, the improvement plans will be implemented. At the Check stage, verification will be executed to assure the corrective actions are being properly carried out. Lastly, at the Act stage, the firm may standardize all processes or maintain the current improvement activity (García-Alcaraz et al. 2016).

5 Implementation of Kaizen

According to García-Alcaraz et al. (2016), implementation of Kaizen can be separated into three different stages:

5.1 Kaizen Planning Stage

In this stage, the company must focus on planning the implementation of a Kaizen system or continuous improvement (Recht and Wilderom 1998). At this stage, raw materials are purchased, sales are organized, and human resources receive training. Moreover, research and analysis of available information must be executed to identify any possible problem in the production process. The three important elements in this process, consist of **Managerial commitment, Organizing the work team,** and **human resource training**. In **managerial commitment**, organizational changes must be carried on continuously to achieve ongoing improvement. An acquisition plan for all resources, policies, procedures, regulations, culture, customer feedback, and a process for detecting the failure must be taken into account in order to adopt changes. During **work team organization**, communication is a key to the success of continuous improvement since goals and objectives must be clearly transmitted to the employees. Motivation, enthusiasm, and contribution are also important to the success of Kaizen. Lastly, in **human resources training**, it must be ensured that workers on the plan have appropriate skills to use tools and methods with the support of relevant training and education.

5.2 Kaizen Execution Stage

After setting a detailed plan, a continuous improvement system is applied. At this stage, changes in working groups, training systems, and disciplines will be implemented. Different techniques and practices of Kaizen will be utilized depending on the status of the subject company. There are three critical factors involved in this stage, including **Successful implementation of proposals, Human resource integration,** and **Customer focus**. In the **successful implementation of proposals**, all stakeholders of the company must commit to the proposals, to successfully execute the plan. Techniques to implement Kaizen strategy vary based on the geographical condition, organizational structure, and business status, In human **resources integration**, the assurance of incentives, efforts and achievements, recognition, self-discipline, and well-defined organizational culture must be attainable to succeed in adopting Kaizen. Lastly, in **customer focus**, the company must ensure the main achievement of adopting Kaizen is to enhance the product quality and gain customer satisfaction.

5.3 Kaizen Control Stage

Adequate control and administrative processes are necessary to achieve Kaizen. All parties must be aware of the need to reject unnecessary costs and implement savings on information and documents in different improvement projects (Maarof and Mahmud 2016) and ensure an appropriate organizational culture. There are three different variables in the Kaizen control phase. They are the **Communication process**, **Documentation and Evaluation Process**, and **Organizational Culture**. In the **communication process**, requirements and essential needs must be effectively exchanged between superiors and peers in order to achieve successful improvement changes. In the **documentation and evaluation process**, performance and results should be assessed with proper instruments to fine-tune the process, detect failure, and enhance quality. Lastly in **organizational culture**, according to Caswell (1998), the aim of Kaizen must be to achieve an ideal outcome by balancing the interests of the individual with the common interest and well-being.

6 Benefits and Drawbacks of Kaizen

6.1 Benefits of Kaizen

According to García-Alcaraz et al. (2016), and Lina and Ullah (2019), Kaizen is popularly applicable due to certain benefits such as waste reduction, increase in output quantity and quality, and optimization in operation and production. As a result, businesses that adopt the implementation of Kaizen can achieve their economic advantage, competitive advantage, and human resource advantage.

In terms of economic benefit, the common achievement can be the reduction of material picking and storing, employees, motion and time in the process, cycle time, lead time as well as better inventory management. Kaizen can reduce waste, specifically ignoring and discarding non-added value and defective items, thus cutting irrelevant results and costs. It also improves the time and motion within the operation with an appropriate arrangement of materials and machinery. Activities, which are unnecessary, will also be diminished to optimize the working process. Moreover, Kaizen absolutely maximizes the productivity of each factory where it is applied. Firstly, inventory management can be efficiently quantified with desired amounts. Subsequently, the cost of transportation and storage are minimized, thus optimizing the profitability. With the application of advanced manufacturing technology, production and logistic systems are upgraded, thus complying with the quantities and delivery time.

Moving to the competitive benefit, the improvement in product quality can support the competitiveness of businesses in the market. With proper training and instruction, employees can gain more skills and knowledge, thus participating in complex tasks and increasing the utilization rate. Moreover, the appropriate maintenance, changeover, and check-ups schedule of machinery and equipment can ensure the quality of final products. With modern technology implemented, product samples and designs can attain higher quality. Besides, the adoption of a holistic perspective induces the creation of multidisciplinary teams to deal with problems, hence meeting the interest of the business with the promotion of process-oriented thinking.

Another competitive advantage of Kaizen is the rapid adaptation to market change. The production flexibility can be maximized for the agility of the supply chain, specifically the communication and information technology. Therefore, businesses can fulfill the customers' needs and satisfaction in the circumstances of market trends or demands.

Lastly, the human resource benefit is a crucial element in the Kaizen methodology. Better communication and cooperation within the business can drive the success of the whole organization. Proper managerial leadership can supervise, instruct and promote continuous improvement in the process. Collaboration between departments can ameliorate the participation of employees. As a result, along with effective training plans and rewards, employees are encouraged by motivation, responsibility, self-esteem, attitude, and professional experience. Additionally, a higher working standard with occupational and health safety is an incentive for workers. Therefore, businesses can retain a skilled workforce and reduce the attrition rate from customers as a consequence of the added values that are translated into customer satisfaction and experience.

6.2 Drawbacks of Kaizen

According to Goshime et al. (2019), the implementation of the Kaizen methodology has three major barriers, which are a managerial barrier, operational barrier, and financial barrier.

In terms of managerial barriers, the lack of awareness, culture, and knowledge of lean philosophy cannot mitigate the risks that businesses are facing. Moreover, the failure of adopting Kaizen can worsen the current situation. Besides, the unavailability of appropriate tools, equipment, and real-time data may postpone the full accomplishment of Kaizen. Additionally, the lack of communication between the management team and workers can unsuccessfully transmit the idea and perspectives of Kaizen as well.

Moving to operational barriers, one of the major disadvantages of Kaizen is the resistance to change (Berhe 2021). Implementing Kaizen will adjust the current management system, as a result, affecting the business operation at the primary stage. The current workforce and administration can be reluctant to apply Kaizen as the existing flows of thinking and freedom will be stifled (UXKaizen n.d.). Moreover, adjusting the current working habits is challenging as workers and managers observe no benefits, motivation or enthusiasm to adopt the full concept. Consequently, it can lead to the loss of employees, dissatisfaction, and negative attitude within the organization as well as lower efficiency as compared to the previous situation. Therefore, businesses must ensure that Kaizen methodology is communicated as a tool to enhance efficiency and productivity, thus possessing a higher wealth for the organization as a whole.

Ultimately, financial barriers are preventing many enterprises from implementing the Kaizen methodology. Businesses that are not financially well-prepared may suffer from frustration since major changes are applied to their current operation (Lill 2016). Accordingly, training and specialized programs for the workforce require a large number of fees to enhance professionalism, knowledge, and experience. Moreover, some enterprises require high costs for adoption, thus making it difficult to transform back to the prior management system. The lack of resources, including material, human resources, or machinery may hinder the capability of Kaizen.

7 Kaizen in Fashion and Textile Industry

The fashion and textile manufacturing industries provide employment to about one-third of the population in developing countries and largely contribute towards the gross domestic product (GDP) of the country. However, in many developing countries, manufacturing is still labour-intensive, in spite of some development of automation technologies (Nayak 2017). The traditional technologies produce a large amount of waste, consume higher energy and operate with low efficiency, which is discussed in the following section.

7.1 Use of Traditional Technology

Although there is a wide scope for the fashion and textile industries for the use of modern technologies, there has not been widespread modernization due to high cost, the complicated modernization processes, and the availability of cheap labour. Despite various advantages of implementing the new technologies, traditional technologies are still put in operation by many labour-intensive industries in developing countries. This can be attributed to: (a) high cost of fragmented developments in fashion and textile industries compared with the other sectors, (b) overwhelming number of cheap labour available to work in many developing countries, (c) high skill requirements and complexities of new technologies, and (d) frequent style changes creating challenges for the implementation of new technologies. The major problem is the high cost of implementing modern technologies. To overcome the problem or quality and productivity, the manufacturing industries in developing countries should focus on investing in new technologies in addition to the implementation of lean manufacturing tools such as Kaizen.

7.2 How Kaizen Works in Fashion and Textile Industries

7.2.1 Line Balancing

As stated by Ongkunaruk and Wongsatit (2014), line-balancing problems stem from the standardized products produced by mass production with a cost-efficient strategy. Line balancing is the process to level the operations in the manufacturing process so that output from one section matches the input from the other. In fashion manufacturing, there are a large amount of waiting time due to bottlenecks in various processes. Line balancing find the bottlenecks in processes and tries to find the root cause for improvement. Generally, garments are manufactured in a progressive bundle system (PBS) in garment industries, where several workers work in one line but with different machines. Each operation has a standard time and the productivity per hour is calculated from this. The lines with lower than the standard productivity are the bottlenecks, which should be improved. The use of line-balancing as per the kaizen makes the process smoother by removing the bottlenecks or delays from the processes, which leads to smoother operations with fewer waiting times.

The first example that textile industries receive benefits from the implementation Kaizen is highlighted through a printing company's improved productivity (Chan and Tay 2018). With the application of lean tools such as Kaizen, the textiles printing operation enhanced its productivity. However, the existence of labour-intensive production processes, their skillsets, working environment, fatigue, etc. has posed major challenges for such line-balancing tools to work out successfully. Identically, the impact of Kaizen on problem-solving had been proved through two departments—the handicraft and the packaging department. With the implementation of the lean tool combination (line-balancing tool, facility layout, and standardized work), Kaizen helped the industry to reduce cycle time, work in progress (WIP), and product advancement, then, reduced the waste and increased the throughput in the printing industry.

7.2.2 Facility Layout

Layout-related problems are considered, by Koopmans and Beckmann (1957), as common industrial problems in the fashion and textile industries. To support this statement, Heragu (2008) also claimed in their study that layout and setups, along with choosing a team for layout and facility management are recognized as weak points in many industries. According to Koopmans and Beckmann (1957), the main objective for resolving those issues is to minimize the transportation cost when moving the materials between different workspaces or workstations. The improper layout (i.e., arrangement of departments and work sections in sequence) is a major problem in many fashion and textile industries, which leads to unnecessary or extra transportation of material and motion of workers. However, reducing material handling costs is not the only result that an effective facility layout can generate. In fact, with a well-designed layout, it can also help reduce time utility, including throughput time and WIP time.

Chase et al. (2006) suggested that facility layout should be based on the nature of a department in which similar machines could be used in the same area for best performing their functions. Nevertheless, it still depends on other factors to determine how to set up the most efficient layout. For example, in the textile printing industry, when it comes to the post-press process, they only need smaller and easy-to-move machines, making it more flexible for layout design (Chase et al. 2006). In developing countries working with fashion and textiles, where finding a favourable working space with fewer constraints is a challenging task, the areas are usually designated for specific departments which are commonly separate from each other. This can possibly lead to long-distance in the operational department, taking more time for a process cycle (Chan and Tay 2018).

To solve those above-mentioned issues, Kaizen could be applied for manufacturers to attain Lean Manufacturing's goal. When considering the main purpose of Kaizen which is to reduce Muda (waste) in the workplace, the concept appears to be highly applicable in eliminating waste generated by ineffective facility layouts in the fashion and textile industries. Noticeably, as Ohno (1988) indicated the seven types of waste that are likely to occur in the manufacturing processes, the first four wastes namely transportation, movements between workstations, inventory, and waiting time, could be reduced by designing a good layout architecture. Other studies conducted by De Carlo et al. (2013) also demonstrated the same result that the approach of Lean Manufacturing is the most optimal choice for cost saving by removing wastes including reducing the transportation and handling time due to better layout.

7.2.3 Standardized Work

According to Whitmore (2008), unsystematic task allocation and working sequence based on personal preferences of the workers would lead to unpredictable and unmanageable outcomes, making it hard to obtain flow and pull in the processes. Therefore, Shang and Pheng (2013) proposed standardization as a noticeable tool of Kaizen, which shapes both products and processes with a high level of standardization. This highlights the crucial role of Kaizen in production process management, which is to ensure the process follows the standard operating procedures and only functions within prescribed tolerances. The standardised work leads to standard lead time, hence the standard Takt time for delivery of products. Takt time plays a crucial role in achieving standardization since it estimates the rate of customer demand for a product or a service within the negotiated time frame by dividing the net available time by the quantity demanded (Kumar and Thavaraj 2015). Furthermore, the work sequence must be systemized to ensure consistent workflows amongst the production process (Chan and Tay 2018). In addition, standard WIP must be established to track manufacturing inventory and eliminate wasteful items and defects, therefore with the development of WIP, the problem becomes visible from the detected sources (Whitmore 2008).

7.3 Challenges

With increased consumer awareness about high quality and sustainable fashion, the fashion and textile industries are facing several challenges in maintaining stable growth. Such challenges include financial constraints, the well-being of the labour force, and consumer concerns about environmental issues, which are discussed in this section.

7.3.1 Financial Constraints

A French fashion designer, Kenzo Takada once said: "Fashion is like eating, you shouldn't stick to the same menu" (Eli Motivation Personal Classic Style 2019). The quote has become a ubiquitous mindset for worldwide shoppers, who intrinsically purchase more and more clothes to satisfy a need for constant change. However, the shoppers always demand good quality clothes at a cheaper price. The fashion and textile industry has been described as highly volatile, low-predictable with frequent errors and a short product life cycle in the manufacturing processes of fabric and sewing lines (Ashraf et al. 2017). These trends pose major challenges for fashion-related market players, especially those fashion and textile industries that rely on producing fashion garments to implement lean manufacturing concepts. The same factors also lead to a single major inventory controlling issue, imposing a financial constraint on fashion companies, which is considered severe. The financial obstacle has always been an identical problem worldwide within this industry, not only for large enterprises but also for small and medium-sized enterprises (SMEs) (Ayyagari et al. 2006). So, the interdependence of implementing lean manufacturing to solve the inventory controlling and waste management problem and the resultant financial constraints create a never-ending cycle in this industry.

7.3.2 Concerns for Workers in the Fashion and Textile Industry

An increase in the number of sophisticated consumers has led to more purchasers at cheaper prices, but it is becoming critical for the health and safety aspects of the workers in the workplace manufacturing goods like fashion, apparel, and beauty products (Fan and Lo 2012). Occupational Health and Safety Management System (OHSMS) becomes a prime requirement for leading fashion brands and retailers in order to avoid scandals related to labor abuse, non-payment of wages, unsafe working conditions, and use of forced labour in the fashion and textile supply chain. Hence, the necessity of acquiring OHSMS certification (such as OHSAS 18001) for the fashion and textiles industries has become of utmost importance and is another factor directly affecting the financial performance of a company (Hoque and Shahinuzzaman 2021). Hence, the primary importance of the companies is to solve the OHSMS problem rather than implementing lean manufacturing.

7.3.3 Damage to the Environment and Human Health

Rising concerns about environmental pollution have led fashion and textile industries to face problems stemming from detrimental environmental impacts from various fashion production stages (Slater 2003). As the demand for fast fashion is growing, consumers are purchasing more trendy clothes. This means production is boosted to satisfy growing demand and results in increased textile-related waste (Scofield 2017). This has caused a surging carbon dioxide (CO_2) emission of over 850 million tons, related mainly to manufacturing, transportation and use of goods (e.g., washing, drying, and ironing) (Gavranović 2020). The fashion and textile manufacturers are listed within the top 5 most polluting industries in many countries. Moreover, advancements in the treatment of textiles, such as coating, and functional finishing have also been reported to have deleterious effects on human health. Specifically, textile contact dermatitis is becoming an alarming disease for the laborers in the industry (Alinaghi et al. 2019). In this context, benefits obtained from Kaizen may ameliorate the current problems, which beset the fashion and textile industry. The Kaizen framework of a sustainable operating model involving waste reduction, qualified labour skills, disciplinary and optimal operation, may have the potential to resolve these issues through its structural improvement processes.

7.3.4 Large Amount of Waste Generation

Many of the fashion and textile industries produce copious quantities of solid hard waste, wastewater, or effluent and toxic greenhouse gases (GHGs). The use of manual processes can be one of the major causes of waste generation in addition to the inherent nature of garment manufacturing. Whether it is spinning, weaving, fabric manufacturing, chemical processing or garment manufacturing, several processes are still done manually, which may lead to more wastage. Furthermore, the use of traditional technologies in many fashion and textile industries located in developing countries leads to the generation of wastes. The use of Kaizen concepts even in manual processes can result in the minimisation of waste generation.

7.3.5 Usage of Excessive Energy

The major reason for the fashion and textile industries to use high energy is the use of traditional manufacturing methods. Modern technologies are energy efficient and result in higher productivity and better quality. Several industries are also suffering from the problem of large waiting time. Due to the lack of synchronisation, the machines have to wait for raw materials to be processed. Further, the financial constraints faced by several industries lead to the use of traditional technologies. Switching to new technologies and implementation of Kaizen can provide some solutions in terms of energy saving.

8 Case Studies

8.1 *Textile and Apparel Industries in Bangladesh*

This section discusses some case studies from the textile and garment industries located in Bangladesh. Bangladesh is the fourth largest manufacturer of fashion and textiles in the global supply chain. The case studies in this section are from fabric and garment manufacturing industries.

8.1.1 Fabric Fault Analysis

Fabric constitutes the main raw material used in the apparel industry and constitutes 60–70% of the total cost of a manufactured garment. The failure to early detection of faults in the fabric has always posed difficulties in the garment. Most faults which mandate rejection of the final product are apparent in the initial stages of fabric manufacture. The problem is compounded by various factors including inaccurate checking, lack of adequate lighting, human error in identifying defects, lack of discipline regarding workpieces, skipping inspection due to lack of time, worker stress and fatigue (Ashraf et al. 2017). However, the implementation of Kaizen helped to reduce the number of fabric faults in a Bangladesh fabric manufacturer. Progress was made by applying the Kaizen framework as shown in Table 2. Various tools such as an improved table arrangement (Fig. 6), a defect board (Fig. 7) with brighter lighting facility to support workers in identifying defects in the fabric, and most importantly, a performance report board (Fig. 8) were used to support the workers to identify the percentage of rejected fabrics. The entire process of implementing Kaizen took about three months for process analyses, determination of problems, suggestion of solutions, and other paperwork, without any improvement. Further, a pilot run was staged over 2 months for efficiency checking. After 10 months of observation and working with Kaizen, fabric faults were decreased by about 50% as compared to the previous process, and that has been a milestone for the industry (Table 2).

Table 2 Monthly rejection rate due to fabric fault

Month	Year	Total production	Total fabric fault	% of fabric fault
Oct	2014	400,386	517	0.13
Nov	2014	523,644	723	0.14
Dec	2014	476,763	558	0.12
Jan	2015	528,210	580	0.11
Feb	2015	469,758	744	0.16
Mar	2015	565,707	1621	0.29
Apr	2015	478,625	508	0.11
May	2015	465,209	203	0.04
Jun	2015	475,363	213	0.04
Jul	2015	491,407	201	0.04

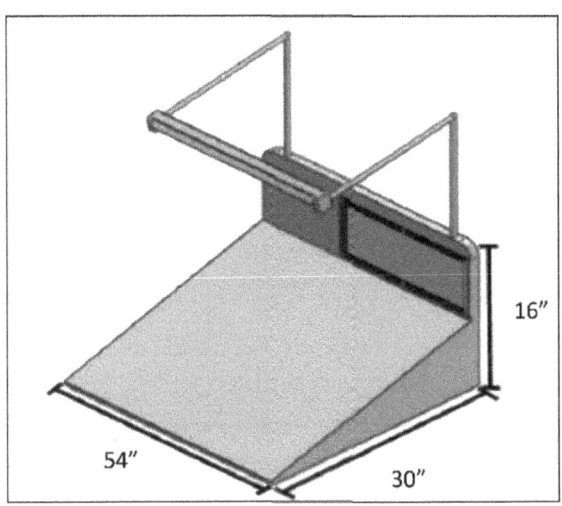

Fig. 6 Work part for improvement in table

8.1.2 Invention of a Shoulder Pad "Gadget" in a Sewing Line for an Ergonomic Standard and Labours' Comfort During Work

An identical case study illustrating the enhancement of workers' comfort through application of Kaizen principles was the implementation of a shoulder pad "gadget". Before Kaizen was introduced, the ready-to-be-joined shoulder pads used to be kept in transparent bags on the left side of the workers during the production of suits, jackets, and overcoats. The inappropriate arrangements of shoulder pads result in a time-consuming working and difficult size selection process and were also space-consuming, which was a waste in the process. The rate of production was slow, which affected the productivity and morale of the workers. To solve the problem as

Fig. 7 Defect board used in cut panel

PERFORMANCE REPORT				
CUTTING TABLE NO: 0				
DATE:				
Sewing Line NO:	Buyer	Order NO.	Input	NO. of fabric fault

Fig. 8 Performance report board

per Kaizen principles, a shoulder pad "gadget" was invented. The frame was 27 inches long, 40 inches high, and 12 inches wide and was made of steel. Each frame was divided into three box segments, which were allocated by small, medium, and large sizes. The total capacity was 420 pairs of shoulder pads. An air-filled cylinder was fitted onto the frame to balance the weight of the pads and the pressure of the air. With this, the next pair of pads automatically appeared under the air pressure whenever the one on top of the gadget was taken. This naturally reduced the operational steps for the workers, thereby increasing their efficiency, productivity as well as their job satisfaction.

8.2 Productivity Improvement in Peru

"Industrias EKL" is a small and medium-sized enterprise (SME) manufacturing industry with expertise in manufacturing men's garments (Tapia-Leon et al. 2019). Located in Lima, the capital of Peru, Industrias EKL mainly produced T-shirts for global fashion brands. The company was struggling financially. Customers were dissatisfied and switching their sourcing elsewhere often because the company was unable to fulfill production orders on time. The company's investment in Kaizen turned the results around and improved the working conditions. A Research team in the company collected information and data before and after applying the Kaizen methodology which demonstrated a comprehensive turnaround. Three phases were proposed to improve the performance of the company.

- **Phase 1**: This phase involved process mapping, through the application of value stream mapping (VSM). The preparation stage in garment manufacturing was shown to be disproportionately longer, compared to other stages such as marker making, cutting, finishing, and packaging, and was slowing down the whole process.
- **Phase 2**: This phase involved the analysis of waste and numerous non-value-added activities, which were contributing to unnecessary costs. In particular, the activity of searching and collecting the right garments in the manufacturing process took the longest time and cost, which could be described as inefficient waiting time and inventory management, which was established using the Pareto analysis.
- **Phase 3**: This phase involved work distribution, using the 5S method to accomplish a continuous improvement to solve the above problems. As a result, preparation time was more efficiently controlled. The manufacturing cycle and delivery time were reduced by the implementation of 5S. Management of inventory was optimized, allowing a 10% reduction, and more effective organization of existing stock. Significantly, timely fulfillment of production orders increased by 15%, along with a 25% increase in equipment usage. Customer satisfaction was also restored by the implementation of 5S.

8.3 Reduced Lead Time in India

A final case study illustrating how the Kaizen application benefits the Indian garment industry. From an industrial perspective, the traditional garment industry in India confronts several shortcomings including longer lead time, low productivity, poor line balancing, rejections and rework, and inflexibility in the manufacturing changeover (Kumar 2019). In the context of a need for improvement in this sector, the implementation of Lean Manufacturing techniques, including Kaizen principles has been shown to bring the advantages that they can bring. More specifically, the capacity of Kaizen applications to improve cycle times by modifying the layout for productivity improvement is highlighted.

In a normal sewing process, key activities are usually carried out on the floor including designing products, sourcing raw materials, trims, and then proceeding on to the manufacturing, including cutting the fabric, sewing, inspecting, packaging, and shipment. The case study involves the manufacturing of a Colombia 1058 jacket at Gokaldas exports. Figure 9 shows a detailed layout of the sewing section from the beginning to the final stages. In the manufacturing of the Colombia 1058 jacket, each item is prepared with 154 workstations over 130 min cycle time during which the waste occurred for 30 s per jacket. For a total of 200 jackets prepared per day, resulting in roughly 1 h and 40 min of wasted time. The reason behind most of the wasted time was because of improper arrangement of the workstations and unwanted movements by checkers during the checking stage.

Figure 10 demonstrates the proposed layout of a sewing section that would help to address the disadvantages in Colombia 1058 jacket manufacturing. Significantly, the major changes appear in the adjustment of the fashion tab, bringing it to the front ready instead of directly loading, then checking, then zip to collar. That was expected

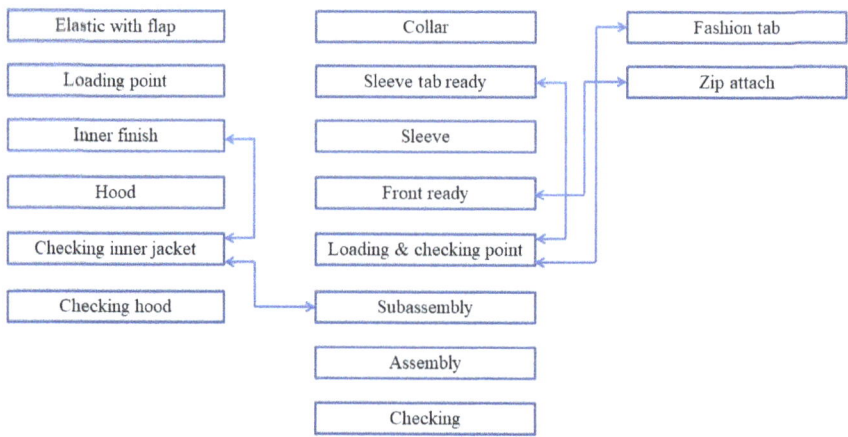

The current layout of the sewing section

Fig. 9 The current layout of the sewing section

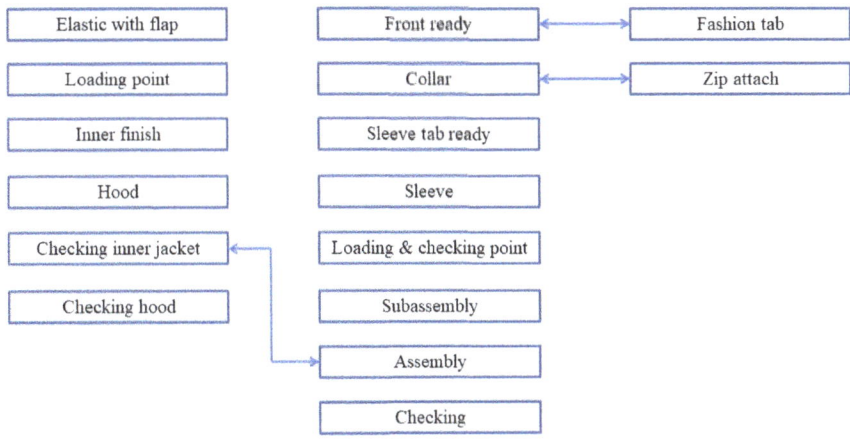

Proposed layout of the sewing section

Fig. 10 New layout of the sewing section with Kaizen application

to result in not only time-saving but also boosting the producing sequence so that the cycle time was reduced by up to 50%. In short, there were two aspects that could be improved from just a small change in the layout of this producing system (by applying Kaizen principles). These were time savings and a productivity increase. Other modifications of the layout to achieve higher productivity and save more time are specified as follows.

- Moving front ready before collar and sleeve sectors so that the fashion tab and zip attach could be near front ready and collar areas.
- Moving the checking table in between the hood and inner finished eliminating time wastage (around 30 s) per 10 jackets on an average.
- The new design of workstations as in the proposed design helped to save about 1 min overall for an average of 4 jackets.

9 Conclusions

In this chapter, we have provided a comprehensive analysis of the Kaizen concept and illustrated its advantages and disadvantages in the fashion and textile industry. In the first part of the chapter, the concept of Kaizen has been thoroughly examined and seven aspects of waste management by using Kaizen have been explained including the definition, history, principles, frameworks, implementation, techniques, and benefits mainly from an organizational standpoint. Due to its long evolution, widespread coverage, numerous applications, and an extended relationship to other managerial methodologies and techniques (TQM, Lean Manufacturing, Job Methods, etc.), Kaizen is gaining increasing attention from many manufacturing industries including fashion and textiles. Kaizen is an essential tool for fashion and textile industries which are characterised by a large amount of waste generation, effluent problems, waiting time, rejection, and rework.

Kaizen can be viewed in many ways when analysing its beneficial aspects in fashion and textile industries. Kaizen is meant to create continuous improvement by the involvement of all parties concerned, from every facet, and accounting for every action, even the smallest, to gain an optimal level of organizational productivity and Gemba waste being eliminated. The application of Kaizen in the fashion and textile industry is illustrated in the second half of the chapter. A context, as well as current shortcomings in the industry, is provided from both theoretical and practical perspectives. The authors have also demonstrated a wide range of benefits Kaizen can bring to the fashion and textile industry. Practical aspects of Kaizen advantages including productivity enhancement, improvement of employee well-being, and customer satisfaction were highlighted by four different case studies. The case studies from some of the major manufacturing countries of fashion and textiles have shown the improvement in productivity and efficiency, reduction of the lead time, and reduction of the wastes.

References

Alinaghi F, Bennike NH, Egeberg A, Thyssen JP, Johansen JD (2019) Prevalence of contact allergy in the general population: a systematic review and meta-analysis. Contact Dermat 80(2):77–85. https://doi.org/10.1111/cod.13119

Ashmore C (2001) Kaizen and the art of motorcycle manufacture. Manuf Eng 80(5):220–222

Ashraf RB, Mahmud I, Akram A, Rashid ARMH, Rahman T, Tanvir MH (2017) The practice of kaizen tool in the apparel industry of Bangladesh for process improvement and development of ergonomics standard. Int Res J Eng Technol (IRJET) 4(10):1747–1754. https://doi.org/10.13140/RG.2.2.33341.15844

Ayyagari M, Demirguc-kunt A, Maksimovic V (2006) Firminnovation in emerging markets: the role of finance, governance, and competition. J Financ Quant Anal 46(6):1545–1580

Barnwell N (2007) Japanese management: its emergence into Western consciousness and its long term impact. In: Asia Pacific economic and business history conference. Economic History Society of Australia and New Zealand, Sydney

Berhe HH (2021) Application of Kaizen philosophy for enhancing manufacturing industries' performance: exploratory study of Ethiopian chemical industries. Int J Qual Reliab Manag. https://doi.org/10.1108/IJQRM-09-2020-0328

Bessant J, Caffyn S (1997) High-involvement innovation through continuous improvement. Int J Technol Manage 14(1):7–28. https://doi.org/10.1504/IJTM.1997.001705

Brunet AP (2000) Kaizen: from understanding to action. Inst Electr Eng 1–45. https://doi.org/10.1049/ic20000198

Brunet AP, New S (2003) Kaizen in Japan: an empirical study. Int J Oper Prod Manag 23(12):1426–1446. https://doi.org/10.1108/01443570310506704

Burton T, Boeder S (2003) The lean extended enterprise: moving beyond the four walls to value stream excellence. Boca Raton

Carvalho CP, Goncalves LWN, Silva MB (2017) Kaizen and 5S as lean manufacturing tools for discrete production systems: a study of the feasibility in a textile company. Int J Res Stud Sci Eng Technol 4(7):1–12. ISSN 2349-476X

Caswell JA (1998) Valuing the benefits and costs of improved food safety and nutrition. Aust J Agric Resour Econ 42(4):409–424. https://doi.org/10.1111/1467-8489.00060

Chan CO, Tay HL (2018) Combining lean tools application in kaizen: a field study on the printing industry. Int J Product Perform Manag 67(1):45–65. https://doi.org/10.1108/IJPPM-09-2016-0197

Chase B, Jacobs F, Aquilano J, Agarwal K (2006) Operations management for competitive advantage. 11th ed. Tata McGraw-Hill Publishing Company Limited. NewDelhi

De Carlo F, Arleo MA, Borgia O, Tucci M (2013) Layout design for a low-capacity manufacturing line: a case study. Int J Eng Bus Manag 5:5–35

Deming WE (1986) Out of the crisis. MIT Press

Elgar T, Smith C (1994) Global Japanization: The transnational transformation for the labour process. Routledge, London

Eli Motivation Personal Classic Style (2019) Fashion is like eating you shouldn't stick to the same menu: funny fashion designer lined notebook. J Str Fash Jpn Unique Spec Inspir Birthd Gift Idea 110 Pages. ISBN13 9781692413743

Fan D, Lo CK (2012) A tough pill to swallow?: The impact of voluntary occupational health and safety management system on firms' financial performance in fashion and textiles industries. J Fash Mark Manag 16(2):128–140. https://doi.org/10.1108/13612021211222798

García-Alcaraz JL, Oropesa-Vento M, Maldonado-Macías AA (2016) Kaizen planning, implementing and controlling. ProQuest Ebook Central

Garcia-Alcaraz JLG, Oropesa-Vento M, Maldonado-Macias AA (2017) Kaizen planning, implementing and controlling. Springer International Publishing. https://doi.org/10.1007/978-3-319-47747-3_1

Gavranović A (2020) Textile market–clothing industry facing new challenges. Text Leather Rev 3(2):103–106. https://doi.org/10.31881/TLR.2020.PP03

Goshime Y, Kitaw D, Jilcha K (2019) Lean manufacturing as a vehicle for improving productivity and customer satisfaction: a literature review on metals and engineering industries. Int J Lean Six Sigma 10(2):691–714. https://doi.org/10.1108/IJLSS-06-2017-0063

Hamel MR (2010) Kaizen event fieldbook: foundation, framework, and standard work for effective events. Society of Manufacturing Engineers

Hellsten U, Klefsjö B (2000) TQM as a management system consisting of values, techniques and tools. TQM Mag 12(4):238–244

Heragu S (2008) Facilities Design. CRC Press, Taylor & Francis Group, Florida

Hirano H (2009) JIT implementation manual—the complete guide to just-in-time manufacturing: volume 1—the just-in-time production system. Productivity Press

Ho SK, Fung CK (1994) Developing a TQM excellence model. TQM Mag 6(6):24–30

Hoque I, Shahinuzzaman M (2021) Task performance and occupational health and safety management systems in the garment industry of Bangladesh. Int J Workplace Health Manag. https://doi.org/10.1108/IJWHM-09-2020-0169

Huntzinger J (2002) The roots of lean training within industry—the origin of Japanese management and kaizen. AME 18(1):13

Imai M (1986) Kaizen: the key to Japan's competitive success. McGraw-Hill Education

Imai M (2007) Gemba kaizen. A commonsense, low-cost, approach to management. In: Boersch C, Diest FV (eds) Das Summa Summarum des management. Gabler, pp 7–15. https://doi.org/10.1007/978-3-8349-9320-5_2

Jones C (2017) Software methodologies: a quantitative guide. ProQuest Ebook Central

Juneja SS, Doctor G, Azir V (2012) Innovative Kaizen implementation in textile industry. Nirma Univ J Bus Manag Stud 7(1 & 2):75–82. ISSN 2249-5630

Kiran DR (2017) Total quality management: key concepts and case studies. Butterworth-Heinemann. https://doi.org/10.1016/C2016-0-00426-6

Koopmans TC, Beckmann M (1957) Assignment problems and the location of economic activities. Econ J Econ Soc 25(1):53–76

Kumar D (2019) Productivity improvement with kaizen tool in garment industry. Int J Curr Eng Sci Res (IJCESR) 6(2):95–100. ISSN 2394-0697

Kumar S, Thavaraj S (2015) Impact of lean manufacturing practices on clothing industry performance. Int J Text Fash Technol (IJTFT) 5(2):1–14

Lareau W (2003) Kaizen office. American Society for Quality ASQ

Leseure M (2010) Key concepts in operations management. SAGE Publications Ltd.

Lill D (2016) The advantages and disadvantages of kaizen to business. Talk Business. https://www.talk-business.co.uk/2016/06/21/advantages-disadvantages-kaizen-business/

Lillrank P (1995) The transfer of management innovations from Japan. Organ Stud 16(6):971–989. https://doi.org/10.1177/017084069501600603

Lina LR, Ullah H (2019) The concept and implementation of Kaizen in an organization. Glob J Manag Bus Res Adm Manag 19(1):9–17. ISSN 2249-4588

Maarof MG, Mahmud F (2016) A review of contributing factors and challenges in implementing kaizen in small and medium enterprises. Procedia Econ Financ 35:522–531. https://doi.org/10.1016/S2212-5671(16)00065-4

Macpherson W (2008) How the Japanese interpret Kaizen: an exploration of Japanese spirit. SSRN Electron J. https://doi.org/10.2139/ssrn.2364270

Malloch H (1997) Strategic and HRM aspects of Kaizen: a case study. N Technol Work Employ 12(2):108–122. https://doi.org/10.1111/1468-005X.00028

Murata K, Katayama H (2010) Development of Kaizen case-base for effective technology transfer–a case of visual management technology. Int J Prod Res 48(16):4901–4917. https://doi.org/10.1080/00207540802687471

Nayak R, Van Thang LN, Nguyen T, Gaimster J, Morris R, George M (2021) Sustainable developments and corporate social responsibility in Vietnamese fashion enterprises. J Fash Mark Manag Int J

Nayak R, Panwar T, Nguyen LVT (2020) Sustainability in fashion and textiles: a survey from developing country. Sustainable technologies for fashion and textiles, pp 3–30

Nayak R, Akbari M, Far SM (2019) Recent sustainable trends in Vietnam's fashion supply chain. J Clean Prod 225:291–303

Nayak R, Padhye R (eds) (2017) Automation in garment manufacturing. Woodhead Publishing

Nayak R, Padhye R (2015) Introduction: the apparel industry. Garment manufacturing technology. Woodhead Publishing, pp 1–17

Ohno T (1988) Workplace management. Productivity Press

Ongkunaruk P, Wongsatit W (2014) An ECRS-based line balancing concept: a case study of a frozen chicken producer. Bus Process Manag J 20(5):678–692

Recht R, Celeste W (1998) Kaizen and culture: on the transferability of Japanese suggestion systems. Int Bus Rev 7(1):7–22. https://doi.org/10.1016/S0969-5931(97)00048-6

Saleem M, Khan N, Hameed S, Abbas M (2012) An analysis of relationship between total quality management and Kaizen. Life Sci J 9(3):31–40

Scofield J (2017) Faster fashion cycle accelerates. Naked capitalism. https://www.nakedcapitalism.com/2017/04/faster-fashion-cycle-accelerates.html

Shang G, Pheng LS (2013) Understanding the application of Kaizen methods in construction firms in China. J Technol Manag China 8(1):18–33. https://doi.org/10.1108/JTMC-03-2013-0018

Shingo S (1988) Non-stock production: the Shingo system of continuous improvement. Productivity Press

Singh J, Singh H (2015) Continuous improvement philosophy–literature review and directions. Benchmarking Int J 22(1):75–119. https://doi.org/10.1108/BIJ-06-2012-0038

Slater K (2003) Environmental impact of textiles: production. Processes and protection. Woodhead Publishing Limited/The Textile Institute, Cambridge, Reino Unido

Suárez-Barraza MF, Ramis-Pujol J, Kerbache L (2011) Thoughts on Kaizen and its evolution: three different perspectives and guiding principles. Int J Lean Six Sigma 2(4):288–308. https://doi.org/10.1108/20401461111189407

Tapia-Leon R, Vega-Neyra X, Chavez-Soriano P, Ramos-Palomino E (2019) Improving the order fulfillment process in a textile company using lean tools. In: 2019 Congreso Internacional de Innovación y Tendencias en Ingeniería (CONIITI), pp 1–5. https://doi.org/10.1109/CONIITI48476.2019.8960698

Tozawa B, Bodek N (2012) The idea generator: quick and easy Kaizen (Workbook). PCS Press

Tsutsui WM (1996) W. Edwards Deming and the origins of quality control in Japan. J Jpn Stud 22(2):295–325

UXKaizen (n.d.) Advantages and disadvantages of Kaizen. https://www.uxkaizen.com/post/advantages-and-disadvantages-of-kaizen

Vento MO, Alcaraz JLG, Macías AAM, Loya VM (2016) The impact of managerial commitment and Kaizen benefits on companies. J Manuf Technol Manag 27(5):692–712. https://doi.org/10.1108/JMTM-02-2016-0021

Vieira L, Balbinotti G, Varasquin A, Gontijo L (2012) Ergonomics and Kaizen as strategies for competitiveness: a theoretical and practical in an automotive industry. Work 41(Supplement 1):1756–1762. https://doi.org/10.3233/WOR-2012-0381-1756

Webley P, Cartwright J (1996) The implicit psychology of total quality management. Total Qual Manag 7(5):483–492. https://doi.org/10.1080/09544129610595

Whitmore T (2008) Standardized work. Manuf Eng 140(5):171–179

Wittenberg G (1994) Kaizen—the many ways of getting better. Assem Autom 14(4):12–17. https://doi.org/10.1108/EUM0000000004213

Womack JP, Roos D, Jones DT (1990) The machine that changed the world: the story of lean production—Toyota's secret weapon in the global car wars that is now revolutionizing world industry. Simon and Schuster

Kanban Applications in Fashion and Textile Industries

Majo George, Le Phan Thanh Truc, Vuong Nguyen Dang Tung, Le Khac Yen Nhi, Nguyen Minh Ngoc, and Rajkishore Nayak

Abstract This chapter will provide a thorough scrutiny of the Kanban concept, its tools and methods as well as its implementation from different perspectives, especially the application of Kanban in the fashion and textile industries. Derived from a Japanese notion, Kanban means "visible sign", originated with the efforts of the Toyota automotive company in applying the concept to the manufacturing process. However, Kanban has eventually become a popular methodology that is used widely not only in the automobile industry but also in any other sectors such as fashion and textiles. It has a close correlation with the Lean Manufacturing and Just-in-time systems with the purpose of maximizing productivity and minimizing workplace waste, which has been discussed in this chapter. The Kanban system is designed to help make the workflow as agile and efficient as possible through the utility of signal visualization, thus, to some extent, the traditional Kanban system has been considered even more useful than a computerized system. Implementation of Kanban system has several advantages and disadvantages in the fashion and textile industry to reach an optimal point of production, which are also briefly highlighted. Various case studies focusing on the fashion brands and manufacturers are included in this chapter. Overall, Kanban implementation in fashion and textile industries can help to increase the productivity, control inventory, ensure supplier and employee participation, remove the bottlenecks and improve the quality of the products within the production lines, which are also discussed in this chapter.

Keywords Kanban · Fashion and textiles · Productivity · Efficiency · JIT and Lean

---

M. George (✉) · L. P. T. Truc · V. N. D. Tung · L. K. Y. Nhi · N. M. Ngoc
School of Business and Management, RMIT Vietnam, Ho Chi Minh, Vietnam
e-mail: majo.george@rmit.edu.vn

R. Nayak
School of Communication and Design, RMIT Vietnam, Ho Chi Minh, Vietnam

© The Author(s), under exclusive license to Springer Nature Singapore Pte Ltd. 2022
R. Nayak (ed.), *Lean Supply Chain Management in Fashion and Textile Industry*,
Textile Science and Clothing Technology,
https://doi.org/10.1007/978-981-19-2108-7_8

1 Introduction

1.1 Definition of Kanban

Originally, Kanban (看板) is a Japanese lexeme, literally meaning "visible sign" or "visible record" (Surendra et al. 1999). It is well-known as the subsystem of the Toyota Production System (TPS) having the purpose of controlling inventory levels, the flow of production and distribution, and management of raw materials. Gupta et al. (1999) mentioned Kanban is recognized as a fundamental component of the just-in-time (JIT) system. This derives from Kanban's objective that aims to achieve an agile production stream, aligning with the purpose of minimizing waste through the JIT system. In specific, the Kanban system is designed to help simplify production schedules, lower the burden on operational processes, smoothen the process of locating and dispatching manufacturing items, as well as reducing the use of paperwork. Another definition from Graves et al. (1995) and Tardif and Maaseidvaag (2001) explains Kanban as a Material Flow Control mechanism (MFC). This notion concentrates on the function of Kanban methodology instead of its objective. Thus, the authors reckoned that Kanban is a technique used to mainly control the proper quantity as well as a proper duration of time in the production lines.

As stated by Cuellar (2011), the concept of Kanban makes it difficult to formulate a detailed definition regarding its principles and functional processes, and since Kanban is such a broad concept, its presence also varies. Especially in discussions by Monden (1993) and Suzaki (1987), Kanban appears as many different types, such as production Kanban, withdrawal Kanban, supplier Kanban, signal Kanban, express Kanban, and tunnel Kanban. In short, Gupta et al. (1999) described Kanban as an information delivery system and certain kinds of Kanban will relate to the information carried, but usually, such indispensable information that is illustrated through Kanban, including the type of Kanban, the name and number of manufacturing components, and the location or destination of the items needed.

The Kanban system, according to Rahmana et al. (2013), is perceived as an efficient tool to be used in Lean Manufacturing (LM) to minimize the amount of inventory stocked in the production and distribution chain. Thus, for companies that pursue the LM scheme, Kanban appears as an operational strategy when discussing its advantages in organizational management, especially for manufacturing companies to have product available to meet the demand. As mentioned in earlier chapters, LM or Lean Production (LP) is defined as a production strategy that deploys several tools and techniques to attain the goal of cost and waste reduction and create more added value in the production process (Selçuk 2013). Among LP methods, Kanban is the most prominent production management technique that is used in the pull approach. It was utilized to assure the right quantity of products is produced at the right time. Thus, the correlation between Kanban implementation and LM and JIT manufacturing is seen to be close (Savino and Mazza 2015). The goal is to help enhance productivity and flexibility, hence improving the quality of production and customer responsiveness

(Taj et al. 2011). In doing so, what must be attained is small-lot production, schedule stability, shorter setup times, and layout efficiency (Savino and Mazza 2015).

Accordingly, it is widely accepted that Kanban plays a crucial role in the LM and JIT systems. In other words, Mojarro-Magaña et al. (2018) stated that Kanban's implementation can be considered as the major practice in setting up LM principles. Following that concept, Wakode et al. (2015) described Kanban as more like a methodology that has to be attached with a pull approach to implement JIT in a variety of domains such as manufacturing, assembly lines, or supply chain systems. In short, Kanban, from a wider perspective, is viewed as a scheduling system used to maximize operational productivity by minimizing idle time, futile motion, defects, overproduction as well as waiting time (Wakode et al. 2015). Notwithstanding, when viewed at micro-levels, Kanban is normally categorized by the use of cards. Sivakumar and Shahabudeen (2008) noted that the traditional Kanban system refers to the deployment of cards in order to manage the delivery and production stream of manufacturing items and raw materials. Thus, a great number of publications about the term Kanban, say it can be used to indicate the meaning of either "cards" or "the system". This relationship will be considered further in the final section discussing Kanban techniques.

1.2 History of Kanban

The beginning of Kanban goes back to the 1940s when the Toyota Motor Corporation (Toyota), a Japanese multinational automotive manufacturer, developed a strategic breakthrough in its methods of management and production. At that time, the company was confronting issues with its automobile production. The problem was the way Toyota practiced its push approach and tried to produce as much as it was possible to achieve at every stage of vehicle component production, and there were disparities in the quantity of components manufactured in a given period of time. Toyota's production flow was being disrupted, making it difficult for the company to control the operational process and blurring the pathway to achieve the goal of achieving an optimal point of production.

In seeking a solution, Toyota took notice of the way contemporary Western grocery stores replenished their products. The system turned out to be very simple but more effective in the stock replenishment. When a product was taken off a shelf, a blank spot was created immediately and that blank spot worked as a visualized signal, informing the grocery staff to refill it. From that starting point, Toyota found the answer that is believed to have initiated the pull approach they still follow. The way the Western grocery stores refilled their shelves was considered as practicing a pull process, which is the fundamental of Kanban system. It's astonishing efficiency emerged as the key for Toyota to improve its previous production strategy in managing the inventory by using signals. Therefore, the Kanban system also became known as the "Supermarket method" since Taiichi Ohno (an engineer at Toyota) gained inspiration from the American grocery stores (Kirovska and Koceski 2015). Toyota's Kanban system was designed to support its JIT production and reach the maximum level of worker and inventory utilization (Sugimori et al. 1977).

Under the Kanban system, Toyota was able to increase both efficiency and productivity in production and lower its inventory costs (Sugimori et al. 1977). This application of Kanban also formed part of the implementation of the TPS, in which the technique starts with the final desired delivery schedule (Henderson 1986). The company has each of its specific items scheduled at a specific time. Yet, when referring to Kanban as a philosophy of agile movement in the operation process, Henderson (1986) stated that the concept of Kanban methodology evolved after Henry Ford pioneered the continuous production flow. Then, when the Great Depression began in 1929, the event pushed other Japanese industrial enterprises to use this methodology (making the production process *Lean*) to gain a competitive share in the global market at that time. But it was not possible for most of the companies to fully achieve Kanban implementation in a methodical way. Hence, the discovery and deliberate investigation and employment by Toyota of this concept is undoubtedly a milestone in the history of Kanban.

1.3 Aim of the Kanban System

The aim of Kanban was clarified by Sugimori et al. (1977). By applying the Kanban System, reliance on electronic computers was no longer compulsory for Toyota (Fig. 1). Particularly, certain reasons for implementing the Kanban System instead of a computerized system are pinpointed. Firstly, the cost of information processing is minimized due to the reduction in the cost of implementation of a complicated system requiring a production schedule for the whole execution dealing with suppliers, any modifications, or different approaches by real-time control. Secondly, the Kanban system speedily informs a business about the system changes. In detail, managers can identify consistent alternatives through a process of monitoring capacity, rate of operation, and manpower, regardless of a computer. Finally, the Kanban system mitigates the surplus of existing shops by multiple processing stages in automotive industries.

1.4 Description of Kanban System

Kanban can either be referred to as a whole system with the purpose of managing workflow in the manufacturing and production process, or as the cards used in such a system. Kanban is a tool for pull production in LM that controls the material flow among workstations by using Kanban signals. The signal may be a physical card in the traditional Kanban system, or, since the information and technology revolution, it is more likely to be a technological sign (Abbadi et al. 2018).

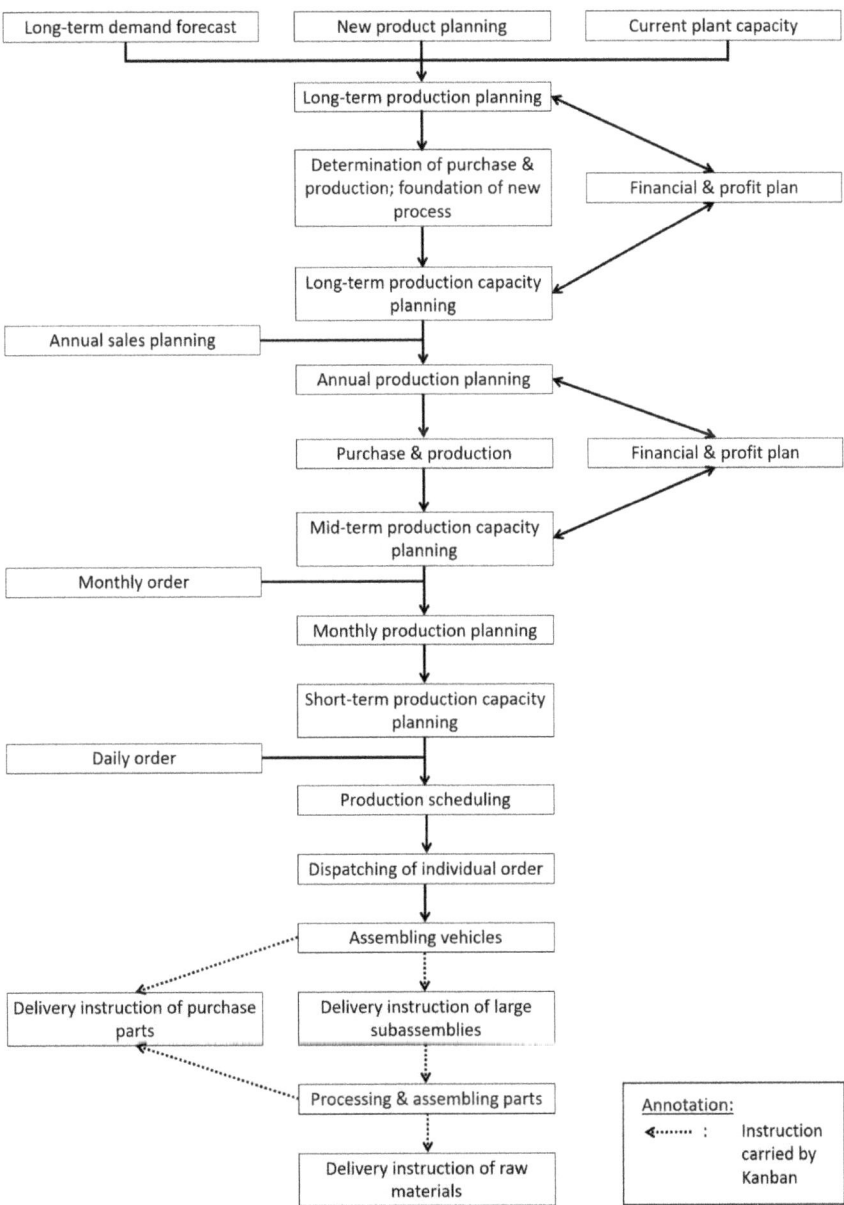

Fig. 1 Structure of Toyota production system

The Kanban system focuses its core concentration on customer demand for a specific item or product (which is why it works most efficiently in the pull approach). More specifically, a customer could be an external, actual consumer in the market for a final product or a production worker at another workstation in an internal manufacturing/production line. From that core focus, Surendra et al. (1999) recognize the premise of the Kanban system in which the material needed will not be produced or transferred until a signal is sent from the customer. The cards thus go backward from the final order to the upstream production points (Iannone et al. 2009).

The cards are Kanban cards, which are commonly used as order cards in a production process. This element includes "production Kanban" card and "withdrawal Kanban" card, which is also called "conveyance Kanban" or "move Kanban" (Ohno 2018; Monden 1993). These two types of cards must work together to set up a pull system, in which customers stay in the downstream position whilst the company remains in the upstream position (Sugimori et al. 1977). According to Naufal et al. (2012), production Kanban gives the instruction to produce a particular product and the required amount, whilst withdrawal Kanban represents authorization signals for the withdrawal process in accordance with the customers' order. To fulfil the missions of these cards, which is producing the right parts at the right amount for the customers, there are various requirements for outlining a card:

- Customer information: customer name, product name, and type of model.
- Product information: name of the required part, its picture, and quantity per packing.
- Process information: production process, address, and storage area

Sugimori et al. (1977) demonstrated that the Kanban pull system starts downstream. Following an order from the client, the product will be taken to the customer from the assembly, and simultaneously, the withdrawal Kanban is removed and put back with the containers holding parts. Accordingly, the withdrawal Kanban is now working as the new demand and will accompany the container until its demand is noticed and processed. Its existence presents the required amount of withdrawal parts, therefore, a production Kanban is extracted and returned to the production ordering centres to notify a new production. As long as no parts are produced or withdrawn without a Kanban, a complete pull system is sustained.

In short, in the Kanban system the appearance of those cards will follow the containers throughout the whole production process to accentuate and assure in-time response to demands at each stage in the production line. From there, it will help control the workflow, making it more efficient and minimizing inventory and waste in the in-process manufacturing (Singh et al. 1990). To better visualize the ongoing process in the manufacturing, a Kanban board is usually used (Corona and Pani 2013). It is an important tool that helps to manage the workflow where all personnel including the manager can observe all the activities undergoing in the production process. A simple design of a Kanban board has three columns: "To do", "In progress" and "Done"; nevertheless, there are more columns that can be added if the production and distribution process is more complicated (Bieniusa et al. 2018). As well as the two Kanban cards of production and withdrawal, there are also Kanban cards which reflect different needs, namely supplier Kanban, procurement Kanban, or subcontract Kanban (Huang and Kusiak 1996).

2 Principles of Kanban

Ohno (2018) first mentioned the six rules of Kanban implementation in his book "Toyota Production System Beyond Large-Scale Production". Later, after experiencing the process of development and incorporating other factors in the modern era, the concept of Kanban methodology is now being more widely used in various industries and various countries, the six rules have been improved and restated by Cimorelli (2013) in the book "Kanban for the Supply Chain: Fundamental Practices for Manufacturing Management". The six rules are discuses in the following section.

2.1 Level Production

There should be a point of balance between the number of items produced during every stage of production and the demand. Since the level or capability of production varies between each stage, limiting the number of requests and accepting the appropriate number of orders to achieve a harmonized equilibrium of production can boost productivity and consistency in manufacturing. It can therefore be easier to control and keep track of the quantity produced, prevent fluctuation, minimize wastes, and contribute to a lean flow.

2.2 Smooth Information Flow in a Factory Floor

Information sharing is one of the most important keys in the Kanban system. Therefore, once a process has consumed all its inventory, it must deliver that information and make a corresponding request to its suppliers. Signal sending is among the top critical requirements in the process; thereby, assuring a smooth signal conveyance will help to keep the production flow on the track, everyone involved is informed and the production line can carry on continuingly, making the items and products in time.

2.3 Do not Produce or Move Parts Without a Kanban

Nothing should be made or moved without a request. Since the Kanban system closely relates to the pull approach (or customer demand), request and order play a significant role in this process. Thus, the previous stage only produces the quantity required to make sure the following stage will receive what it needs, and eventually, enough products will be produced. At every stage, Kanban cards are used consistently and continuously to keep delivering the right information in relation to necessary items to be produced. Obeying this rule not only helps to reduce the time for production but also reduces the cost of storing excessive inventory.

2.4 Always Attach a Kanban to a Product

The Kanban card that represents a request or demand must always be attached to the items or products. Making sure this association is made will help utilize the function of that card in specific and of the whole system in general. This rule is relevant to the first rule when strengthening the information flow appears to be the key requirement in the Kanban system. Again, this will bring more efficiency generated in the manufacturing process as information and requests are sent and received constantly and rhythmically.

2.5 Avoid Delivering Faulty Parts to the Succeeding Stages

All defective items must be eliminated immediately. That means whenever there is a defect in the product at any stage, it must not be sent out to the next workstation without removing the defect. This will help guarantee that every final product will be free from deficiency, thus assuring the quality of the product as well as brand name of the company. Moreover, avoiding defects in time also helps keep the production flow going smoothly and effectively by eliminating waste time, bottlenecks, waste items, and customer complaints.

2.6 Produce the Precise Quantity of Parts Withdrawn

Each stage of production will produce in response to the exact quantity and follow the sequence of incoming requests. In other words, the later process must refer back to the previous stage in order to produce the number of production components that are needed. This rule is necessary to avoid overproduction by creating an intersection between product quantity produced and customer demand. Therefore, a stable flow of production must be ensured to prevent disruption at one stage from affecting others.

Implementing the Kanban in fashion and textile context carries a great value as the system is know as excessive production and waste of materials (Nayak 2020). The six principles as mentioned above can be used to eliminate waste products from a process being used in the subsequent process. The normal practice in garment manufacturing of the final inspection can be eliminated by kanban and in process inspection can help to detect the occurrence of the fault from the early stage. Furthermore, the excessive production can be eliminated from the manufacturing system by Kanban. The fashion and textile industries are well-known for excessive production, which can be reduced by the implementation of Kanban.

3 Kanban Systems (Push and Pull Approach)

3.1 Types Based on Approach

According to Sugimori et al. (2007), a form of order card is used to apply the Kanban system. There are two types of order cards, withdrawal Kanban and production Kanban (these types will be explained further in the next section). These cards work as an information flow and must be attached to the containers that stock the produced parts, creating a mechanism of conveyor to connect every workstation within a manufacturing location. With different patterns of material handling systems, Kanban is separated into three control systems, including Single Kanban system, Dual Kanban system, and Semi-dual Kanban system (Huang and Kusiak 1996). Table 1 provides the comparison between the three control systems.

3.1.1 Single Kanban System

The single Kanban system applies solely to production Kanban to block the material handling process depending on the type of WIP. In this system, the part mix and the finished products buffer vary in size. At each stage, the block of production is determined by the total production size. The quantity of containers in the output buffer should not exceed the capacity and the containers will contain the parts that need replenishing. To efficiently operate the single system, there are several conditions such as the low WIP, the fast WIP and Kanban's turnover, small space for buffer, short distance between consecutive workstations and synchronization is necessary between production and WIP movement.

Table 1 Comparison of Kanban systems

Items	Single	Dual	Semi-dual
Work-in-progress (WIP) between two stages	Small	Small	Large
Distance between two stages	Small	Moderate	Large
Turnover of WIP	Fast	Moderate	Slow
Turnover of kanban	Fast	Fast	Slow
Synchronisation of WIP movement and production	Necessary	Necessary	Unnecessary

3.1.2 Dual Kanban System

In this system, withdrawal and production are applied simultaneously with distinguishing inbound and outbound buffers (Kumar and Panneerselvam 2007). Specifically, the material handling area based on the type of WIP, and the station will be implemented at the same time with the presentation of withdrawal Kanban in the buffer area. For WIP, there will be a buffer to deliver the finished parts from the previous stage to the following stage. The dual Kanban system is said to be unsuitable for manufacturers that do not adopt rigid regulations for the control of buffer inventory. There are conditions when applying dual Kanban system: the necessary synchronization between the material handling process and the production rate, WIP is required in the buffer, an external buffer is needed for the production, the distance between stations is moderate, and Kanbans acquire a fast turnover.

3.1.3 Semi-Dual Kanban System

In this system, production Kanban and withdrawal Kanban are changed at intermediate stages of the process. Semi-dual Kanban system acquires several characteristics: the slow turnover of Kanban's and WIP, preceding stages and succeeding stages is operated in a large distance, the large WIP is required in the preceding stages, and synchronization is unnecessary between the material handling process and the production rate.

3.2 Types Based on Operation Stages

The main objective of the Kanban system is to implement JIT in the manufacturing workplace and provide information about allowed production quantities and amount of semi-manufactured products to the preceding stages between manufacturing workstation and supply chain components for replenishment and finalization (Huang and Kusiak 1996). The Kanban system is able to achieve three major functions, including visibility function, production function, and inventory function. Based on the functions, Kanban is classified into five different types.

3.2.1 Primary Kanban

This type of Kanban represents the flow from a stage of production to another stage within the same manufacturing area or production preparation station. Specifically, there are two separate kinds of primary Kanban. They are withdrawal Kanban (conveyor Kanban) and production Kanban. Withdrawal Kanban determines the time to transfer WIP, in the preceding stage that is in charge of producing that part, to the succeeding stage between different areas with the production line or the supply

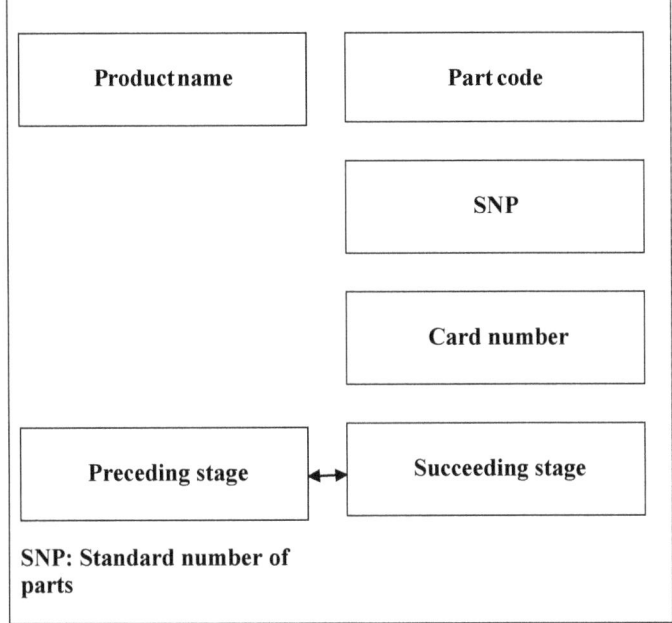

Fig. 2 Production Kanban framework

chain of materials. The key information of withdrawal Kanban is the location to receive and stock, part number, description of the item, and quantity (Cimorelli 2013). Production Kanban is applied to order what production is allowed, the time and quantity within the given manufacturing cell or workstation at the succeeding stage (Baradaran 2018). An example of production Kanban is shown in Fig. 2. Key information required for production Kanban includes the number and description of components to be manufactured, process location, description, quantity, and stocking area of final products (Cimorelli 2013).

3.2.2 Supply Kanban

Supply Kanban authorizes the travel of suppliers to deliver parts and components from a storage facility or warehouse to the manufacturing cell or workstation (Huang and Kusiak 1996). Key information for supply Kanban includes the number, description, and quantity of WIP that are delivered, supplier information, and stocking area after delivery (Cimorelli 2013). Figure 3 provides an example of supply Kanban.

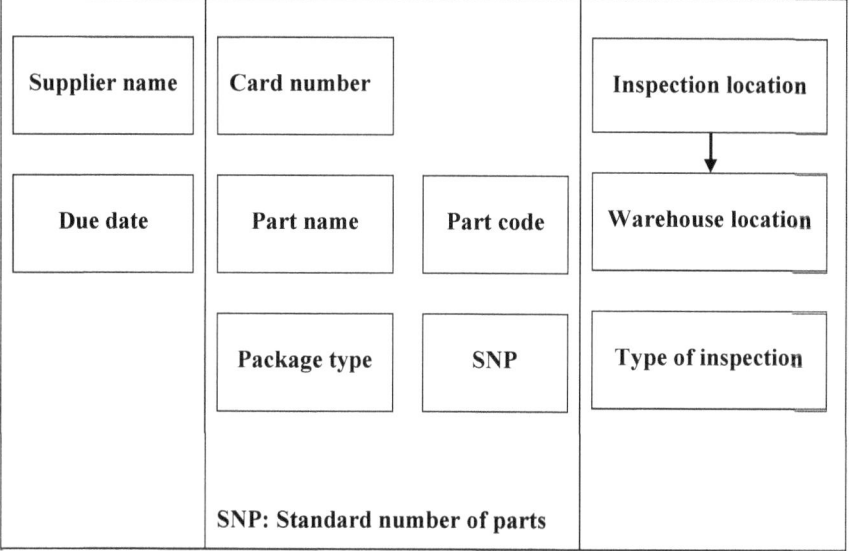

Fig. 3 Supply Kanban framework

3.2.3 Procurement Kanban

Procurement Kanban is related to the WIP parts that are stocking in the input buffers (Gurgur and Altiok 2007). Specifically, it is a process of parts traveling from outside supply to the manufacturing location. Similar to supply Kanban, key information of procurement Kanban includes number and description of parts, storage location, or supplier information. Figure 4 provides a comprehensive example of procurement Kanban.

3.2.4 Subcontract Kanban

The supply of orders obtained from subcontractors can ensure the influence of Kanban on inventory levels in the introductory phase as well as the balance between manufacturing processes within the workstation (Fukukawa & Hong 1993). According to Huang and Kusiak (1996), subcontract Kanban authorizes travel between different subcontracting units.

3.2.5 Auxiliary Kanban

Auxiliary Kanban is utilized to support extra work or backlogs. These kinds of Kanban can be issued depending on the order of the owners or on a weekly-basic work plan. As a result, auxiliary Kanban is issued separately to manage the planning and

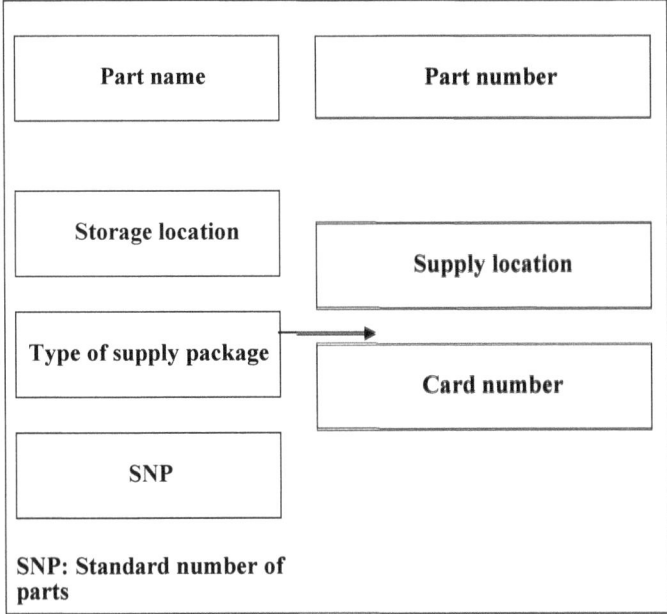

Fig. 4 Procurement Kanban framework

measure performance of additional work (Jang and Kim 2007). Auxiliary Kanban can be issued in the form of emergency Kanban, express Kanban, or any specific type based on the requirement or situation (Huang and Kusiak 1996). There are some examples of the equipment that can be applied in the auxiliary Kanban such as Kanban box (collecting Kanban after withdrawing), Kanban management account (managing Kanban implementation), supply management account (managing raw materials supply), and dispatching board (displaying schedule of production).

4 Benefits and Drawbacks of Kanban System

4.1 Benefits of Kanban

According to Jyothi and Rao (2012), the implementation of the Kanban system is a major source of achieving competitiveness, which provides the following advantages:

- Enhancement of customer satisfaction and retention rate for business.
- Promotion of international trade.
- Improvement of quality, efficiency, and effectiveness of the operation.
- Improvement of employees' attitude, motivation, and awareness.
- Development of adaptiveness and flexibility.

- Prevention of overproduction and elimination of waste.
- Reduction of cost.
- Increase of productivity and profit margin.

Furthermore, Gross and McInnis (2003) provided slightly different research on the advantages of the Kanban system. Specifically, they identified eight different benefits, as discussed below.

- Cutback in inventory.
- Enumeration of the operation flow.
- Prevention of overproduction.
- Enhancement of control at the operation level (material handling level).
- Development of process management and visual scheduling.
- Improvement of adaptiveness and flexibility.
- Minimization of inventory obsolescence risk.
- Enhancement of supply chain management.

4.2 Drawbacks of Kanban

In spite of several advantages, there are certain disadvantages of Kanban implementation in the manufacturing industries, which are discussed below.

- Kanban is less effective where the resources are shared or sudden change in the demand in the downstream.
- Product mix or demand changes may cause problems.
- The Kanban system does not eliminate variability.
- Production flow problems with poor quality products or products with a wide variation such as fashion and textiles.
- Cannot be used independently as it needs other tools such as JIT or 5S.
- Time consuming process as the impacts are realised after a long time.

5 Implementation of Kanban

5.1 Fashion and Textiles Industry Applications

There are key determinants to evaluate the effectiveness of the Kanban system in the manufacturing industries such as fashion and textiles (Nayak 2020). For example, in the fashion and textile industries, measuring the effectiveness of increase output from qualified inventory management; correlation of supplier activities and production lines; ensuring quality improvements and quality control; and maintaining employee devotion and employer commitment is rather a difficult task. However, these are the highly relevant factors defining the success of an industry after the implementation of Kanban.

5.1.1 Increase Output from Qualified Inventory Management

Without appropriate inventory management, a cost-efficient strategy can never be acquired (Heizer and Render 2005). For these authors, inventory management can be divided into four areas of management, namely: raw material, work-in-process (WIP), finished products, and processing (maintenance or repair) inventory. With a close correlation to the holding costs and storage in manufacturing plants, inventory management can be impacted by the volatility of real-life demand by external parties (Kobbacy and Liang 1999). Due to the application of Kanban systems with a high level of flexibility, an industry, especially in the inventory management process can respond quickly to market changes, assess each task comparatively and concentrate on an ultimate goal to gain a competitive edge. As the fashion and textile industries use a large quantity of inventory, the Kanban system can help in the effective inventory management.

5.1.2 Ensure Supplier Participation Correlate with the Production Lines

The application of the Kanban system only demands a certain inventory capacity that adapts the production numbers. Specifically, compared to the just-in-time (JIT) management, the Kanban system performs in a wider scope, in which not only the production and supplier scheduling are fine-tuned, but also the inventory is always kept to a minimum level with closely monitored production and WIP (Donald 2003). That given, supplier commitment in fashion and textile industries is important to ensure efficient and smooth production lines, and therefore, five criteria of opting for a qualified supplier include up-to-standard quality, willingness to cooperate, and competency in technical, geographical, and pricing aspects.

5.1.3 Ensuring Quality Improvements and Quality Control

Low-standard operational quality in a business may result in lower costs but will also cause a higher level of defects (Balram 2003). The fashion and textile industries suffer from the problem of good quality products in several instances. The implementation of Kanban system can ensure a better cost-saving process by managing inventories and quality improvements in output. JIT is important to evaluate the total quality management (TQM) system (Flynn et al. 1995). A JIT is considered effective when a business achieves a certain level of quality standards before being confirmed in a continual operation intended to reach customers, which ensures a competitive advantage that high quality affords the business' processes and consumers without any prevention, appraisal, and failure costs bias (Bernstein 1984). Hence, by ensuring quality improvement and control through minimizing waste from incorrect work, over-processing, and space-consuming execution, the Kanban system brings greater efficiency to a business as a whole.

5.1.4 Maintain Employee Devotion and Employer Commitment

By provision of improved visibility and problem solving, the Kanban system assists employees to visualize task divisions and allocate responsibilities appropriately, quickly identify problems and offer solutions. From those advantages, a business can enhance employee participation and management to achieve the final objective of providing incentives and thereby ensure a loyal, cooperative, flexible, and dedicated workforce. This approach is beneficial in the fashion and textile industries to improve the employee devotion as well as employer commitment.

5.2 How Kanban Support Fashion and Textile Industries

5.2.1 Case Study 1: Kanban Application by ZARA

In this case study, ZARA (one of the biggest fashion global fashion retailers from Europe) implemented the Kanban system for its inventory management process. Specifically, the brand had chosen to concentrate on a low stock-based approach (i.e., reduced inventory level), relying on the pull system to quickly adapt to customer trends with new designs introduced monthly. The digital Kanban system was utilized to closely monitor the production efficiency of ZARA's partners in other countries, in which the data was better tracked by catching signals from multi-coloured lights (red refers to a machine stopping or breaking down, purple for changeover, blue for maintenance, and green for availability) and the data flow was monitored overall. With the application of Kanban in the manufacturing and supply chain of ZARA, the inventory turnover of the brand was maximized.

ZARA also adopted the Kanban system to mitigate its inventory holding costs and control shortages of inventory. In detail, the system was implemented in all the warehouse levels of the brand located in Europe and other locations. As a fast-changing brand adapting quickly to customer fashion preferences, ZARA emphasized the Kanban role in manufacturing, sourcing, design, and distribution to the retail process almost all over the world. Specifically, the system has been effectively applied in over 300 small factories across Morocco, Portugal, Turkey, and even Asia. Eventually, certain benefits were achieved by ZARA as a success as discussed below.

- Better monitoring of production flow

With the application of multi-coloured lights in the digital Kanban system, defects were easily detected, with detailed identification of where they occurred in the process. The supervisors and workers can easily view the different coloured light and can take action accordingly. For example, a worker can observe the red light, when there is a machine stoppage and quickly fix the problem so that the machine can be brought to working to improve the productivity and reduce the waiting time.

- Improved production flow with on-time information

Another benefit stemming from the digital Kanban system was that the performance in a diversity of production flows were all controlled and ensured the same performance rate. This on-time information availabiliy also meant that any malpractice noted from the flow was quickly examined and corrected so that the flow could be turned back to its normal performance stage.

Through these two benefits, the efficiency of the brand (ZARA) was increased to a level of 100%; a statistic that many industries, especially the fashion and textiles industries, could strive to emulate. The implementation of Kanban can bring the efficiency to a 100% level.

5.2.2 Case Study 2: Kanban Application in a Shirt Manufacturing Industry in India

The shirt manufacturing industry of an Indian domestic retailer acquired inhouse manufacturing for clothing types varying from shirts, trousers, blazers to suits. With a production capacity of up to 2500 pieces per day, the industry acquired 10 departments executing the steps as shown in Fig. 5. The various departments included production planning and control (PPC) department, fabric and trim store, CAD (computer aided design) and pattern making room, cutting room, fusing room,

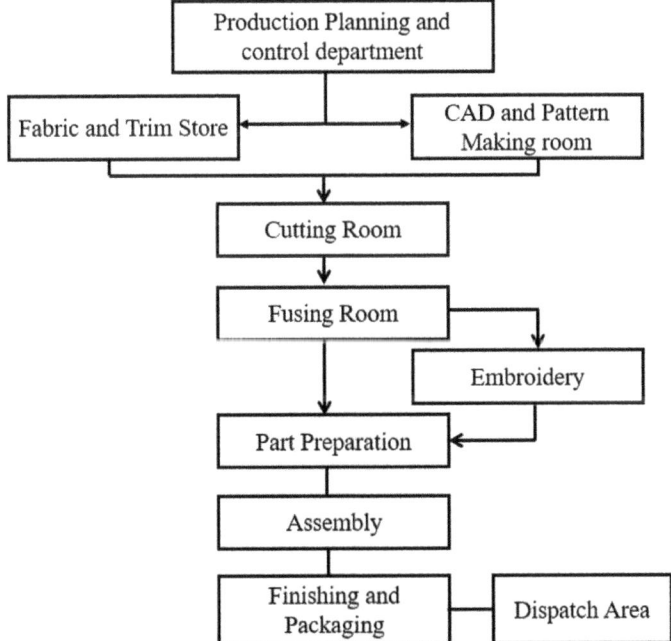

Fig. 5 Process flow chart in the shirt manufacturing industry in India

embroidery, part preparation, assembly, finishing and packaging, and dispatching (shipment) department.

Before, the industry was following the traditional manufacturing practices, which lead to several problems, higher lead time, excessive inventory and delaying of orders. Specific problems were identified at each of the above stages and are detailed below.

- **Delayed orders**: Because of a lack of communication between management and supervisors when deciding on the prioritization level for each order, some orders were totally ignored. This resulted in delays in production as well as customer dissatisfaction when orders were not available on demand.
- **Excessed inventories and WIP**: Since the whole production process was not unified, the industry suffered from an increase in WIP and inventory.
- **Unpredictable production**: Since the departments and management decisions were not unified, the production floor lost track of upcoming and processing orders. This created less predictability and involvement in production. Specifically, as each line only paid attention to producing more pieces regardless of particular orders, the management board found it more difficult to predict accurate statistics.
- **More concentration on the push system**: The industry was paying more attention to pushing its products to consumers rather than pulling what the customers were really demanding.

After these problems had been identified the Kanban system was implemented, especially at the cutting, sewing, and finishing stages. The object was to enhance the production process, improve communication, and involve staff at all levels in the overall management of the business. Specifically, the system provided information regarding weaknesses in each process, orders that were being produced, and orders that were in most demand with a certain required quantity. From that information, excessive inventories, lead time, scrap, and rework were considerably reduced. This in turn increased flexibility, made the production process of the business much more transparent, visual, and predictable.

Implementation of the Kanban system moved the business to a more demand-driven production process. As a result, the production process was allocated within a more structured framework, and output was increased by 20%. Inter and intra-department communication was also enhanced and more effective, which reduced WIP by 30%, minimized idle time in operations with less waiting time. The space occupied was also released by up to 50% with better inventory management.

6 Conclusions

In conclusion, this chapter has discussed the fundamental principles of Kanban system and its implementation in the fashion and textile industries for enhancing efficiency. Kanban systems are based on the use of signals in manufacturing industries to control the movement of goods in the production floor. In kanban system, a pull system is followed to produce as much as quantities which are actually in demand. Kanban system uses the JIT approach which means each process in the manufacturing line pulls the number of components in the next process just at the right time.

As a major tool in the lean manufacturing process, the Kanban system has been designed to help the workflow as agile and efficient as possible through the utility of visualised signals such as different coloured lights. Compared to the traditional system, Kanban system has been considered more efficient in reducing the bottlenecks by visual signalling. Depending on different types of material handling, the Kanban system may be separated into three control systems, namely Single Kanban system, Dual Kanban system, and Semi-dual Kanban system.

In fashion and textile industries, the Kanban system has undoubtedly brought about a positive impact in several developing countries. Some case studies have been presented for the fashion brand (ZARA) and the shirt manufacturer from India. It has been shown that the improved management of inventories (raw materials, semi-finished products, WIP and finished goods), the system has helped businesses to mitigate waste, and aim for a zero-defect production line, in which the quality is strictly maintained right from the initial source to delivery of the final product to the customer. Kanban implementation specifically, as a lean manufacturing tool, perform as a great support to the fashion and textile industries.

References

Balram B (2003) Kanban systems: the stirling engine manufacturing cell. University of Manitoba, Department of Mechanical & Industrial Engineering
Baradaran V (2018) Muti-objective optimization and simulation model to design the withdrawal Kanban systems. Prod Oper Manage 9(16). https://doi.org/10.22108/JPOM.2018.92445.0
Bernstein J (1984) GM exec discusses commitment to Kanban; system has tremendous potential. Automotive News 48.
Bieniusa A, Zeller P, Barke S (2018) Collaborative work management with a highly-available Kanban board. In: Müller P, Schaefer I (eds) Principled software development. Springer, Cham, pp 58–72. https://doi.org/10.1007/978-3-319-98047-8_4
Cimorelli S (2013) Kanban for the supply chain: fundamental practices for manufacturing management, 2nd edn. Productivity Press
Corona E, Pani FE (2013) A review of lean-Kanban approaches in the software development. WSEAS Trans Inf Sci Appl 10(1):1–13
Cuellar R (2011) Kanban for help desks: managing the unplannable. Cutter IT J 24(3):16–23
Donald W (2003) Inventory control and management, 2nd edn, John Wiley & Sons Ltd.
El Abbadi L, El Manti S, Samah Elrhanimi MH (2018) Kanban system for industry 4.0 environment. Int J Eng Technol 7(4.16):60–65. https://doi.org/10.14419/ijet.v7i4.16.21780
Flynn BB, Sakakibara S, Schroeder RG (1995) Relationship between JIT and TQM: practices and performance. Acad Manage J 38(5):1325–1360. https://doi.org/10.5465/256860
Fukukawa T, Hong SC (1993) The determination of the optimal number of Kanbans in a Just-In-Time production system. Comput Ind Eng 24(4):551–599. https://doi.org/10.1016/0360-8352(93)90197-6
Graves RJ, Konopka JM, Milne RJ (1995) Literature review of material flow control mechanisms. Prod Planning Control 6(5):395–403. https://doi.org/10.1080/09537289508930296

Groos JM, McInnis KR (2003) Kanban made simple-demystifying and applying Toyota's legendary manufacturing process. American Management Association
Gupta SM, Al-Turki YA, Perry RF (1999) Flexible Kanban system. Int J Oper Prod Manage 19(10):1065–1093. https://doi.org/10.1108/01443579910271700
Gurgur CZ, Altiok T (2007) Analysis of decentralized multi-product pull systems with lost sales. Naval Res Logistics (NRL) 54(4):357–370. https://doi.org/10.1002/nav.20210
Heizer J, Render B (2005) Flexible version: operation management, 7th edn. Prentice Hall
Henderson BD (1986) The logic of Kanban. J Bus Strategy 6(3):6
Huang CC, Kusiak A (1996) Overview of Kanban systems. Int J Comput Integr Manuf 9(3):169–189. https://doi.org/10.1080/095119296131643
Iannone R, Miranda S, Riemma S (2009) The search for the optimal number of Kanbans in unstable assembly-tree layout systems under intensive loading conditions. Int J Comput Integr Manuf 22(4):315–324. https://doi.org/10.1080/09511920802206427
Jang JW, Kim YW (2007) Using the Kanban for construction production and safety control. In: Proceedings of the 15th annual conference of the international group for lean construction (IGLC), pp 519–528
Jyothi VE, Rao KN (2012) Effective implementation of agile practices-incoordination with lean Kanban. Int J Comput Sci Eng 4(1):87–91
Kirovska N, Koceski S (2015) Usage of Kanban methodology at software development teams. J Appl Econ Business 3(3):25–34
Kobbacy KA, Liang Y (1999) Towards the development of an intelligent inventory management system. Integr Manuf Syst 10(6):354–366. https://doi.org/10.1108/09576069910293022
Kumar CS, Panneerselvam R (2007) Literature review of JIT-KANBAN system. Int J Adv Manuf Technol 32(3–4):393–408. https://doi.org/10.1007/s00170-005-0340-2
Mojarro-Magaña M, Olguín-Tiznado JE, García-Alcaraz JL, Camargo-Wilson C, López-Barreras JA, Pérez-López RJ (2018) Impact of the planning from the Kanban system on the company's operating benefits. Sustainability 10(7):2506. https://doi.org/10.3390/su10072506
Monden Y (1993) Toyota production system: an integrated approach to Just-In-Time, 2nd edn. Industrial Engineering and Management Press
Naufal A, Jaffar A, Yusoff N, Hayati N (2012) Development of Kanban system at local manufacturing company in Malaysia–case study. Procedia Eng 41:1721–1726. https://doi.org/10.1016/j.proeng.2012.07.374
Nayak R (ed) (2020) Supply chain management and logistics in the global fashion sector: the sustainability challenge. Routledge
Ohno T (2018) The Toyota production system: beyond large-scale production. Routledge
Rahman NAA, Sharif SM, Esa MM (2013) Lean manufacturing case study with Kanban system implementation. Procedia Econ Finance 7:174–180. https://doi.org/10.1016/S2212-5671(13)00232-3
Savino MM, Mazza A (2015) Kanban-driven parts feeding within a semi-automated O-shaped assembly line: a case study in the automotive industry. Assem Autom 35(1):3–15. https://doi.org/10.1108/AA-07-2014-068
Selçuk B (2013) Adaptive lead time quotation in a pull production system with lead time responsive demand. J Manuf Syst 32(1):138–146. https://doi.org/10.1016/j.jmsy.2012.07.017
Singh N, Shek KH, Meloche D (1990) The development of a Kanban system: a case study. Int J Oper Prod Manag 10(7):28–36. https://doi.org/10.1108/01443579010140498
Sivakumar GD, Shahabudeen P (2008) Design of multi-stage adaptive Kanban system. Int J Adv Manuf Technol 38(3):321–336. https://doi.org/10.1007/s00170-007-1093-x
Surendra MG, Yousef AY, Ronal FP (1999) Flexible Kanban system. Int J Operat Prod Manage 19(10):1065–1093
Sugimori Y, Kusunoki K, Cho F, Uchikawa S (1977) Toyota production system and Kanban system materialization of just-in-time and respect-for-human system. Int J Prod Res 15(6):553–564. https://doi.org/10.1080/00207547708943149

Sugimori J (2007) Celebrating the enigma: the continuing puzzle of the Toyota production system. Int J Prod Res 45(16):3545–3554. https://doi.org/10.1080/00207540701223386

Suzaki K (1987) The new manufacturing challenge: techniques for continuous improvement. The Free Press

Taj S, Motlagh CK, Hazen MA, Morosan C (2011) Dependencies within dimensions of lean manufacturing: evidence from the Chinese manufacturing plants. Int J Oper Quant Manage 17(4):279–302

Tardif V, Maaseidvaag L (2001) An adaptive approach to controlling Kanban systems. Eur J Oper Res 132(2):411–424. https://doi.org/10.1016/S0377-2217(00)00119-3

Wakode RB, Raut LP, Talmale P (2015) Overview on Kanban methodology and its implementation. IJSRD-Int J Sci Res Develop 3(02):2321–613

Other Lean Tools in Fashion and Textile Manufacturing

Hiep Cong Pham⊙, Irfan Ulhaq⊙, Paul Yeow⊙, and Mohammadreza Akbari⊙

Abstract This chapter discusses how waste reduction, process efficiency and safety in textile manufacturing can be managed using lean manufacturing tools: namely Muda, six Sigma and statistical process control, in addition to ergonomics. Muda approach identifies seven waste groups in post-implementation manufacturing process. The chapter highlights four Muda techniques to perform the waste identification including building lean thinking and cultures, hybrid approach of combining multiple decision-making methods, integrating lean and green management, and value stream mapping to prioritize responses to wastes. Statistical process control focuses on process deviation in manufacturing and provide a real-time analysis to enhance the future production predictability. Six Sigma is a process-based strategy consisting of five stages (DMAIC-Define, measure, analyze, improve, and control) used as a data-driven process to eradicate defects in manufacturing process. Finally, ergonomics which focuses on interactions among humans and other elements of a system to optimize human well-being and overall system performance, is recommended to combine with lean manufacturing to improve quality and well-being of workers in textile industry.

Keywords Lean · Muda · Six sigma · Statistical process control · Quality management tools · Ergonomics

1 Introduction

Fashion and textile industries create an immense global market that influences the economy of almost every country, either directly or indirectly. Despite its extensive size in the global business, the industrial sector still faces many challenges. It

---

H. C. Pham · I. Ulhaq · P. Yeow · M. Akbari
RMIT University Vietnam, Ho Chi Minh City, Vietnam

M. Akbari (✉)
James Cook University, 1 James Cook Dr, Douglas QLD 4811, Townsville, Australia
e-mail: reza.akbari@rmit.edu.vn

© The Author(s), under exclusive license to Springer Nature Singapore Pte Ltd. 2022
R. Nayak (ed.), *Lean Supply Chain Management in Fashion and Textile Industry*,
Textile Science and Clothing Technology,
https://doi.org/10.1007/978-981-19-2108-7_9

traditionally experiences low productivity, labor intensive, manual work practice, high product defects, and inaccurate demand forecast due to fast changing apparel styles (Karthikadevi 2014). The industrial sector is under pressure to focus more on effective and efficient manufacturing processes for survival in a highly competitive market.

Minimizing production wastes, product defects and labor idle time are some key conditions of reducing cost, increasing productivity, and adequately focusing on high value-added activities (Akbari and Do 2021; Akbari and Ha 2020). Wastes which can be any excess in materials, components, or equipment in the manufacturing process can be traced to various stages. Lean manufacturing comprises a number of principles and techniques aimed to eliminate wastes and reduce low value-added activities at various stages of production or services to achieve high quality and competitive products or services (Fercoq et al. 2016; Hodge et al. 2011; Pettersen 2009). The focus of lean manufacturing is considered as a managerial approach rather than a set of tools to improve business performance based on the integration of socio-technical system (Angelis and Fernandes 2012; Bortolotti et al. 2015).

This chapter presents three main approaches in lean manufacturing namely: Muda as a waste identification technique, six Sigma and statistical process control for process performance improvement and their applications in fashion and textile manufacturing. Besides, the issue of ergonomics in the manufacturing is also covered as it can have a significant impact on the improvement of working quality and well-being of workers in the industrial environment.

2 Muda in Lean Manufacturing

2.1 Overview

Muri, Mura and Muda are the three waste reduction techniques proposed in 1950s by Toyota Production System (Monden 1983). Out of the three, Muda focuses on the identification of production wastes after the process has been put in place, helping businesses to accurately identify either socio or technical aspects that need improvements and develop appropriate strategies to adjust and optimize production processes (Shamah 2013; Vaněček et al. 2018).

Muda focuses on seven major wastes of business operation, including transportation, inventory, motion, waiting, over production, over processing, and defects (Fercoq et al. 2016; Behnam et al. 2018). Although initially discovered and applied in the Japanese automotive industry, Muda approach can be adapted for different business types to better identify potential wastes or factors that cause wastes of the business operations or processes (Vaněček et al. 2018). Furthermore, types of wastes can be added to provide a more complete representation of the business waste management (English 2005; Sutherland and Bennett 2008). The next section will expand the types of Muda wastes and how the Muda waste technique has been applied in several industries.

Waste identification haven been explored in different industries, such as in managing information (English 2005), in supply chain (SC) management (Akbari et al. 2017; Sutherland and Bennett 2008), in clothing manufacturing (Behnam et al. 2018) and fast food industry (Morales-Contreras et al. 2020). The general definition of seven deadly waste elements of Muda can be explained as below.

- Transportation: Time, resource wasting and other costs that are generated from unnecessarily moving products or materials.
- Inventory: The excess of materials or products that are not processed.
- Motion: Wasted human efforts and time for unnecessary works.
- Waiting: The time spent on waiting between processes' steps.
- Over production: Making more products than demanded or required.
- Over processing: Wastes from higher requirement of quantity and quality of works than needed.
- Defects: Wastes from a product/service/process's failure.

The nature of each waste element can vary subject to industry characteristics or types of product/service provided. For example, overproduction waste in information management can include redundant systems, duplicate records, or hidden information storage; whereas in SC management it relates to higher order quantity than needed; in clothing manufacturing it can be the number of unsold products after a sale-period; or in fast food industry it indicates over number of cooked meals or finished goods inventory. Waste sources can be either eliminated or identified and reduced depending on business nature. When discovering waste factors in managing information, English (2005) added two new elements, including repairs/rejects and wrong decisions, while Sutherland and Bennett (2008) added space element as a supplement waste element in SC management. On the other hand, Sutherland and Bennett (2008) eliminated over processing waste in SC management and Morales-Contreras et al. (2020) dropped the over-processing and motion wasting element in the fast-food industry. The name of the waste element can be also modified to better fit with the business context: defects was called process of failure in information managing (English 2005), or errors in SC management (Sutherland and Bennett 2008).

Multiple methods have been developed to apply Muda in lean management to identify waste sources in business operations or processes. Muda methods can include: (1) building lean thinking and cultures (Sutherland and Bennet 2008; Shamah 2013) to promote and adopt Lean management approach; (2) hybrid strategies such as Pareto analysis or 80/20 principle, decoupling points, separation between "base" and "surge" demands to integrate lean and Agile approach to deal with uncertain business environment (Christopher and Towill 2001); (3) integrating lean and green management approach to minimize wastes by combining 3R (reduce, reuse and recycle) hierarchy with deadly wastes of lean management (Fercoq et al. 2016); and (4) using value stream mapping (Behnam et al. 2018), and (5) Analytical Hierarchy Process (AHP) and Fuzzy AHP (Gnanavelbabu and Arunagiri 2018) to more fully and accurately identify sources of wastes and prioritize responses. Each of these methods will be further explained in the following sections.

Acquiring lean thinking and cultures has been identified as a core competence in effectively implementing lean approach and application of the Muda (Shamah 2013). Christopher and Towill (2001) emphasized the important role of senior executives in the Lean culture development. To ensure that everyone is aware of the various forms of wastes and solutions to eliminate the wastes, senior executives need to regularly observe the employees' working activities, develop the problem-solving capability by asking the "Five Whys" questions. Womack and Jones (1996) proposed five principles that firms should follow to develop lean culture and achieve the overall value creation: (1) identifying customer value, (2) managing the value stream, (3) developing the capacity to flow production, (4) applying a pull system to support the flow of materials to constrained operations, (5) detecting excellence by eliminating all forms of waste in the production system.

Similarly, Shamah (2013) suggested six fundamental tasks to promote sustained and organization-wide efficiency and waste reduction: (1) planning for change, (2) designing suitable strategies and procedures to support Lean operation, (3) implementing the suggested plans/measuring employees', departments' or organizations' performance. (4) analysing defects/lean from leaders or feedback to create a knowledge base to achieve optimum performance by eliminating waste, (5) exceeding customer expectations and (6) adding value to stakeholders. These Lean principles and tasks help managers or even employees in tracking the degree of leanness in all business parties and activities. However, this process needs to be implemented with sensible consideration of negative impacts on employees' morale, or work and life balance. It might lower the employee performance if they perceive it as unnecessary, time-consuming, and feeling stressed because it might indicate redundance in an employee's position. Thus, suitable actions and decisions from managers or senior executives are important in successfully generating and maintaining leanness among employees' and business' activities.

The separation between the Agile and lean management is debatable. Some prior studies indicated that there is a fundamental difference between these two concepts, whereas others showed possible integration and common effects of both methods on the business operation. Christopher and Towill (2001) indicated three hybrid strategies in order to integrate lean approach within Agile SC management. The first strategy is the Pareto curve approach with the rule of 80/20, in which "80 percent of total volume will be generated from just 20 percent of the total product line" (Christopher and Towill 2001). Lean approach should be applied to the top 20% of products by volume, which is predictable, whereas the other 80% which are less predictable are where the Agile management approach should be applied. Secondly, another suggested hybrid strategy is the decoupling point or strategic inventory approach (Christopher and Towill 2001). This approach suggests that companies can apply lean method to effectively manage the holding materials, which is predictable, and only complete final assembly when customer requirements have been confirmed, which requires Agile management approach. The transforming point between these two methods is called as decoupling point. Another lean and Agile hybrid strategy proposed by Christopher and Towill (2001) is the base and surge demand separation, in which the business can put customers into base demands, which are more likely to be forecasted and managed by the lean approach, and surge demand, which are less predictable and managed by Agile management approach. Based on these demand differences, organizations can distribute the manufacturing capacities to both reduce costs and ensure the flexibility to match customers' needs.

Each of these hybrid strategies is suitable for different types of market conditions. For example, the Pareto analysis or 80/20 strategy is better used in an environment that has high levels of variation in customer demands, while the decoupling point strategy is suitable with market conditions of modular production or intermediate inventory, and surge and base demand separation strategy is suitable when the base demand can be predicted confidently for applying a lean approach and adapting local manufacturing available for applying Agile approach. In other words, under a dynamic environment, lean management can reduce wastes (i.e. Muda) through the implementation of the Agile management.

Another tool designed to minimize the seven deadly wastes in Muda is the Lean/3R matrix by Fercoq et al. (2016). In this matrix, each of the seven wastes from Muda is linked with the 3R items, including Reduction, Recycle, and Reuse, to better identifying the wastes and deciding the solutions to eliminate them. The 3R hierarchy provides a process that includes three steps of eliminating wastes. The first step is to identify and quantify the waste sources. At this stage, the seven Muda concepts from the lean approach can be applied to provide a full and accurate overview of waste identification. Second step involves analyzing the effects of the identified wastes on the business performance. In the final step, the managers decide whether to reduce, re-use, or recycle the wastes and formulate a detailed solution to implement an appropriate waste minimizing method. The Lean/3R matrix showed a positive result in minimizing the wastes, following the seven Muda, and improve the environmental performance of the business operation. The seven Muda elements server as a checklist that allows managers to assess wastes that might exist from the business operation, whereas 3R hierarchy provides directions for analyzing the potential causes, effects and minimizing solutions of identified wastes based.

Another focus of Muda is to accurately identify the seven deadly wastes in business operation by classifying the chain of operational activities into separate stages, then identifying potential wastes at each operational stage or activity under different working scenario (i.e. stable or uncertainty working environment), which can be done through value stream mapping. Behnam et al. (2018) proposed the use of value stream mapping as an efficient tool to detect the seven wastes. This mapping tool allows the managers to obtain an overview of the production process, from importing raw materials to transporting finished products to customers. For example, study of Behnam et al. (2018) broke the shirts sewing production into 20 separate activities then classified them into five stages, including cutting stage, sewing stage, enzyme stage, quality control and ironing stage, and barcoding, folding and warehouse transportation stage. Then, the potential wastes in each activity were identified and categorized according to the seven deadly wastes. For instance, at the cutting stage, several wastes have been identified such as cutting wastes (classified to "Defects" element), cutting workers' movements (classified to "Motion" element), delay in the fabric delivery from supplier and delay in fabric cutting (classified to "Waiting" element) and cut fabric stock (classified to "Inventory" element). Based on this mapping tool, managers can review and identify wastes at each small stage in a large production process. Furthermore, each of the identified waste will be classified by their consequence in either delaying time or financial expenses. Thus, Muda can provide the managers an accurate overview of the wastes in the production processes to develop an efficient waste minimizing program. The wastes in the entire production process, therefore, can be fully discovered, helping managers to provide appropriate and effective waste minimizing programs or strategies.

Gnanavelbabu and Arunagiri (2018) proposed AHP and Fuzzy AHP algorithm in segregating, measuring the weights of seven deadly wastes and prioritizing responses. The aim of both methods is to break down a lean problem into smaller components. AHP helps the managers to focus on a limited number of lean wastes using weighted rankings in each stage. Fuzzy AHP can consider the uncertainty in the environment and the responses' deviation, hence providing more accurate prioritization among lean wastes. Both methods showed a good result in helping managers to assess waste elements and to identify the major wastes. However, the Fuzzy AHP method showed a better performance in measuring the lean wastes and solving the multi-criteria decision-making problem.

2.2 Muda Application in Fashion and Textile Industry

Several studies examined applying seven deadly wastes of Muda approach to manage and eliminate wastes in the fashion and textile production systems. For example, Behnam et al. (2018) proposed the value stream mapping approach as a tool to identify and prioritize production system's Muda in clothing manufacturing companies. Because Muda from different process stages might relate to each other, hence eliminating of all wastes one-time can be difficult. Value stream mapping tool allows the manager to breakdown the whole production chain into separate stages and identify the potential wastes in each production stage. The managers, then, can measure the impact of time and financial expenses of each waste separately and decide waste reduction responses. In their study, 40 potential wastes have been identified from 20 production activities in five major stages. Three wastes in Waiting stage accounted for 2.79% of total wasting expenses, six from Motion (0.58%), four from Defect (74.5%), two from Over Processing (1.41%), one from Over Production (2.04%), six from Transportation (2.14%) and 18 from Inventory (16.54%). The managers, thus, can decide which wastes they should focus to eliminate to bring the most efficient improvement in production process. For example, 20 out of 40 identified wastes accounted for 97% of the production inefficiency whereas the remaining 20 contributed only 3%. Managers should focus their resources on addressing high impact wastes, instead of spending resources on minimizing less important wastes.

3 Six Sigma and Statistical Process Control

3.1 Overview

Over the years, firms have started adopting quality management tools at operational and strategic level to enhance their supply chain capabilities and overcome the performance dilemmas. Six Sigma and statistical process control (SPC) methods are a part of total quality management (TQM) improvement programs, and its techniques are widely embraced to monitor and discover process variations in manufacturing and supply chain process performance (Neyestani 2017). Moreover, these tools become a complete integrated system of information sharing and knowledge management tool (Aini et al. 2017).

To the same degree, SPC is also helpful in delivering right results required by the end market and customers, supporting management to decide on supplier's performance and ranking them as preferred suppliers, and developing a systematic approach conclusive evidence to supply chain level strategies as part of organisation wide strategies. Shahin et al. (2010) conducted an exploration to develop an integrated roadmap of quality management by combining the traditional seven quality tools and new quality tools as well. However, majority literature suggests six sigma and SPC application largely limited to large enterprises, as well as more explorations are required in SMEs (small and medium-size enterprises) and in developing countries.

3.2 Statistical Process Control

Dr. W. Edwards Deming endorsed the idea of statistical rationale as part of organisation process management strategy. Today Deming's approach is established as part of quality management strategies in many firms to enable process insights in reducing variations (Saidy et al. 2020). SPC is defined as a managerial toolkit using statistical methods to monitor and control manufacturing processes data to ensure that the outputs conform to specifications. SPC is a practice based on Shewhart's conception of process variability, which is widely applied not only in manufacturing processes but also in service operations. Compared to inspection method, Madanhire and Mbohwa (2016) establish and extend SPC as a lean tool by detecting and minimizing errors whilst developing a product according to conformance standards. SPC is further designated as a major tool for sequential process monitoring to ensure operational adequacy such as production lines, customer services issues, improper use of raw material, and operational related errors (Das 2013).

In the 1960's Kaoru Ishikawa advocated the idea of quality management using statistical tools named as seven quality tools (Mach and Guáqueta 2001). The aim of these tools was to enhance the team and workforce capabilities by utilizing staff skills in business process improvement. Several tools could be attributed as process quality control tools including Pareto Chart, Cause-and-Effect diagram (or Ishikawa diagram), Scatter Diagram, Flow Chart, Histogram, Check sheets and Control Charts (Muhammad 2015). However, in fashion and textile industry, majority of studies have commonly used only three major tools namely Pareto analysis, cause and effect diagram and control charts (Ata et al. 2020; Das 2013; Das et al. 2007; Hossen et al. 2017; Majumdar et al. 2012). The following section provides a brief elaboration of commonly used quality control tools in fashion and textile industry.

SPC tools and methods could be applied in two phases. During the first phase often control charts are developed by plotting data to uncover and understand whether process is under control. By analysing the control charts (inbound and out bound) one could identify the potential causes of variations by incorporating other tools such as Pareto principles and brainstorming sessions are conducted to make process adjustment (Madanhire and Mbohwa 2016; Mason and Antony 2000). Secondly, understanding the process capability is key to quantify the capability of the studied process to produce products that could meet required product specifications. The second phase covers the process capability by calculating capability indices of the process as mentioned by Mason and Antony (2000). The pre-defined specifications defined customers as the lower specification limit (LSL) and important tasks as the upper specification limit (USL), for an organisation to comply with the limits that brings returns and benefits (Tuna 2018).

Pareto analysis or ABC analysis, often named as 80/20 rule is a common method deployed to categorize the performance or outcomes to understand the variances from different angles. Pareto principle is based on the idea that 80% of all problems (production or supply chain in this case) could be solved by removing the causes related to 20% of all problems (Mach and Guáqueta 2001; Wang and Zhang 2008). Studies in the fashion and textile industry shows positive impacts of Pareto analysis. As shown in the Ata et al. (2020) study, Pareto analysis helped the firm to quantify the problems in denim washing process and found that out of all the causes only five categories, namely chemical repair, blue floor, chemical intensive, light floor and deep floor defects accounted for 78.6% of the total 53 washing defects. In another study involving a yarn manufacturing industry in Bangladesh, the application of 80/20 rule, established that among all production and manufacturing losses, stoppage losses contributed the most at 67% compared to breakdown losses of 22% (Hossen et al. 2017).

Cause-and-effect or Ishikawa or fishbone diagram is one of the systematic approaches to brainstorm causes (Liliana 2016). This diagram represents causes related to a specific process or event. The diagram became popular in lean management in identifying root causes of a variety of problems (Hossen et al. 2017). Fishbone diagrams can shed insights and support in brainstorming to identify possible causes of a problem and in sorting ideas as feasible solutions, thus being a useful tool to explore process abnormalities from multiple factors (Liliana 2016). In a study conducted on textile industry washing defects by Ata et al. (2020), cause-and-effect diagram facilitated to categorize causes of defects into six different categories in denim fabric washing. This study also found majority of causes emerged due lack of color experienced staff at different stages during denim fabric washing. Moreover, chemical defects could be associated with defects in the color category (Ata et al. 2020).

Process control charts or P Chart is a graphical diagram supporting process control by drawing and pointing in/out of control of a cause' values (Muhammad 2015; Mach and Guáqueta 2001). Control chart provides controlling graphics of all attribute values that indicate if process is kept under control, or the values are exceeding upper control limits (Ata et al. 2020). Upper Control Limit (UCL) shows the maximum value is acceptable variation from the average value on the process, Lower Control Limit (LCL) is the minimum value acceptable variation from the average value of the process and Central Line (CL) is average value on the control charts. P charts could help firms to predict the performance of a process in the long run. Therefore, process control charts can serve as a preventive tool to further research and develop necessary corrections if a process is out of control (Das 2013).

Check sheet is a simple form that allows employees and operators to record data and information in a systematic manner. A check sheet captures specific events of a process frequency in a consistent and economical manner (Neyestani 2017). Data captured from these events could be rearranged for future reuse but check sheets are not seen as a prominent method of problem analysis. One type of check sheet for example is tally sheet which is a very basic tool in the form of counting the currencies of occurrences of a specific event.

3.3 Six Sigma Concepts

Six Sigma is a systematic strategy to quality management and was coined and formalized in 1980s by Motorola (Hoerl 1998; Mach and Guáqueta 2001). Since then, Six Sigma is found one of most effective management approaches which has helped many Fortune-500 firms to reap benefits by increasing market share, healthy cost reductions, and dramatic efficiency enhancements by optimizing business processes (Harry and Schroeder 2000). Following the TQM philosophy (approach to zero defects), six Sigma's major goal is to diminish inconsistencies and continuously meeting customer expectations over time (Mach and Guáqueta 2001). Sigma (standard deviation from the mean) measures the variation of a process from its mean value. Since a major aim of six sigma is striving for zero defects, a value of 3.4 defects per million opportunities is allowed to achieve the lowest probability of an error (Hoerl 1998).

3.4 Six Sigma- A Five Stage Process

Six Sigma is a process-based strategy consisting of five stages: define, measure, analyze, improve and control (DMAIC). During the "define" stage, a firm need to establish the goals and scope of the improvement activity at project and operational quality levels. In addition, during this stage a project charter is also developed, and project teams are identified (Hossen et al. 2017). The "measure" stage requires a project team to identify the relevant parameters and past information to measure the existing performance of a process (Hoerl 1998). For example, machine usage; current product quality levels; fabrics shade, dyeing requirements and length of the fabrics etc. are some of the parameters (Das et al. 2007). It is necessary to establish the metrices and KPIs (key performance indicators) to monitor the future goals (i.e., developing the baselines values and target sigma level). During the "analyze" stage, current process or system related root causes are traced. Often in this stage statistical tools are used for root cause analysis (Das et al. 2007; Mach and Guáqueta 2001). Furthermore, a project team could use the brainstorming sessions and develop the cause-and-effect diagrams to visualize the defects from different perspectives (Chang et al. 2012; Hoerl 1998). The main goals of this stages are to develop a shared meaning, researching the root cause of actual issue, and to establish key actions on the decision-making (Das et al. 2007).

The "improve" stage requires the project team to operationalize the decisions towards a solution for the root causes identified during analysis phase (Mach and Guáqueta 2001). During this stage key improvements are made through feedback and prototypes development. The solutions are finalized once the Sigma level established in the start of the project are achieved (Mach and Guáqueta 2001). These changes could be from simple to more innovative implementations including process change, change of dyeing process or quality level of the color (Das et al. 2007). Control stage is the final stage of a Six Sigma project. The goal of this stage is to sustain the achieved results (Mach and Guáqueta 2001). Das et al. (2007) exemplified three main points to be addressed during this stage: development of a change control plan and structure; new knowledge transfer of process changes to staff; and document and reporting of unattended issues during the project for future improvements.

Six Sigma is conducted on a project-based approach to remove waste and process improvement (Chang et al. 2012). As majority of studies employing six Sigma as a data-driven process to eradicate defects, SPC tools are often found to be a subset of six Sigma. The real essence of six Sigma is employing DMAIC method taking all SPC related tools at different stages along the process (Chang et al. 2012). Six Sigma uses a systematic cycle problem solving methodology, SPC used statistical procedures at each phase during the manufacturing process to identify and tracking of errors and variations. Both approaches help firms in coping with the quality variations on operations in supply chains and support managers to further design experiments to explore critical factors affecting process and product quality that could lead to best possible performance. Evans and Lindsay (2002) highlighted three statistical principles of continuous improvement.

- All work occurs in a system of interconnected processes.
- Variation exists in all processes, and
- Understanding and reducing variation are keys to success.

3.5 Six Sigma and SPC in Fashion and Textile Industry

Process parameters in supply chain and manufacturing could deviate from their base values due to some chance variations. SPC tools could offer a great real-time analysis and enhance the future predictability by analysing data in supply chain and logistics related processes (Das 2013). However, developing and adopting these methods comes with a greater risk from different perspectives.

Although several studies found positive impacts of SPC in the manufacturing industry, its applicability in fashion and textile industry requires more academic work (Madanhire and Mbohwa 2016). Previous studies show that SPC offers an advantage on assessing the detection and corrective action against the operational problems in supply chains (Saidy et al. 2020; Wang and Zhang 2008). Aini et al. (2017) developed a quality control system for apparel production SMEs in Indonesia. This study concluded that the SPC as an information system with mobile platform could effectively contribute to feedback and knowledge transfer for fashion and textile experts. In the garment industries SPC techniques including flowchart, Pareto analysis and fishbone diagram help to analyze current process and with historical data helped to understand the nature of the mistakes during the production process (Tuna 2018).

Shafira and Mansur (2018) applied six Sigma in production quality improvement analysis of grey cambric in a textile setting. Using a Waste Assessment Model (WAM) approach they explored the frequency of different types of production waste results. They found that production defects, inventory, motion, transportation, over production were the highest among other categories ranked. The analysis revealed a level of 3.3 sigma lowered than the average Japanese firm. FMEA (Failure Mode Effects Analysis) and AHP was also used to understand malfunctioning of machines. More and Pawar (2015) survey on twenty-six textile companies in India found that textiles companies only use ISO 9000 Quality Management System. Moreover, this study stated that Six Sigma is not given importance as a quality tool in several industries. Using a case study method in the South Indian textile firms, Karthi et al. (2013) also found the ISO 9000 implementations are common among the firms, however, companies seldom try Six Sigma as a tool of their quality drive.

Taner (2012) investigated the critical success factors of six Sigma adoption in Turkish textile SMEs. Using a survey-based approach this study found involvement and commitment of top management linking quality initiatives to employees; information technology and innovation are found to be important success factors for Turkish textile SMEs. In addition, leadership and commitment of top management, strategic vision, data collection and measurement, are found to be the most CSFs (critical success factors) for the successful initiation of six Sigma programs. Furthermore, the lack of knowledge of the system to start the initiative and the presence of ISO 9000-certification in SMEs are found to hinder its implementation of six Sigma. Results of this study also concluded that personnel and incompetency with new technologies are among the failure of implementation of six Sigma in SMEs.

The above review has shown that SPC and six Sigma could help fashion and textile firms to overcome several quality and waste management issues. But recent research highlights that majority of firms apply these tools to rectify defects and issues emerging during the production processes. SPC charts, and other basic quality tools are commonly applied to understand and explore process anomalies in the supply chains. Moreover, review of studies finds that fashion and textile firms are good at applying quality control systems such as ISO 9000 standard along with common SPC tools. However, SMEs are lacking its appreciation in their supply chain processes (Taner 2012). A study from Hill et al. (2018) on the implementation of six Sigma in an MRO (Maintenance, Repair and Operations) case found that common issues on the project were: feedback from employees, commitment of the operational staff, and lack of strong leadership drive to push project forward. During the implementation, the study found the people with expertise at operation were away to manage operational and supply chain tasks. Furthermore, continuous training, development of staff and effective communication throughout the project is essential to six Sigma as part of quality management programs (Hill et al. 2018).

4 Ergonomics Concepts and Tools

4.1 Overview

Ergonomics is defined by the International Ergonomics Association (IEA) as "the scientific discipline concerned with the understanding of interactions among humans and other elements of a system, and the profession that applies theory, principles, data, and methods to design in order to optimize human well-being and overall system performance" (IEA 2021). It takes account of physical, social, technical, cognitive, organizational and environmental aspects when humans interact with systems which include other humans, processes, tools, machines and technologies. The goal of ergonomics is very much related to the "Muri" concept of lean manufacturing that means avoiding physical strain, bending to work, lifting heavy weights, repeating tiring action or overburdening (Arezes et al. 2015). Additionally, "Muri" concept includes elimination of wastes not only related to physical burden but also related to mental burden, particularly addressing the motivation and welfare issues of the workers, e.g., workers' autonomy, participation and engagement (Arezes et al. 2015; Eswaramoorthi et al. 2010). As such ergonomics and lean manufacturing are complementary to each other.

4.2 Ergonomics in Lean Management

There is evidence that ergonomics is integrated in the lean manufacturing approaches. For example, an ergonomics risk assessment tool was incorporated in the original Toyota car assembly lines which is called "Toyota Verification of Assembly Line" (Benders and Morita 2004). In this tool, ergonomics risk factors, e.g., physical workload and demanding tasks were identified, and solutions were developed by the assembly workers themselves using Kaizen (continuous improvement) techniques. Solutions developed included the "comfort car seat" where the worker was seated in a location in the car with close reach of the assembly locations without difficult physical body extensions. This intervention reduced musculoskeletal discomfort, increased productivity, and improved quality. Another example is a study of 41 lean manufacturing projects in various industries (including textile) in Portugal (Alves et al. 2019). Like the "Toyota Verification of Assembly Line", another assessment tool, i.e., the Ergonomics Workplace Assessment (EWA) (Ahonen et al. 1989) was used to identify the ergonomics risk factors. Out of 14 ergonomics risk factors in the EWA, 5S addressed 10 risk factors, Standard Work nine risk factors, Visual Management six risk factors, SMED (Single-Minute Exchange of Die) six risk factors, and Poka-Yoke four risk factors. These five lean manufacturing approaches covered 13 out of 14 ergonomics risk factors such as worksite, general physical activities, lifting tasks, postures and movements, accident risks, job content, job restrictiveness, communication and personal contacts, decision-making, work repetitiveness, attention, lighting conditions, and thermal conditions. Only noise risk factor was not addressed, possibly because there was no workplace that has elevated level of noise.

4.3 Ergonomics Tools in Fashion and Textile Industry

Ergonomics is particularly important because fashion and textile industry is labor intensive that may require intensive manual handling; thus, leading to poor body posture, and repetitive movements. There are several studies that validated the effectiveness of ergonomics integrated with lean manufacturing in terms of improvement of quality and well-being of workers in textile industry. For example, Vayvay and Erdinc (2008), and Erdinc and Vayvay (2008), introduced Quality Improvement Through Ergonomics (QUITE) methodology in a textile factory in Turkey, which integrated the Kaizen's continuous improvement approach, and an ergonomics assessment tool. The tool identified the main cause of 36 sewing operators' poor-quality performance, i.e., poor postures due to bending of neck and body during the monitoring of machine stitching operation. Ergonomics intervention of correcting the posture, i.e., by tilting the single-needle straight stitch machine by 4.2 degrees. This has resulted in reduction of defects from 7% to 3.4% (Erdinc and Yeow 2011).

There are other researchers who integrated ergonomics and lean manufacturing in other labor-intensive industry. For example, Axelsson (2000) introduced Select,

Interpret, Measure, Progress, Learn and Echo (SIMPLE) methodology in an assembly plant in Sweden. It is a combination of Kaizen's continuous improvement, QC Tools particularly the Pareto principle and an ergonomics assessment tool. Through the tool, it was discovered that the 17 out of 40 tasks with poor body postures contributed 70% of the plants' quality problems. Ergonomics interventions such as improving work body postures, lighting conditions, accessibility of materials and the comfort of assembly procedures have reduced the rejection rate from 8.9% to 5% and improved the well-being of the workers.

This section introduces some ergonomics assessment tools that can be incorporated in lean manufacturing approach in textile industry, i.e., the tools that evaluate ergonomics risk factors. Chiasson et al. (2012) has identified eight popular tools, i.e., the Quick Exposure Check (QEC), Finnish Institute of Occupational Health's EWA, Job Strain Index (JSI), Hand Activity Level (HAL), the Occupational Repetitive Actions (OCRA) Index, the Rapid Upper Limb Assessment method (RULA), the Rapid Entire Body Assessment method (REBA) and European standard for safety in human physical performance (EN 1005–3). Below is a brief description of the tools together with their strengths and weaknesses.

The QEC (David et al. 2008) measures musculoskeletal disorder (MSD) factor at the neck, back, arms upper extremities, and overall body. It takes into consideration of movement frequency, force applied, posture, and duration. The EWA (Ahonen et al. 1989) has 14 areas, i.e., physical workload, workstation design, lifting, accident risk, working posture and movements (multiple body areas), task restrictions, task content, decision-making, personal contact and communication, repetitiveness, attention required, noise, thermal environment, and lighting. Both the QEC and the EWA are particularly useful tools for initial ergonomics assessment because they can be easily learnt and can be conducted quickly (Chiasson et al. 2012). The workers' perception is included, which can be both an advantage and a disadvantage. Workers know best about the problems they face in their workplace and their discomfort and difficulty in body movements; however, their opinion may be biased because of the searchlight effect, i.e., they know they are being observed; thus, they might try to provide polite answers rather than the truth of the actual situation.

JSI (Moore and Garg 1995) and HAL (American Conference of Governmental Industrial Hygienists 2002) measure the MSD risks at the hands and wrists. Both tools require a time-motion study, and they are suitable for evaluation of workstations with short-cycle time (Chiasson et al. 2012). They consider the perceived exertion and its duration within the overall cycle time. Additionally, they consider the workers' perception. However, both tools have a different measurement for repetitive movement (Chiasson et al. 2012).

The OCRA (Colombini 1998) measures MSD risk factors particularly repetitive movements at the shoulders, arms, wrists, and elbows. Like JSI and HAL methods, it considers the risk of exposure within the overall cycle time. The OCRA and the QEC have more than 70% agreement when evaluating workstations with high-risk level (Chiasson et al. 2012). The OCRA covers many work characteristics; however, it takes 2.6 times longer to complete an assessment compared to using the QEC tool. Thus, it is costly to use particularly for large sample size (Chiasson et al. 2012).

The RULA (McAtamney and Nigle Corlett 1993) measures MSD risk factor at the upper limbs which includes the shoulders, upper and lower arms, wrists, trunk, and neck. It takes into consideration the posture, load and repetition movement. The REBA (Hignett and McAtamney 2000) measures MSD risk factor at the entire body including legs, trunk, neck, shoulders, arms and wrists. It also has the final score like the RULA. Both tools have a final score that determines if an ergonomics improvement is required. They are easy to learn and use and can be performed quickly. Based on Chiasson et al. (2012), they take about the same time to complete. However, both tools take a cross-sectional assessment of a work cycle; thus, they may not represent the whole cycle. Loo and Yeow (2015) recommended the observation of several cycles through video recording to determine the work case scenario (e.g., the worst postures) when recording the scores. The RULA and the REBA have high agreement with the QEC tool. According to Chiasson et al. (2012), more than 90% workstations identified as high risk by the QEC tool were also identified the two tools.

Lastly, EN1005-3 tool measures the acceptable force applied during a work. For example, it stipulates the maximum acceptable force applied to the shoulder joint based on the 15th percentile of workers. The acceptable force also takes into account of the duration, frequency and speed of the action performed. Like OCRA method, it considers many work characteristics. However, it is costly to use (as it takes much time) and difficult to learn. Both OCRA and EN1005-3 tools are not suitable for preliminary screening studies (Chiasson et al. 2012).

Most of the above assessment tools make use of observational methods, i.e., the assessor observes the work performed and estimate the measures (e.g., posture, angle, load, repetitions, and bending.) on various body locations such as upper limbs, and lower limbs, based on a set of standards. Scores are assigned according to the standards and then calculations can be done to get the final scores (e.g., the RULA and the REBA). Based on the scores, the risk factors for specific body locations are identified. For those with high scores (which indicate high risk), ergonomics improvements will be conducted through redesigning of workstation, work process and tool. The purpose is to reduce or remove the risk factors. After the improvements, the assessment tool will be used again to validate their effectiveness in removing the ergonomics risk factors.

One example of the use of tools is the application of the RULA in brazing operations of air-handler production (Loo and Yeow 2015). The RULA was used to identify the arm, wrist, neck and posture risk factors in the brazing task. The tool identified a final high-risk score i.e., 7 which required urgency to investigate and make changes to remove the risk factors. The tool identified bending of neck, raising of arms, twisting of wrist and bending of back during the brazing task throughout an eight-hour shift. The workstation was redesigned by raising it to avoid bending of back. The single-brazing gun was modified into a twin-brazing torch so as to avoid the need of twisting the wrist and bending of neck when the operator works the flames around the joint area. This movement was replaced with straight-line movement perpendicular to the joint area with the use of the twin-brazing torch. After the ergonomics improvements, RULA was used again which resulted to a lower final score of 3. The results also improved quality by 59%, increased productivity by 140%, reduced waste of filler material (silver alloy) significantly and saved $79,000 USD/year. Additionally, it improved the well-being of the workers through the reduction in pain in the neck, shoulders, wrist, and lower back.

In conclusion, this section explains what ergonomics is, why and how ergonomics can be integrated with lean manufacturing through the similarity in their goals. Justification was also given of the importance of ergonomics in fashion and textile industry, i.e., its labor-intensive nature and evidence of ergonomics integrated with lean manufacturing through QUITE methodology (Erdinc and Yeow 2011) together with the effectiveness in reducing quality defects and improving worker's wellbeing. Eight ergonomics assessment tools were introduced and discussed in terms of their strengths and weaknesses. Lastly, an example of application a tool was given together with its effectiveness.

The limitation is this section only focuses on the most common ergonomics tools that are useful for lean manufacturing in fashion and textile industries. There are many other ergonomics tools, methods, and techniques that can be used in lean manufacturing, for example, the mathematical modelling of ergonomics in hybrid assembly lines for safe assembly work (Botti et al. 2017) and integration of multiple ergonomics assessment tools and questionnaires, ergonomics improvement methods in the context of human–robot-collaborations (Colim et al. 2021). However, due to the constraint of space, and their limited use in fashion and textile industry, they have been excluded from this chapter.

5 Conclusion

Fashion and textile manufacturing possesses are often quite inefficient, due to multiple complex stages, labor intensive work and high changing demand. As the industry gets more competitive, fashion and textile firms are under pressure to optimize production, achieve better forecasting accuracy, reducing product defects and increasing work safety. The key challenges in achieving competitive textile manufacturing practice are to identify sources of wastes, reduce product defects and reduce injury-prone activities due to its highly manual and intensive work nature.

The chapter introduces three lean manufacturing approaches as solutions to cost reduction and resource optimization in the fashion and textile manufacturing. First approach, Muda, a waste identification technique, identifies seven waste elements in any established supply chain (Shamah 2013; Vaněček et al. 2018). These wastes once clearly recognized can help firms to eliminate inefficiency, reducing defects or prioritize resources on most productive activities. Second and third approaches including statistical process control and six Sigma, respectively being part of the TQM suite, focus on monitoring and identifying process variations from a baseline model in manufacturing and supply chain process performance (Neyestani 2017), which can help understand the nature of the mistakes during the production process (Tuna 2018).

Another key issue of the fashion and textile manufacturing is work safety or ergonomics due to its intensive manual labor, poor body postures and repetitive movements of the workers. The chapter introduces eight ergonomics assessment tools that can be incorporated in lean manufacturing in fashion and textile manufacturing to evaluate ergonomics risk factors. The main objectives of applying ergonomics integrated with lean manufacturing are to reduce quality defects and improve worker's wellbeing. It was established through the case studies that the fashion and textile industries can be benefitted by applying the lean manufacturing approaches and the ergonomic tools.

References

Ahonen M, Launis M, Kuorinka T (1989) Ergonomic workplace analysis. Ergonomics Section, Finnish Institute of Occupational Health, Helsinki

Aini N, Kusumaningrum R, Mustafa, Hidayat E (2017) Statistical process control systems in apparel production. In: Proceedings of 2017 international conference on information technology systems and innovation, ICITSI. IEEE, pp 134–138. https://doi.org/10.1109/ICITSI.2017.8267931

Akbari M, Do NAT (2021) A systematic review of machine learning in logistics and supply chain management: current trends and future directions. Benchmarking Int J (In-Press). https://doi.org/10.1108/BIJ-10-2020-0514

Akbari M, Ha N (2020) Impact of additive manufacturing on the Vietnamese transportation industry: an exploratory study. Asian J Shipp Logist 36(2):78–88. https://doi.org/10.1016/j.ajsl.2019.11.001

Akbari M, Clarke S, Maleki Far S (2017) Outsourcing best practice-the case of large construction firms in Iran. In: Proceedings of the informing science and information technology education conference, Ho Chi Minh City, Vietnam. http://proceedings.informingscience.org/InSITE2017/InSITE17p039-050Akbari3237.pdf

Alves AC, Ferreira AC, Maia LC, Leao CP, Carneiro P (2019) A symbiotic relationship between lean production and ergonomics: insights from industrial engineering final year projects. Int J Ind Eng Manag 10(4):243–256. https://doi.org/10.24867/ijiem-2019-4-244

American Conference of Governmental Industrial Hygienists (2002) Hand Activity Level (HAL). Threshold limit values for chemical substances and physical agents & biological exposure indices. ACGIH, Cincinnati, pp 112–114

Angelis J, Fernandes B (2012) Innovative lean: work practices and product and process improvements. Int J Lean Six Sigma 3(1):74–84. https://doi.org/10.1108/20401461211223740

Arezes PM, Dinis-Carvalho J, Alves AC (2015) Workplace ergonomics in lean production environments: a literature review. Work 52(1):57–70. https://doi.org/10.3233/wor-141941

Ata S, Yildiz MS, Durak I (2020) Statistical process control methods for determining defects of denim washing process: a textile case from turkey. Tekstil Ve Konfeksiyon 30:208–219

Axelsson J (2000) Quality and ergonomics: towards successful integration. Thesis (PhD). Linkoping University

Behnam D, Ayough A, Mirghaderi SH (2018) Value stream mapping approach and analytical network process to identify and prioritize production system's Mudas (case study: natural fibre clothing manufacturing company). J Text Inst 109(1):64–72. https://doi.org/10.1080/00405000.2017.1322737

Benders J, Morita M (2004) Changes in Toyota Motors' operations management. Int J Prod Res 42(3):433–444. https://doi.org/10.1080/00207540310001602883

Bortolotti T, Boscari S, Danese P (2015) Successful lean implementation: organizational culture and soft lean practices. Int J Prod Econ 160:182–201. https://doi.org/10.1016/j.ijpe.2014.10.013

Botti L, Mora C, Regattieri A (2017) Integrating ergonomics and lean manufacturing principles in a hybrid assembly line. Comput Ind Eng 111:481–491. https://doi.org/10.1016/j.cie.2017.05.011

Chang SI, Tsai T-R, Lin DKJ, Chou S-H, Lin Y-S (2012) Statistical process control for monitoring nonlinear profiles: a Six Sigma project on curing process. Qual Eng 24(2):251–326. https://doi.org/10.1080/08982112.2012.641149

Chiasson ME, Imbeau D, Aubry K, Delisle A (2012) Comparing the results of eight methods used to evaluate risk factors associated with musculoskeletal disorders. Int J Ind Ergon 42(5):478–488. https://doi.org/10.1016/j.ergon.2012.07.003

Christopher M, Towill D (2001) An integrated model for the design of agile supply chains. Int J Phys Distrib Logist Manag 31(4):235–246. https://doi.org/10.1108/09600030110394914

Colim A, Morgado R, Carneiro P, Costa N, Faria C, Sousa N, Arezes P (2021) Lean manufacturing and ergonomics integration: defining productivity and wellbeing indicators in a human–robot workstation. Sustainability 13:1931. https://doi.org/10.3390/su13041931

Colombini D (1998) An observational method for classifying exposure to repetitive movements of the upper limbs. Ergonomics 41(9):1261–1289. https://doi.org/10.1080/001401398186306

Das A (2013) 3—testing and statistical quality control in textile manufacturing. In: Majumdar A, Das A, Alagirusamy R, Kothari VK (eds) Process control in textile manufacturing. Woodhead Publishing, pp 41–78

Das P, Roy S, Antony J (2007) An application of six sigma methodology to reduce lot-to-lot shade variation of linen fabrics. J Ind Text 36:227–251. https://doi.org/10.1177/1528083707072360

David G, Woods V, Li G, Buckle P (2008) The development of the quick exposure disorders. Appl Ergon 39(1):57–69. https://doi.org/10.1016/j.apergo.2007.03.002

English L (2005) I.Q. and Muda: information quality eliminates waste. DM Rev 15:8, 40

Erdinc O, Vayvay O (2008) Ergonomics interventions improve quality in manufacturing: a case study. Int J Ind Syst Eng 3(6):727–745

Erdinc O, Yeow PHP (2011) Proving external validity of ergonomics and quality relationship through review of real-world case studies. Int J Prod Res 49(4):949–962. https://doi.org/10.1080/00207540903555502

Eswaramoorthi M, John M, Rajagopal CA, Prasad PSS, Mohanram PV (2010) Redesigning assembly stations using ergonomic methods as a lean tool. Work 35(2):231–240. https://doi.org/10.3233/wor-2010-0975

Fercoq A, Lamouri S, Carbone V (2016) Lean/Green integration focused on waste reduction techniques. J Clean Prod 137:567–578. https://doi.org/10.1016/j.jclepro.2016.07.107

Gnanavelbabu A, Arunagiri P (2018) Ranking of MUDA using AHP and Fuzzy AHP algorithm. Mater Today 5(5–2):13406–13412. https://doi.org/10.1016/j.matpr.2018.02.334

Hignett S, McAtamney L (2000) Rapid entire body assessment (REBA). Appl Ergon 31(2):201–205. https://doi.org/10.1016/S0003-6870(99)00039-3

Hill J, Thomas AJ, Mason-Jones RK, El-Kateb S (2018) The implementation of a Lean Six Sigma framework to enhance operational performance in an MRO facility. Prod Manuf Res 6(1):26–48. https://doi.org/10.1080/21693277.2017.1417179

Hodge GL, Ross KG, Joines JA, Thorney K (2011) Adapting lean manufacturing principles to the textile industry. Prod Plan Control 22(3):237–247. https://doi.org/10.1080/09537287.2010.498577

Hoerl RW (1998) Six Sigma and the future of the quality profession. IEEE Eng Manage Rev 26:87–94

Hossen J, Ahmad N, Ali SM (2017) An application of Pareto analysis and cause-and-effect diagram (CED) to examine stoppage losses: a textile case from Bangladesh. J Text Inst 108(11):2013–2020. https://doi.org/10.1080/00405000.2017.1308786

IEA (2021) What is ergonomics? International Ergonomics Association. Retrieved from https://iea.cc/what-is-ergonomics/

Karthi S, Devadasan SR, Selvaraju K, Sivaram NM, Sreenivasa CG (2013) Implementation of Lean Six Sigma through ISO 9001: 2008 based QMS: a case study in a textile mill. J Text Inst 104(10):1089–1100. https://doi.org/10.1080/00405000.2013.774945

Karthikadevi M (2014) Lean based manufacturing to increase the productivity, quality and reduce waste of textile industries. Int J Ind Eng (SSRG-IJIE) 1(1):23–32

Liliana L (2016) A new model of Ishikawa diagram for quality assessment. In: IOP conference series: materials science and engineering. IOP Publishing, p 12099

Loo HS, Yeow PHP (2015) Effects of two ergonomic improvements in brazing coils of air-handler units. Appl Ergon 51:383–391. https://doi.org/10.1016/j.apergo.2015.06.007

Mach P, Guáqueta J (2001) Utilization of the seven Ishikawa tools (old tools) in the Six Sigma strategy. In: Proceedings of the 24th international spring seminar on electronics technology. IEEE, pp 51–55. https://doi.org/10.1109/ISSE.2001.931009

Madanhire I, Mbohwa C (2016) Application of statistical process control (SPC) in manufacturing industry in a developing country. Procedia CIRP 40:580–583. https://doi.org/10.1016/j.procir.2016.01.137

Majumdar A, Das A, Alagirusamy R, Kothari VK (2012) Process control in textile manufacturing, process control in textile manufacturing. Elsevier

Mason B, Antony J (2000) Statistical process control: an essential ingredient for improving service and manufacturing quality. Manag Serv Qual Int J 10(4):233–238. https://doi.org/10.1108/09604520010341618

McAtamney L, Nigle Corlett EN (1993) RULA: a survey method for the investigation of work-related upper limb disorders. Appl Ergon 24(2):91–99. https://doi.org/10.1016/0003-6870(93)90080-s

Monden Y (1983) Toyota production system. An integrated approach to just-in-time. Springer. US. https://doi.org/10.1007/978-1-4615-9714-8

Moore JS, Garg A (1995) The strain index: a proposed method to analyze jobs for risk of distal upper extremity disorders. Am Ind Hyg Assoc J 56(5):443–458. https://doi.org/10.1080/15428119591016863

Morales-Contreras MF, Suárez-Barraza MF, Leporati M (2020) Identifying Muda in a fast-food service process in Spain. Int J Qual Serv Sci 12(2):201–226. https://doi.org/10.1108/IJQSS-10-2019-0116

More SS, Pawar MS (2015) Implementation of six sigma with ISO in the Indian textile industry for improvement in performance. In: Proceedings of 2015 international conference on technologies for sustainable development (ICTSD), 30 Apr 2015, Mumbai, India

Muhammad S (2015) Quality improvement of fan manufacturing industry by using basic seven tools of quality: a case study. J Eng Res Appl 5(4):30–35. Retrieved from http://www.ijera.com/papers/Vol5_issue4/Part%20-%204/G504043035.pdf

Neyestani B (2017) Seven basic tools of quality control: the appropriate techniques for solving quality problems in the organizations. SSRN, Retrieved from https://zenodo.org/record/400832#.XdPBdJMzaUl

Pettersen J (2009) Defining lean production: some conceptual and practical issues. TQM Journal 21(2):127–142. https://doi.org/10.1108/17542730910938137

Saidy C, Xia K, Kircaliali A, Harik R, Bayoumi A (2020) The application of statistical quality control methods in predictive maintenance 4.0: an unconventional use of statistical process control (spc) charts in health monitoring and predictive analytics. Smart innovation, systems and technologies. Springer, pp 1051–1061

Shafira YP, Mansur A (2018) Production quality improvement analysis of grey cambric using six sigma method. In: The 2nd international conference on engineering and technology for sustainable development (ICET4SD 2017), Yogyakarta, Indonesia, 13–14 Sept 2017

Shahin A, Mazaher G, Arabzad SM (2010) Proposing an integrated framework of seven basic and new quality management tools and techniques: a roadmap. Res J Int Stud Issue 17:183–195. Retrieved from https://www.researchgate.net/publication/264236503.Stojanovic

Shamah RAM (2013) Measuring and building lean thinking for value creation in supply chains. Int J Lean Six Sigma 4(1):17–35. https://doi.org/10.1108/20401461311310490

Sutherland J, Bennett B (2008) The seven deadly supply chain wastes. Supply Chain Manag Rev 12(5):38

Taner MT (2012) A feasibility study for six sigma implementation in Turkish textile SMEs. South East Eur J Econ Bus 7(1):63. https://doi.org/10.2478/v10033-012-0006-6

Tuna S (2018) Keeping track of garment production process and process improvement using quality control techniques. Period Eng Nat Sci (PEN) 6(1):11–26

Vaněček D, Pech M, Rost M (2018) Innovation and lean production. Acta Universitatis Agriculturae et Silviculturae Mendelianae Brunensis 66(2):595–603. https://doi.org/10.11118/actaun201866 020595

Vayvay O, Erdinc O (2008) Quality improvement through ergonomics (QUITE) methodology: conceptual framework and an application. Int J Prod Qual Manag 3(3):311–324

Wang W, Zhang W (2008) Early defect identification: application of statistical process control methods. J Qual Maint Eng 14:225–236. https://doi.org/10.1108/13552510810899445

Womack JP, Jones DT (1996) Beyond Toyota: how to root out waste and pursue perfection. Harv Bus Rev 74(5):140–172. Retrieved from https://hbr.org/1996/09/how-to-root-out-waste-and-pursue-perfection

Digital Technologies for Lean Manufacturing

Majo George, Le Khac Yen Nhi, Nguyen Minh Ngoc, Vuong Nguyen Dang Tung, Le Phan Thanh Truc, and Rajkishore Nayak

Abstract This chapter aims to highlight the digital technologies that are designed to complement the operation of lean manufacturing. Firstly, Industry 4.0 has been explained that helps to complement Lean Manufacturing to gain continuous improvement, better customer satisfaction and improved operational efficiency. The results of combining digital technologies with Lean Manufacturing yield the concept of "Lean Industry 4.0". Secondly, blockchain occurs as a disruptive innovation to resolve the problems of lack of an integrated lean management system across the supply chain network, which is also discussed. Thirdly, the Radio Frequency Identification (RFID) system is analyzed, and its ability to offer high levels of accurate, real-time information, decrease time-consuming activities and labor cost while increasing product visibility and operational speed is covered. Fourthly, Artificial Intelligence (AI) and robotics are also discussed with the ability to deal with complexity, increase productivity and efficiency with the automatic system, and decrease production costs. Finally, other non-common yet useful tools are mentioned to give a comprehensive view of the application of digital technologies with Lean Manufacturing, including automated guided vehicles (AGVs), virtual stimulation (VS), and cybersecurity. To consolidate our findings, two case studies are presented to give realistic viewpoints of digital technology adoption from two giant firms in the textile and apparel industry, namely Uniqlo and H&M. The findings of a survey based on Vietnam's fashion and textile industries on the use of technology such as RFID is also included in this chapter.

Keywords Industry 4.0 · Blockchain · RFID · Artificial intelligence · Robotics · Automated guided vehicles

---

M. George (✉) · L. K. Y. Nhi · N. M. Ngoc · V. N. D. Tung · L. P. T. Truc
School of Business and Management, RMIT Vietnam, Ho Chi Minh, Vietnam
e-mail: majo.george@rmit.edu.vn

R. Nayak
School of Communication and Design, RMIT Vietnam, Ho Chi Minh, Vietnam

© The Author(s), under exclusive license to Springer Nature Singapore Pte Ltd. 2022
R. Nayak (ed.), *Lean Supply Chain Management in Fashion and Textile Industry*,
Textile Science and Clothing Technology,
https://doi.org/10.1007/978-981-19-2108-7_10

1 Introduction

Ever since the first industrial revolution, the manufacturing sector has always been changing and improving compared to its original version. What causes those changes are undoubtedly various technologies. Sanders et al. (2016) noted that, with the same purpose of making the production process more efficient, while the Western companies choose to apply advanced technology such as automation or computational technology into the process to make it more streamlined, the world has witnessed another capable manufacturing principle called Lean Manufacturing derived from Japan with the huge success from the Toyota Motor Corporation (Chiarini et al. 2018). With those two methods being the fulcrum (automation and Lean Manufacturing), companies and industries around the world have attempted to apply either one or both approaches to achieve competitive advantage in the more intense competitive market as in today.

In light of the fourth industrial revolution, the concept of Lean Manufacturing, although still maintaining its core objective of enhancing processes' productivity by simultaneous methodologies including waste elimination, product quality improvement, lead time reduction and cost contraction (Fullerton et al. 2014), the approaches towards achieving those goals are evolved to some extent. The reasons behind this shift can be explained by the new production stand in this Industry 4.0 era (Valamede et al. 2020). Particularly, Industry 4.0 (some may refer to the term smart manufacturing) requires new ways for production in response to the strong rise of smart digital technologies. Thus, the existing physical manufacturing production and business operation can be now developed to a new level to be "smarter" and "connected" with the integration of technology, data and network applications.

According to Varela et al. (2019), the new age of production considers Lean Manufacturing and Industry 4.0 as the modern fundamental pillars for companies to attain business productivity and sustainability. The effectiveness of using advanced digital technologies to reinforce and accrete Lean Manufacturing should be put into more investigations for drawing a final conclusion. However, this chapter will introduce and demonstrate some of the promising technologies that can be applied in firms to improve their manufacturing performance, thus aligning with the objectives of Lean Manufacturing.

This chapter has discussed the application of various digital technologies such as Industry 4.0, Blockchain, RFID, AI and robotics in lean manufacturing. These digital technologies are designed to complement the operation of Lean Manufacturing for improving the effectiveness to achieve the objectives of lean. Furthermore, this chapter includes the findings of the research work on the implementation of RFID in the fashion and textile manufacturing industries in Vietnam, the third-largest global clothing exporter (Nayak et al. 2019). Due to significant role of Vietnam in global fashion and textile industries, we conducted a survey on a total of 11 Vietnamese fashion and textile enterprises. A qualitative interview was done on the status of implementation of RFID technology in fashion and textile industries and the views expressed by the manufactures have also been included. This qualitative research will help to improve the understanding of the readers on the implementation of technology in real-life conditions.

2 Digital Technologies for Lean Manufacturing

2.1 Industry 4.0

2.1.1 The Evolution of Industry 4.0

The manufacturing sector, has passed through four stages of evolution (from Industrial 1.0 to 4.0) that also come along with the transformation of nearly all facets of manufacturing including changes in the way products are produced, consumer expectations, and especially how firms improve their productivity to adapt or innovate in the changing of competition landscape (Fujimoto 2012). A brief summary of the four industrial revolutions is discussed below and listed in Table 1.

In 1760, along with the invention of the steam engine and hydropower, the first industrial revolution happened that indicated the huge development in the way products are created. In fact, the steam engine worked like a milestone that transformed the economy from mainly agricultural and farming-based into a manufacturing epoch. Also, in this phase of manufacturing, coal was used as the main energy source, helping two industries, textile and steel to become dominant and among the largest sectors for economic growth. In the 1901, the second industrial revolution started, which is being highlighted with the discovery of the internal combustion engine. Taking this as the base, widespread industrialization occurred with the introduction of more modern production lines such as the automatic assembly lines and more importantly, the mass production that came along with the usage of electricity.

Table 1 Brief history of the four industrial revolutions

Industry	Period	Energy resource	Main technical achievement	Main developed industries	Transport means
1.0	1760–1900	Coal	Steam Engine	Textile, Steel	Train
2.0	1901–1960	Oil, Electricity	Internal Combustion Engine	Metallurgy, Auto, Machine Building	Train, Car
3.0	1961–2000	Nuclear Energy Natural Gas	Computers, Robots	Auto, Chemistry	Car, Plane
4.0	2001–Present	Green Energy	Internet, 3D Printer, Genetic Engineering	High Tech Industries	Electric Car, Ultra-Fast Train

(Reproduced from: Prisecaru 2016)

Around the 1961s, the third industrial revolution began, and the focal point of this stage was the rise of electronics, information technology (IT), specifically telecommunications and computer systems. The third industrial revolution worked as a very solid cornerstone for the fourth industrial revolution (Prisecaru 2016). Officially named in 2011, as the Industry 4.0 started with a fine amalgamation between physical and digital aspects with the crucial involvement of advanced technologies (Ghobakhloo 2020). Xu et al. (2018) asserted in their research that the unique characteristics of the Industry 4.0 in comparison to its predecessor belong to the superiority of technologies, greater velocity and impact of information systems on the process.

At the moment, when the fourth industrial revolution comes into the picture, its intense impact on human productivity cannot be neglected especially when talking about the significance of its facilitation on the speed and efficiency that could be achieved (Lee et al. 2018).

2.1.2 What is Industry 4.0

Coming at the dawn of the twenty-first century, Industry 4.0 is all about adopting advanced digital technologies to improve production, intercommunication and data-driven decision-making processes, mainly by gathering real-time data for analysis (Wang et al. 2016). Those technologies/devices will interconnect to each other, creating a seamless sequence of communicating, analyzing and performing, thus boosting the efficiency of operating tasks.

According to Hakeem et al. (2020), traditional manufacturing can now be turned into a more smart and autonomous process using the Industry 4.0. The key driver, as indicated above, relies on the interconnectivity between the horizontal and vertical spheres consisting of materials, machinery and operators. As a result, a Networked Smart Manufacturing system is created.

2.1.3 The Aim of Industry 4.0

As stated above, with digitalization being the trait of Industry 4.0, it is possible to create a more-than-ever interrelationship between everything and everyone in a system (including processes of logistics, production, capital and human resources), hence, making a cross-technology and linking every element within and without industries together. Fragapane et al. (2020) found that by adding the real world virtual components, Industry 4.0 aims to enhance flexibility, responsiveness, intelligence, and most importantly, the overall efficiency of the manufacturing process through the improvement in data analysis and data transferring.

Hakeem et al. (2020) claimed that Industry 4.0 makes the process become smarter and more automatic by replacing traditional practices with computerization. Apparently, the higher level of interconnection and automation brings better unification in the production process. A more collaborative operational method could be generated

Digital Technologies for Lean Manufacturing

with technology-driven supports. Thus, a higher operational productivity, and efficiency could be achieved in manufacturing process (Alcácer and Cruz-Machado 2019). Taking Industry 4.0 as a foundation, firms are expected to upgrade their manufacturing process into a whole new standard, where smart manufacturing and smart products are incorporated. Industry 4.0 is considered as the promising technology for innovating the whole manufacturing industry in which, continual data stream and effective communication will drive into more efficiency in utilizing resources (Nayak 2020).

2.1.4 Industry 4.0 Tools

At the moment, there has not been an agreed-upon list of technologies that best represent the application of Industry 4.0. The list of noted technologies varies in different research papers. According to Türkeș et al. (2019), the differences on the concept of "digitization" in different countries may be related to how Industry 4.0 performs in the country.

Some featured technologies for Industry 4.0 are Internet of Things (IoT), cloud computing and big data analytics, which can be used to generate a computational-physical system (Lu 2017). There are supplementary systems used to support the aforementioned technologies, namely robotics, Artificial Intelligence (AI), 3D printing (additive manufacturing), and Automatic Vehicles.

According to Alcácer and Cruz-Machado (2019), Industry 4.0 is an inclusion of nine fundamental technologies. This framework is believed to be sufficient for creating a well-rounded integration in both horizontal and vertical directions for an end-to-end engineering synthesis as well as a typical networked manufacturing system (Akinlabi 2020). Thus, when effectively integrating those technologies, an advanced manufacturing system is achieved, called cyber-physical system (CPS). Becoming even smarter, factories are bringing automation to be the main function in a production process that can now make decisions and execute tasks beyond human controls (Azouz and Pierreval 2019; Gräßler and Yang 2016). All the mentioned tools used in Industry 4.0 are summarized in Fig. 1.

The above framework is also aligned with the findings from Kamble et al. (2020). Although the above framework (Fig. 1) is not separating Industry 4.0 technologies into primary and supplementary types, these technologies are vital for the operation of Industry 4.0. The list of Industry 4.0 technologies bases on the application domain has also been given in Table 2.

2.1.5 How Industry 4.0 Helps Lean Manufacturing

In accordance with Demirkol and Al-Futaih (2020), when investigating the relationship between Industry 4.0 and Lean Manufacturing, it can be said that Industry 4.0 works as an applicable set of tools to support Lean Manufacturing. When considering the purpose of Lean Manufacturing, waste reduction and continuous production

Fig. 1 The main tools of Industry 4.0

Table 2 Key technologies representing Industry 4.0

Purpose	Industry 4.0 technologies
Smart data collection, storage, analysis, and sharing technologies	Big data analytics (BDA), cloud computing (CC), Internet of things (IoT), and sensors, simulation, and prototyping
Shop floor technologies	Additive manufacturing (AM), virtual reality (VR), augmented reality (AR), IoT, sensors, and robotic systems (RS)
Integration technologies	Cyber-physical systems (CPS) and cyber-security system (CSS)

(Reproduced from: Kamble et al. (2020))

flow appears as the two most important principles (Thangarajoo and Smith 2015). To reach the optimal performance in Lean Manufacturing, it is necessary to resolve seven categories of wastes, as listed here.

- Defects
- Excessive inventory
- Transportation
- Overproduction
- Waiting time
- Overprocessing
- Movement

In fact, there are many well-designed tools used to address the goal of Lean Manufacturing such as Kaizen principles, Just-In-Time (JIT) production, Six Sigma and 5S, which are discussed in other chapters of this book. Although these tools are more likely to be presented as theoretical methods and frameworks, how manufacturers implement the tools in terms of physical and mechanical execution still depends on certain situations. Therefore, when Industry 4.0 comes with a variety of digital technologies, it also brings up a more straightforward solution to attain Lean Manufacturing. More specifically, when the goals of the two concepts (achieving production efficiency, and higher resource utilization) are considered, Industry 4.0 application can certainly support to achieve the goals of Lean Manufacturing (Wagner et al. 2017).

There are fundamental differences when carefully comparing the two concepts, Industry 4.0 and Lean Manufacturing. One significant difference that must be put into deliberation is that while Lean Manufacturing involves people's inclusiveness, Industry 4.0 tends to center on the role of computer-based and automation systems. In Lean Manufacturing, the harmony in people's interaction with each other and with the production system is essential and people are always put at the core focus in the process (Gaiardelli et al. 2019). In contrast, Industry 4.0 concentrates more on the function of technology with computer-based systems and automation. According to Thoben et al. (2017), the ultimate goal of Industry 4.0 is to create smart manufacturing systems, whereas the goal of Lean Manufacturing is improve the productivity by eliminating unnecessary waste. Additionally, in Industry 4.0, the significant role of data either in data transmission or data-driven decision making is appreciated at a much higher level than in Lean Manufacturing. Thus, the utilization of big data is considered as one of the key pillars in building networked systems.

Overall, the philosophies of Lean Manufacturing and Industry 4.0 can impeccably complement one another. In a nut shell, Industry 4.0 helps in gaining continuous improvement with the application of digital technologies, for better addressing customer expectations and facilitating higher level of efficiency and productivity. In the era of Industry 4.0 a new notion could be generated, sustaining with the quintessence of both concepts, called "Lean Industry 4.0" (Ejsmont et al. 2020).

2.2 Blockchain

2.2.1 What is Blockchain

According to Christidis and Devetsikiotis (2016), blockchain is a distributed system to cryptographically capture and record a consistent and direct chain of transactions between individuals within the same network. When a transaction is completed, a new block is formed under the format of a data package to record the information related to the transaction. The verification of each block follows governance rules in the blockchain network (Wang et al. 2019). Each block after being created is linked together in an irreversible chain to block the transaction (Nofer et al. 2017). This means that after creation, any block in the system (blockchain) cannot be deleted or adjusted by any participants, which explains the choice of terminology, blockchain. In summary, a blockchain includes multiple data sets which are created by a chain of blocks, that possesses information of a transaction (Ghode et al. 2020).

Furthermore, each block has not only the information of the transaction but also a timestamp, a hash value of the prior block (parent block), and a nonce (i.e., a random number to verify the hash). With the association between the blocks, having the hash value of the parent block and the nonce can ascertain the integrity of the whole chain to the very first block (genesis block). Moreover, since hash values are idiosyncratic, they will change accordingly to any attempts to change the block, and consequently, this will notify the system since the hash values are no longer matched, hence an effective function to detect fraud (Nofer et al. 2017).

Nonetheless, it is not true that all transactions are automatically added to the block strings, which are included in a ledger. Instead, each transaction must go through a consensus mechanism of validity of transactions, which requires the agreement of the majority of blocks in the network, and the validity of the pending block, to be supplemented in the string (Nofer et al. 2017). Swanson (2015) mentioned that "Consensus mechanism is a framework that a majority of network validators agree on the rules and procedures set by the ledger, which enables archiving coherent set of facts between participating blocks in the network". It means that instead of being added directly to the chain, this process warrants that the transaction is fully stored in a block for a certain time before being transferred to the ledger.

2.2.2 Characteristics of Blockchain

Blockchain is a type of database architecture created with the Distributed Ledger Technology as a foundation (Schmidt and Wagner 2019). Therefore, unlike traditional database networks, it has some key characteristics such as:

- **Open distributed ledger for public verification**: Blockchain is designed as a decentralized system, therefore, it is distributed and shares the copies of all information to all the participants within the network. Equivalently, this can be understood that blockchain is not under control of specific third-party intermediaries

Digital Technologies for Lean Manufacturing

such as governments and banks. Hence, information can be updated faster by the participants within the network as the new block will be created and linked to the previous block to update the information of the transaction without the need for centralized authority, hence limiting costs and time. Moreover, with the real-time updating of information, blockchain is a useful tool to adopt when involving different parties and transactions (Korpela et al. 2017).

- **Transparency**: The transparency of data and the updating process is a required element for public verifiability. Every transaction is transparent and visible across the network regardless of the need to access all information of the participants in the network (Wüst and Gervais 2018).
- **Cryptographically sealed and immutability of data**: One of the utmost important pillars of blockchain is cryptography. It is needed for digital signatures and data integrity to avoid manipulating blocks after the transaction has been validated and recorded into the blockchain database. This cryptographic mechanism creates the immutability for data stored in blockchain because it cryptographically connects all new blocks to the parent block (Wüst and Gervais 2018).
- **Consensus mechanism and trustiness**: As mentioned above, before being officially added to the blockchain, transaction must be stored in the block and meet the agreement of majority validators using consensus protocols, including Proof of Work (PoW), Proof of Stake (PoS), and the Proof of Elapsed Time (PoET) (McQuinn and Castro 2019). This mechanism can hinder the risk of wrong inputs and fraudulent conducts that could impact the database network.

2.2.3 Blockchain 1.0, 2.0, and 3.0

According to Swan (2015), the current activities and potentials of the blockchain revolution can be classified under three categories:

- Blockchain 1.0—digital currency. Blockchain is used as cash in the form of digital currency. Furthermore, it can function normally with basic features as currency remittance, transfer, and digital payment.
- Blockchain 2.0—smart contract. Blockchain is further employed in the market and financial applications for transactions that are more extensive than cash, including stocks, bonds, loans, futures, mortgages, titles, smart property, and smart contracts.
- Blockchain 3.0—the future. The potential application of blockchain beyond currency, finance, and markets; including the application in logistics and supply chain.

2.2.4 Application of Blockchain on Lean Manufacturing

The Lean manufacturing, created by the Toyota Motor Company in the middle 1970s, aims to minimize waste and non-value-added elements within the production process, such as unnecessary space, non-value-added steps, prolonged lead time, excessive inventory, defects, and overproduction (Bozarth et al. 2008). Lean Management System is an extension of lean philosophy from a single plant to the entire supply chain network (Holweg 2007). According to Liu et al. (2013a, b), it is unlikely to have an integrated Lean Management system across the network due to the differences in preferences and priorities of the decision makers at each stage of the supply chain. Furthermore, from the perspective of Schmidt and Wagner (2019), lean practices in supply chain networks inevitably cope with various type of risks steamed from the relationship between suppliers and buyers, namely opportunistic behavior, negotiating power imbalances, and asymmetry information. Hence, a systematic database is believed to be the solution to this alerting matter, and blockchain is full of potential to perform a solution that can integrate lean practices to the whole supply chain network as discussed below.

Firstly, supply chain is normally visualized as a complex network of systems with multiplex logistics, a wide range of data, and their relationships. Therefore, minor errors can lead to the bullwhip effect, causing massive delays (Hearnshaw and Wilson 2013; Lee et al. 1997). Nonetheless, blockchain as a digital ledger with available access by all the participants within the network. Real-time updating features can enable transactions, records, and processes to be observable across the entire supply chain network, hence, it is easier for stakeholders to monitor and detect problems in real time. Furthermore, since blockchain allows data to be synchronized and shared throughout the network, it saves cost by eliminating unnecessary steps and time that is normally caused by document duplication in the traditional process (Laurence 2019).

Secondly, blockchain is capable of being a digital platform to implement smart contracts, which are self-validating and self-executing agreements between parties in the supply chain network once certain conditions are met (Dolgui et al. 2020). According to Min (2019), these smart contracts enable entities in the supply chain network to exchange shares, money, or even property in a transparent way by eliminating the third-party intermediaries. Ultimately, smart contracts help corporations to benefit from lower transaction costs with the cut-off of intermediaries during the process and facilitate order fulfilment by minimizing delays between delivery and payment.

Thirdly, blockchain can ensure and enhance trust between all stakeholders in the supply chain network, including suppliers, buyers, and focal firms. This is concluded based on the immutability of data feature, since every new transaction after being added to the blockchain will be cryptographically sealed and unable to make changes (Wüst and Gervais 2018). Furthermore, Saberi et al. (2019) added that the tamper-evident nature of blockchain can diminish the opportunity costs that often exist among stakeholders in the supply chain network. From the above discussions it can be concluded that blockchain will be working as an important technology to achieve the objectives of Lean Manufacturing.

2.3 RFID Technology

2.3.1 Introduction to RFID

Radio frequency identification (RFID), refers to the process of exchanging information between stationary and mobile units using radio communication (Landt 2005), which is becoming an indispensable part of our daily life. Through a thick array of data, information regarding the location and entity of properties can easily be traced through wireless scans, which is further applied to imaging, localization, sensing and security (Chu 2015). RFID technology is becoming more popular due to the increased integration with Internet-of-Things, Machine-to-Machine (M2M) and Ambient Intelligence, offering a faster, yet lower cost, structural simplicity and less power consumption demanded for the tag circuits.

2.3.2 History of RFID

RFID technology started it journey from World War II, when radar was discovered by Sir Robert Alexander Watson, a Scottish physicist in 1935. This radar was used to alert approaching aircrafts, identify whether they are from the allies or enemies and inform the army, also known as an identify-friend-or-foe (IFF) approach. The technology was then widely utilized, in which certain innovations were recorded in Germany. The radar could alter its reaction signals, convert them to be more translatable. The British also took advantage of this technology, in which radar played as a trustful transmitter from the ground to detect friendly aircrafts for planes.

From 1950s to 1960s, the existence of various scientific articles proved the effectiveness of RF range in identifying objects remotely. Radar and RF systems were then further developed by the USA and Europe researchers. The first exploration of RFID technology was started by Harry Stokman in 1948, when the specific reflected-power communication problem was solved.

The concept embedded in RFID technology remained as a secret between 1960s and 1970s. Later, the technology continued to be developed into a tracking system for nuclear and sensitive materials by the Los Alamos National Laboratory in the USA. The advantages of the technology were then exploited by investors, educational centers, programmers, and government laboratories. In January 1973, an US patent on active RFID tag was received by Mario W. Cardullo, allowing memory to be rewritten.

The technological developments were reported to be more sophisticated in manufacturing in between1980s and 1990s, while the RFID systems succeeded in scanning up to millions tags in the USA automotive sector. In 1999, an Auto-ID center, specializing in automatic identification, was established by the Massachusetts Institute of Technology (MIT) and received enhancements from the Uniform Code Council, Procter and Gamble, Gillee and EAN International. With significant development of the RFID technology, Auto-ID center was supported by over 100 corporations, including the US Department of Defense and other key RFID vendors during 1999 to 2003. Its research laboratories were then distributed to the UK, Australia, Switzerland, China and Japan with the establishment of two air interface protocols while marking its evolution to the Electronic Product Code (EPC) to promote the EPC standard. The EPC global, with its RFID technology, soon stood at the second standard to receive the most utilization from many industries.

While environmental issues were becoming top concerns, 2010 marked a considerable achievement with decreased cost of tags and increased performance to reach a sustainable standard of the RFID systems. At present, the RFID technology is evolving from the "security technology" to serve the army, to various industries including health, lifestyle, automotive, logistics, transport, aerospace and fashion. This system was also certified by the International Organization for Standardization (ISO) for a high-leveled interchangeability of a technical standard. The RFID varies in frequency bands, from a microwave level (2.4–2.5 GHz) to the recent development of the UHF (Ultra high frequency) Generation 2 (869–969 MHz) (Chu 2015).

The RFID, an application from a wide range of expertise (Weinstein 2005), has been through a long history (Table 3) and is continuing its sustainable development that adapts current requirements of businesses and society.

Table 3 The history of RFID technology

Time frame	Events
World War II (1935)	• Radar was discovered and used as an identify-friend-or-foe (IFF) approach
1950s–1960s	• Radar and RF range's effectiveness was being proved • 1948: Harry Stokman explored RFID by acknowledging certain reflected-power communications problems had been solved (Domdouzisa et al. 2017)
1960s–1970s	• RFID was utilized as the tracking system for nuclear and sensitive materials for Los Alamos National Laboratory in the USA • January 1973: Mario W. Cardullo acknowledged the allowance for memory to be recorded by the US patent active RFID tag
1980s–1990s	• The system became more sophisticated, managing to scan millions of tags in the USA automotive sector • 1999: MIT founded Auto-ID center, specializing in automatic identification • 1999–2003: Auto-ID center received supports from over 100 corporations, US Department of Defense and RFID key vendors
October 2003	• The Auto-ID center was changed into EPC global, with widespread distribution around the world
December 2004	• The system become the second technology suppliers for many industries
2010s	• The application of RFID marked considerable decreased cost of equipment and tags and increase in sustainable performance • The technology is applied in a diversity of industries

2.3.3 Functions of RFID

The functions of RFID are illustrated as in Fig. 2, in which information is delivered through an RFID tag, Antenna, Reader and controller (CPU-Central processing unit). Particularly, when the information is required from an RFID tag, a signal will be delivered by the CPU to the Reader. Through the Antenna, the Reader will transfer the digital information back to the CPU from the RFID tag.

This process is supported by a structural network of wired and wireless communication infrastructure, in which the Reader will sense the radio signals and broadcast it to the Antenna of the tag. When receiving the signals, the data can be read from the chip via radio wave, and with this done, the Reader can simultaneously and accurately process and store data from a large number of tags. This process is considered to be more sustainable in comparison to the current barcode systems allowing only single tag scanned per time.

Fennani et al. (2011) proposed three main components of the complicated digital RFID system. This includes the RFID Tag, RFID Reader and RFID Controller.

- **RFID Tag**

The RFID Tag, known as the transponder, is a small device made from a silicon microchip and the Antenna (Chechi et al. 2012). In a simpler word, RFID Tag is actually a tag that can be found on any selling items or boxes. The data in the RFID Tag can be delivered wirelessly through a chip with different power requirements, hence, RFID Tag is classified in different aspects as shown in Table 4.

An important part of the RFID Tag includes the RFID Antenna. This device is known as the data storage unit for any tag, varying in types from patch, linear polarized, stick, adaptive, gate and omni directional antennas. To perfectly perform, RFID Antennas should fulfil the following criteria.

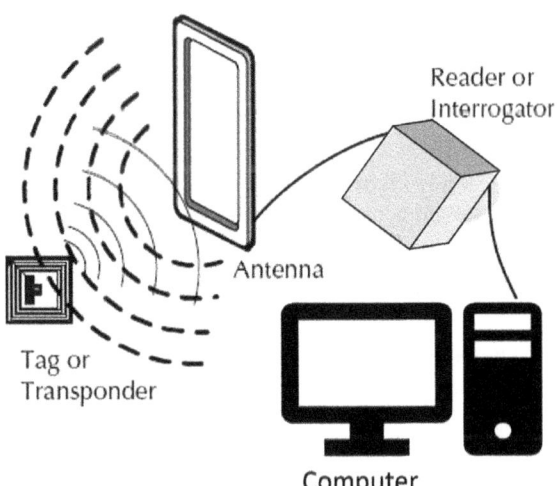

Fig. 2 Functions of RFID systems

Table 4 RFID Tags classification

Classification of RFID Tags	
Passive tag	With the passive RFID Tag, the electromagnetic signal is transmitted forward and backward from the Reader with integrated circuit power (Fennani et al. 2011). This type has limited energy, hence, restricts the distance between the Reader and Tags as well as the capacity of circuit power in the computer. With such limitations, passive Tag is usually cheaper than the other types
Semi-passive tag	In the semi-passive tag, the battery is the power source for the integrated circuit to achieve a more special processing journey but still with the passive transmission module. This is considered an advanced type from the passive tag, despite such restrictions from the waiting time that create unacceptable delivery time from dormant to active state (Moroz 2004)
Active tag	The last and also the most powerful type, comprising better and larger capacity from the passive tag, with further and wider distance allowed to transmit data between the tags and the Readers (which can vary due to the transmission frequency). Above all, the active tag can be incorporated with a sensor with complicated circuits embedded inside, allowing the system to operate advanced features during its operation. The price of an active tag can reach up to hundred US dollars
Memory type of RFID Tag	
Read-only tag	In the Read-only Tag, the memory has already been programmed in the factory. Hence, the static data cannot be changed after the manufacturing process. This tag also is limited in space to store the data, so the price is cheaper
Read–write tag	The concept of storing data is the same as the Read-only Tag, except that the Read–write Tag provides a more flexible way to process the data. Through the usage of this tag, data can be easily altered
Wireless signal type between the tag and the reader	
Induction method	The Induction method utilizes the LF and HF frequency bands for close proximity electromagnetic or inductive coupling-near fields (Chechi et al. 2012)
Propagation method	The Propagation method, however, utilizes the UHF and microwave frequency bands for propagating far-field electromagnetic waves (Chechi et al. 2012)

1. It should be in an acceptable size, usually small.
2. It should acquire an omni-directional or hemispherical coverage.
3. It should deliver signals to the microchip at the highest rate.
4. It should perform robustly.
5. It should be cheap to adapt to the RFID systems.

Alternative types of RFID Antenna are designed to fit different RFID applications. For example, the RFID passive tags with 5.8 GHz frequency can be used with the dual-polarized antenna, while the other ultra-high frequency type (1 GHz) can be used with the meander line antennas (MLA).

Table 5 Classification of RFID Reader

RFID Reader type based on design and technology used	
Read	• The RFID read type reader is only able to read data from the tags by detecting backscatter modulation • The reading system includes micro-controller-based unit, a peak detector hardware, firmware, comparators and wound output coil • There are variable protocols, standards and frequencies exist in this type
Read/write	• The reader is similar to the read type, however, the data collected from the tags can be processed both in read and written form
RFID Reader device fixation type	
Stationary	For the stationary fixation type reader, the RFID device is placed at a fixed location, usually at the entrance/exit gate of the stores
Mobile	For the mobile type reader, the RFID device is movable

- **RFID Reader**

The RFID Reader (interrogator or scanner) delivers RF data forward and backward to the Tag through the antenna. The RFID Reader comprises three main parts: the control section, high-frequency interface and antenna, in which the antenna gain, usage frequency and orientation have certain effects on the read range. The RFID Reader is classified into four types as shown in Table 5 (Ilie-Zudor et al. 2006).

Alongside these classifications, each RFID Reader type operates in different frequencies, and read range as shown in Table 6.

- **RFID Controller**

The RFID Controller is also known as the brain of the RFID technology or the CPU. The CPU is installed with a software (middleware), in a miniature computer (workstation) or a network of various workstations. Asif and Mandviwalla (2005) stated three levels of middleware applications to satisfy the industry context and requirements. The first level ensures the software capability to operate with the connectivity of the industries. The second level ensures that a system of applications can be managed by the software. The third level ensures the strong connectivity between devices, as the RFID shop-floors machine can be considered as an example. The RFID controller can accurately track the items' locations, inventory, sold products and new products, all of which can easily be scanned and recorded via the system. During the

Table 6 RFID reader frequency and reading range

Frequency type	Frequency range	Passive read range
Low frequency (LF)	120–140 kHz	10–20 cm
High frequency (HF)	13.56 MHz	10–20 cm
Ultra-high frequency (UHF)	868–928 MHz	3 m
Microwave	2.45 and 5.8 GHz	3 m

operation, the store manager will be alerted when there are problems with the stock management.

2.3.4 Benefits of RFID Technology

The traditional supply chain without RFID technology suffers from several disadvantages. It is undeniable that the traditional system still faces major challenges in the inventory management and cost control, leading to lost sales, reduced number of customers, decreased productivity, lower efficiency, and customer dissatisfaction due to manual, time-consuming and inaccurate inventory management. The RFID system, through its long development process, has overcome these problems and reached certain achievements while bringing back a more convenient and sustainable operation process for many industries. With the ability to offer high levels of accurate, real-time information, decreased level of time-consuming activities, non-contact technology, reduced labor cost and reduced shrinkage RFID helps in increasing product visibility and operation speed (Whitaker 2007; Tan 2008). This proves the effectiveness of applying RFID system for which more and more industries are using it in the data processing stage.

This chapter also highlights the findings from a study on the benefits of using RFID technology in fashion and textile supply chain in Vietnam. Vietnam is one of the leading fashion and textile manufacturing countries in the South-East Asia, where the manufacturing industries are facing several challenges to achieve the triple bottom line of sustainability. Therefore, this study has attempted to investigate the sustainability benefits achieved by the implementation of RFID technology through empirical research. The research findings showed that the use of RFID technology is in its infancy in Vietnamese fashion and textile enterprises due to several factors, and the major factor is the cost of implementing technology. The findings of the study are discussed in the following section.

2.3.5 RFID and Lean Manufacturing

The application of technology is prominent to reduce losses and achieve the principles of Lean Manufacturing, which is possible due to large database, flexible linkages between functions, and the automation of activities (Aydos and Ferreira 2016). Indeed, RFID has been widely used along with Lean Manufacturing to effectively eliminate wastes and overcome the barriers that hinder the application of Lean philosophy for any producers (Rafique et al. 2016). Patti and Narsing (2008) found that RFID technology is compatible with Lean Manufacturing through electronic Kanbans and the assembly part location tracking. Specifically, a leaner operation is attainable by removing manual data reading processes and adopting automation of scanning (Brintrup et al. 2010; Nayak and Padhye 2017).

Haddud (2011) demonstrated the seven types of wastage reduction by applying RFID to lean manufacturing, particularly: Overproduction, waiting time, inefficient transportation, inappropriate processing, unnecessary inventory (best-eliminated), unnecessary motion, and defects (least-eliminated). Chongwatpol and Sharda (2013) proved that employing RFID technology can help to reduce only some types of wastes. Accordingly, RFID helps to improve the inventory management, and reduce processing time, order time, and backlog orders. Compared to 1D and 2D barcodes, RFID encourages efficient production and increases the productivity.

Rafique et al. (2016) distinguished the impediments in lean implementation into three categories: operational aspect, managerial aspect, and financial aspect. These barriers are derived from the instability of plant operations since production lead times are prolonged, scheduling is wrongly defined, customers' orders are improperly handled, and inventory control is unbalanced (Rose et al. 2013). RFID-based Lean Manufacturing has appropriate features that can surmount these problems, such as real-time traceability, data privacy, data accuracy, and fast responsive manufacturing (Rafique et al.). These features can appeal to managers and employees to adopt lean culture, thus enhancing awareness and increasing the transition towards lean manufacturing.

Su et al. (2009) proposed a modernized Kanban system (RFID-based Kanban management system) that adopted RFID technology to enumerate the convention just-in-time (JIT) production mode. Compared to the traditional method, this system can improve the production management flexibility with real-time data collection, increase JIT response capability, automation degree, and enhance convenience as well as security. Besides, Chen et al. (2012) created a performance monitoring online system, which is known as online RFID-based facility performance monitoring (ORFPM), with the utilization of RFID technology. This system generates a value stream map (VSM) that collects data about lead time and transport time, via the internet with a wireless data transmission technology. This system enhances the flexibility for producers and managers to track the production flow on the shop floor even if they are unavailable at the facility.

2.3.5.1 RFID and Social Benefits

Indeed, many industries nowadays are still attached to the label of "labor-intensive" due to the huge advantage of using a low-cost labor force. The fashion and textile industry can be mentioned as one specific example, where the employers are unwilling to apply advanced technology into the production processes (Negara 2010). It is a well known fact that the traditional labor-intensive operations in fashion and textile industries are involved with many negative social responsibilities including the forced labor or even child labor (Loebbecke et al. 2006). Furthermore, Nakamba et al. (2017) also list some violations of social responsibility such as hazardous working conditions, overtime work, discrimination, and lack of health and safety assurance.

All of the mentioned social concerns can be addressed or mitigated by using RFID technology. Pioneering in this implementation, Walmart and Tesco are the two highlighted names that have practiced RFID partly to prevent such social responsibility issues in their supply chains (Nayak 2019). Therefore, the first social advantage from applying RFID is to help improve the production process not only in its productivity but also in regard to the overall recruitment, treatment and protection of workers.

The second apparent benefit that RFID creates for society is its ability to provide right information on product history to today's consumers leading to transparency and traceability (Zelbst et al. 2019). In simple terms, companies can now make their supply chain more transparent to the end-customers by collecting and uploading relevant information about their products' origin, suppliers, production processes and total environmental impacts in a RFID system or a QR code. This allows consumers to directly access and make their decisions on whether the product is satisfied to them based on other added values and work ethics rather than just the product itself. When the preference standards are met, it is more likely for firms to generate product sustainability, as well as enhance using sustainable materials in their production which can both address sustainability issues and at the same time, generate goodwill for the society (Denuwara et al. 2019a, b).

Shuai et al. (2010) stated in their article, when it comes to social sustainability, the measurement on the quantification is the most challenging aspect of sustainability evaluation. Thus, with the help of RFID technology the social sustainability an be better quantified. RFID can make the whole supply chain become visible to the end customers, providing real-time data and a trusted track list of such goods and services. As claimed by Liu et al. (2010), consumers can be more informative on their consumption and this way of verification can also avoid customers from purchasing counterfeit products in the market. Overall, RFID technology lead to increased consumer satisfaction, which was also established by the study. As explained by one of the stakeholders of the project, RFID can lead to customer satisfaction as explained in Extract 1.

Extract 1: *Thanks to the traceability system, customers' satisfaction can be enhanced since the product quality is fully checked with 100% of device functions tested.*

2.3.5.2 RFID and Environmental Benefits

As the fashion and textile manufacturing industry is exploiting excess resources during manufacturing process while producing large amounts of wastes from various operations. In the entire value chain, the industry is characterized by wastewater generation, solid waste contamination, and water/energy overuse (Denuwara et al. 2019b). Fortunately, the development of RFID systems can uplift its "automation approach" for a more accurate tag navigation and tackling the above mentioned problems with greater efficiency, visibility and forecasting ability (Schindler et al. 2012).

For the increase in sales, the RFID system can identify out-of-season items by only monitoring the tags in-store (Baggchi et al. 2007) and take immediate action of establishing promotions or discounts to push the stocks rapidly (Medeiros et al. 2008). By doing this, the smart retail-shelves monitor and reduce the end-of-season wastage. Moreover, the RFID technology can enrich the life expectancy of clothes with detailed instructions about washing, drying, and ironing of clothes (Azevedo et al. 2012). Amid the manufacturing of fast fashion, short-cycle products emit 400% more carbon dioxide than the traditional clothes used for at least a year or more (Conca 2015). RFID implementation can lead to clothing durability and bring environmental benefits (as stated by one of the interviewees in Extract 2).

Extract 2: *If the average life of clothing was extended by just three months, it would reduce their carbon and water footprints, as well as waste generation by 5 to 10%.*

The other environmental benefit of RFID technology is the speed and accuracy of information (Chowdhury et al. 2008). RFID will generate the most real-time and accurate data on the inventory level and help to reduce the inventory wastage. This can be related to reducing the amount of carbon emission level, and amount of wastewater that helps the business easily point out such pitfalls in the value chain and amend quickly (Mani and Wheeler 1998). Also, the RFID technology can minimize the inaccuracy of inventory management by exactly monitoring the customer purchase in terms of items and timeframe (Marr 2018). Then the RFID produces the exact amount of stock that should be maintained, while eliminating the waste of excess garments at store and cutting the greenhouse gas emissions resulting from inaccuracy in inventory transportation (Karakasa et al. 2007). While the complicated value chain can be better managed, the RFID concurrently proves its accuracy in monitoring the recycling process that can prompt a more sustainable manufacturing process. Whereas the route of recycled clothes is tracked by the readers through mobile platforms (Liu et al. 2013a, b), the RFID encourages more users and investors to recycle garments due to the efficiency. This proposes a brighter opportunity for any business, especially the apparel industry, with the dominance of fast fashion.

2.3.5.3 RFID and Economic Benefits

At the macro-level, the fashion and textile industries have been one of the key financial sectors for some nations, their performance is one of the main drivers for the whole economy (Jabłońska et al. 2020). Therefore, when tracing down to the micro-level, corporate performance, especially the industry leaders, play a crucial role in developing nations' economic growth. The higher possibility of firms achieving economic sustainability means the better for the economy. One of the elements to evaluate whether the corporation has achieved economic sustainability is their profitability generated based on the existing resources. To facilitate profitability, many fashion and textile firms now focus on innovative research and solutions using advanced technology such as RFID to improve their day-to-day business and gain more profit (Shishoo 2012).

First and foremost, RFID technology can improve the supply chain management by tracing the origin of the products, thereby enhancing the visibility of the whole supply chain (Wang 2014). Hence, this implies that any defective products or errors can be easily detected, traced back to the original source, and implement measures to resolve the problem. This is applicable for all products, including those that have gone downstream to the retail. This was also stated by the participants of the RFID project as stated in Extract 3.

Extract 3: *Via the traceability system, if one functional failure happened, there will have people who evaluate and analyze what are the root causes of this failure and which components, link back to the production site to identify how many items and quantities are in different stages or different areas and call back the products.*

Second, RFID technology can enhance production efficiency by assisting corporations to automate their production process and restructure logistics, hence, being able to reduce labor costs (Ustundag 2012). In detail, RFID system can improve goods receiving, product aggregation, sub-item supply tracking, and improve shipping verification in production. Furthermore, RFID can facilitate returnable items' tracking and loading verification during reverse logistics (Denuwara et al. 2019a, b).

Third, RFID technology can improve inventory management (Nayak et al. 2015) by depicting the sales performance of different types of clothing. Combined with its accuracy feature, RFID is capable of detecting and reducing errors as mentioned earlier. Hence, corporations can have a better track of current capital and resources, thereby being able to recycle or eliminate non-value-added assets. This helps in cost-savings, and reducing the lead time (Tong et al. 2010). The statements collected from one of the interviewees in extract 4 justify these benefits.

Extract 4: *Applying automatic machinery, enterprises will save some labor costs, as automatic machinery helps to reduce labor resources in production. RFID technology will help increase productivity so that the company can receive more orders and shorten the production process.*

Finally, it is worth mentioning that the RFID system is a critical tool to improve security and decrease shrinkage in the supply chain and retail owing to its traceability (Ustundag 2012). This implies that firms' authentic products are distinctively differentiated from illegal products or counterfeit products due to RFID system because it can give all necessary information to customers (Nayak et al. 2015). Ultimately, the RFID system can strengthen brands' customer relationship management by providing this information to protect the customers and assisting in personalized services design to enhance customers' shopping experience (Moon and Ngai 2008).

2.4 AI and Robotics

The advancement in Artificial Intelligence (AI) and robotics have encouraged the rise of automatic systems and provided the efficiency within any production, since it can exceed the capability of any human being (Wolla et al. 2019; Nayak and Padhye 2018a, b). These developments help the global economy to grow by creating new products and services leading to increased consumer demand, hence, generating additional revenue streams.

Specifically, in the context of fashion and textile manufacturing, machine learning (ML) and collaborative robotics help to enhance productivity. These technologies have been adopted and applied by many fashion giants, such as Nike or Zara, to optimize the costs, enumerate automation in production and management (Ahmad et al. 2020). ML can help to detect the condition of operation automatically using machines and robotics. Robotics can deal with the complicated manufacturing conditions and collaborate with human workers in a challenging environment.

2.4.1 Artificial Intelligence (AI)

Furman and Seamans (2019) describe AI as a human-like intelligence with a wide range of technologies, such as autonomous robotics, machine learning, and computer vision. Wolla et al. (2019) indicate the main feature of AI is the prediction. It applies current information (big data) and generates assumptions as well as predictions for the upcoming events that we used to intuitively make.

With the prominence of digitalization and globalization, AI has been adopted at numerous stages in the textile and apparel (T&A) industry, such as apparel design, apparel manufacturing, apparel retailing and supply chain management (Kampakaki and Papahristou 2020). Due to the versatility of the T&A industry, the utilization of AI to some extent has occurred for years to help enterprises rapidly respond to the trend and evolvement in consumers' demand.

The fashion and textile industries are well-known as a labor-intensive sector with low automation and an extreme diversity in consumers' demand (Scott 2006). These factors complicate the decision-making process, combined with the strong non-linearity, inflexible production, planning, and control (PPC) phases, and seasonality in product selling. The traditional techniques used in the T&A sector are inappropriate to solve the problems stated here (Guo et al. 2011). Therefore, there have been many AI methods being applied in this sector, which are separated into five distinct categories (Table 7).

Table 7 AI methods that are applied in the T&A industries

Neural networks (NN)	
Genetic algorithm (GA)	
Fuzzy logic (FL)	
Other AI methods	• Artificial immune system (AIS) • Genetic programming (GP) • Multi-agent system (MAS) • Simulated annealing (SA) • Decision tree (DT) • Expert system (ES) • Intelligent system (IS)
Hybrid intelligence (Interactive genetic algorithm) (HI)	• NN + FL • NN + others • GA + FL • Other HI methods

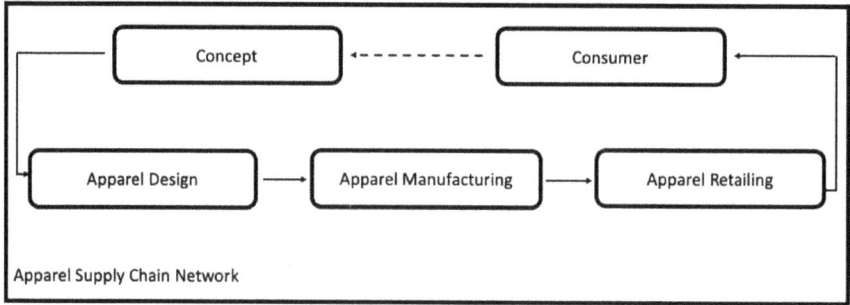

Fig. 3 Processes to develop a T&A product

Prior to the application of AI, the T&A industry dealt with different problems as mentioned earlier due to the high competitiveness of the market. Specifically, four main processes, Concept, apparel design, Apparel manufacturing and retailing (as shown in Fig. 3) have different problems with decision-making. Therefore, industrial manufacturers apply distinguishing AI methods to each process to cope with these problems. As a result, it helps create a clear process within the supply chain, thus minimizing any possible cost or risk and enhancing the productivity of the industry.

Apparel Design

AI is applied in the apparel design phase since it consists of numerous decision-making obstacles, including apparel computer-aided design (CAD), design evaluation in fabric or colour, design of the pattern, and garment evaluation (Guo et al. 2011).

- **CAD system**

The apparel CAD system that is commercially applicable usually concentrates on pattern design, marker making, and grading, thus, significantly enumerating quality and productivity. Designers are provided with a platform, in which they can create a design model with a user-friendly experience, however, it cannot produce a finalized design solution. Therefore, an AI-based apparel CAD system has been developed to generate design solutions for designers. For instance, an iterative GA approach was adopted by Inui (1996), which induced human–machine interaction cycles, helped designers repetitively scout for appropriate design solutions and determine the compatibility of each chromosome.

Kim and Cho (2000) applied 3D technologies and iterative GA approach to enhance system interface and support the design of women's clothes based on customers' preferences and emotions. Additionally, to cope with the impediment of modelling garment drapes to apply 3D technology in T&A design, a drape system that can predict and demonstrate women's clothes made of various styles and materials was proposed. The prediction was made using the data stored in a large database containing a drape image of the garment, then it decided a duet between the garment design and fabric properties as well as feature dimensions.

Noor et al. (2021) developed an intelligent system that provides design solutions that align with customers' demand for different events. However, this system is limited in colour types and only uses five patterns of garments. Another intelligent system that can extract design elements and shape combinations between abstract fashion design and concrete technical design elements based on two principles was also introduced by Noor et al. (2021).

- **Fabric-related problems**

Fabric performance is the key element for garment quality and a smooth production (Noor et al. 2021). There were 11 different indices to evaluate fabric performance that facilitated various stages of garment production, including fabric laying, fabric cutting, and sewing of components (Guo et al. 2011). Additionally, designers can choose suitable fabric in product development based on an AI-based advisory system using 14 different frameworks.

To generate optimal fabric textures, Muni et al. (2006) proposed an interactive texture generation technique. This method was good in creating various textures in fabrics using AI. However, this method was not applicable for generating colour texture. A design advisory system to predict appropriate designs was developed by Guo et al. (2011), which was based on two elements: content-based technique and collaborative filtering technique. However, this was also not much applicable due to the inaccurate prediction of designs.

The mechanical properties of the fabrics, which are evaluated by objective metrics, cannot demonstrate the customers' psychological effects. To address this, Koehl et al. (2008) introduced an FL-based measurement technique to identify the relationship between fabric mechanical properties and customers' psychological effects. However, customers still tend to intuitively assess the fabric performance instead of mechanical properties since it is more convenient to them (Noor et al. 2021).

- **Colour-related problems**

Colour is the most important feature for any fashion and textile products since consumers always purchase a product based on its colour appeal (Guo et al. 2011). A learning-based fuzzy-logic system was introduced to assess the customers' demography (age, gender, or skin colour), and colour taste based on their profiles (Hui et al. 2015). This system can only predict single-colour, which is inappropriate in this dynamic trend nowadays. As a result, an evolutionary fuzzy colour motion model that applied to recommend the colour of clothes based on customers' emotional tendencies was developed by Tokumaru and Muranaka (2008). Besides, Noor et al. (2021) utilized the NN approach to formulate the relationship between 10 different pairs of antonyms, which describe the colour properties, and CMYK colour to anticipate the awareness of customers towards the fabric colour. However, this model was in need of more experimental processes since it was only tested using a scarce sample test (Guo et al. 2011).

- **Pattern design**

The process of converting a 3D design model into a 2D design is a crucial step in the production process of apparels. Conventional pattern design methods cannot exploit the knowledge and experience of fashion experts and designers. Therefore, numerous knowledge-based systems and interactive CAD systems were developed to exploit these elements. Moreover, it also allows the collaboration between these persons under a scheme, where they can provide evaluation and apply interactive functionalities to improve the design quality (Guo et al. 2011). The fitness of component size is important to the appropriateness of the garment. Thus, an intelligent system that scouts for the optimal garment component and predicts the suitability of garments based on customers' body size was introduced (Hu et al. 2009).

- **Garment evaluation**

Since the purchasing decision mostly depends on the sensory and realistic feeling of customers towards the products, it is difficult for them to identify the appropriate clothes. Therefore, many mathematical formalization methods have been built based on customers' demography, preferences, experts' knowledge, and past purchasing behaviours. Thermal comfort is considered by many consumers when choosing a T&A product. As a result, an FL-based method was introduced to anticipate the subjective perceptions based on the data of moisture and thermal sensations. Besides, the assessment of thermal function in non-uniform and dynamic environments using physiological parameters was developed (Luo et al. 2007). However, these methods required a large dataset and multiple criteria to justify the final decisions.

Apparel Manufacturing

At this phase, apparel products are produced, including five main stages: cutting, sewing, inspection, packing, and distribution. Based on the final design the apparel manufacturer will source the raw materials, manufacture the garments and send to the buyer or retailer (Guo et al. 2011). The application of AI supports these stages of apparel manufacturing, specifically, the issues on decision-making during production, planning, and control (PPC); marker making; sewing-related issues; and other issues.

- **Production, planning, and control (PPC)**

PPC helps industrial manufacturers to precisely schedule and execute the production as well as deliver the product on time to the end customers. However, at this stage, there are several issues related to schedule, planning, assembly lines, or the fabric cutting department. Most of the studies just concentrated on dealing with the issues in the short term, while the long-term planning has not been attractive to AI researchers (Guo et al. 2011).

The PPC problems related to the assembly lines consist of sewing assembly line balancing (SALB), machine layout, and operation assignments. Guo and his colleagues developed a GA-based system that solves the issue of SALB in the progressive bundle system (PBS), which can minimize the idle time and optimize the cycle time (Guo et al. 2008). The PBS includes 2 features, workstation revisiting and work sharing. This model is able to decide the suitable operation assignment to the workstation and effectively divide the task for each shared operation. Noor and his colleagues presented a system using the GA approach to deal with the SALB problem in the unit production system (UPS). This system arranges the operator assignment according to the effect of sewing operators' efficiency.

- **Marker making**

The marker making process is important to optimize fabric utilization, increase profit margin, cut production costs, and reduce fabric wastage (Noor et al. 2021). There have been various software applied in this stage, specifically, CAD software is the most popular use in marker making. However, its efficiency is not suitable enough to help industrial manufacturers optimize their productions. Therefore, more tests must be conducted to validate these intelligent systems prior to integrating them into the CAD system.

Heckmann and Lengauer (1998) combined the upper-and-lower-bound method with the SA approach to representing the market layout. Nonetheless, this method was difficult to implement since it was described in a geometric representation and required a large data set as well as complicated features. Therefore, 2D matrices have been applied to represent the shapes of a pattern. There are two advantages for this application, which are the simplicity in selecting the enclosed areas in patterns and detecting overlaps (Noor et al. 2021). In addition, Ozel and Kayar considered the combination of different garment colours and sizes in their production model. This model employed the NN approach to identify cutting time based on the marking lengths, cutting speed, and number of layers (Ozel & Kayar 2008).

- **Sewing-related and Other problems**

There are other problems related to apparel manufacturing, such as issues of modular production systems (MPSs). The size of the carton box for packing towel products is optimized using a model developed by Leung et al. (2008), thus, minimizing cost due to the spatial utilization and cutting unnecessary carton types. Furthermore, a fuzzy-logic-based system was introduced to forecast the consumption of energy based on the change of production inputs, thus stabilizing the manufacturing schedule, power supply and minimizing wastage (Noor et al. 2021). In addition, the implementation of body type analysis into a made-to-measure product process, many industrial manufacturers proposed the intelligent system, using customers' demography, then customizing clothing production by integrating clothing pattern generation, body dimension collection, and cutting of fabric (Guo et al. 2011).

Apparel Retailing

This phase creates a linkage between the T&A industry with final customers since products are delivered to them from retailers (Guo et al. 2011). The problem of decision-making relating to sales forecasting, fashion coordination and recommendation, is solved in this stage in addition to some other problems.

- **Sales forecasting**

An automatic system, developed by Thomassey et al. 2005), can forecast sales of apparel using two complementary principles. The first principal uses the FL approach to quantify the impact of medium-term sales, while the second principal quantifies the short-term sales. Thomassey and Happiette (2007) also introduced the NN sales forecasting system for new items using historical apparel sales data. Sun et al. (2008) proposed a modern NN forecasting system to assess the interrelationship between fashion products' features, such as price, size, or colours, and their sales.

Nevertheless, these systems only concentrate on apparel items' sales. Besides, these items are affected by the uncertain and seasonal demand, thus indicating a short life cycle and high frequency of replacement (Guo et al. 2011). Therefore, firms usually forecast the sales of all items in one or all categories, then select the appropriate items to produce and sell. However, this work requires numerous computing techniques and parameters to accomplish. As a result, Wong and Guo (2010) introduced a system that applied the auto-regressive and moving average method, thus ensuring the accuracy of one-year sales forecasting. Besides, this system can also deal with various data patterns at a time.

- **Fashion coordination and recommendation**

The capability to recommend customers with "fashion-coordination" and "mix-and-match" advice is crucial for any brand to improve its customer service. Numerous systems were developed to give dressing advice to women customers for any occasion. However, the criterion of recommendation varies between geography, cultures, and ages. Therefore, there has not been a perfect system that can provide recommendations for global customers. Besides, a HI + NN system was introduced by Guo et al. (2011) that can give professional advice based on the satisfaction degree of apparel items' attributes, using the coordination satisfaction index.

- **Other problems**

In the era of the internet and technology, e-commerce has become a perfect solution for shopping. Therefore, there have been numerous intelligent models that help customers in choosing and purchasing suitable fashion items via the internet, such as size or colour selection. A sizing system using anthropometric data of women's body type was developed since it is vital for production and retailing (Noor et al. 2021). Lu et al. (2008) proposed a model that guides manufacturers to produce more competitive products based on customers' desirability. Besides, an evaluation system was developed to construct sportswear brand image based on consumer psychology, customer behaviour, and other interests such as experiential marketing.

Apparel Supply Chain Management (SCM)

SCM is the combination of processes within a supply chain, thus creating values for all stakeholders. The apparel supply chain consists of information, technology activities, people, and resources that deliver raw materials from suppliers and distribute final products to end customers (Guo et al. 2011). An effective apparel SCM can help businesses minimize costs and enhance their competitiveness on the market. Many researchers have criticized the decision-making problem of the SCM in the replenishment strategy and control of inventory. An e-fashion SCM system was created by Lo et al. (2008) to convey data and information anywhere and anytime by integrating various information technologies and intelligently adjusting its behaviours. A HI-based system was developed to help industrial manufacturers decide the site for plant location in the global apparel supply chain since there are numerous factors that affect the site selection (Guo et al. 2011).

Industrial manufacturers also face challenges with the replenishment of the apparel supply chain. There are several strategies that are applied in this industry, such as the strategy of vendor-managed replenishment for simple products or the seasonal supply chain to serve the trend of fashion and distribution (Guo et al. 2011). A "GA-based dynamic rolling optimization" system was developed to improve the customer service level and balance the manufacturing capacity (Dong and Leung 2009). Besides, Guo et al. (2011) introduced an AI-based system that can design the supply chain process and create a strategy to replenish quantity. Guo et al. (2011) employed the MAS approach to a management system that can forecast, decide the information replenishment, and control the inventory levels. In addition, Guo et al. (2011) applied the HI approach into a method that helps industrial producers to identify an appropriate completion time and overcome the issue of multi-customers due to date bargaining.

2.4.2 Robotics

The advancement in technologies, such as intelligent robots, has been transforming industrial structures into smart factories and physical production systems that possess more intelligent attributes (Görçün 2018). All industrial processes in this era have converted into automatic operations with the integration between the virtual and physical world. Thus, "smart factory" is the keyword in the fourth industrial revolution. This concept is created by the combination of the physical and digital environments, in which advanced technologies take the control of robotic automation (Gilchrist 2016). The main principle of a smart factory is the interaction and coordination between producers and consumers within all recognizable operations (Görçün

2018). The T&A industry has been transforming into a global business, thus the transition to a "smart factory" can create a positive impact on efficiency and sustainable production (Bertola and Teunissen 2018).

The autonomous behaviour of machines, robotic systems, and automation are the key factors of this system. The current robotic system is functioned to automatically adopt the optimal decisions, thus minimizing the use of manpower in the manufacturing processes. All robots are also able to communicate with raw materials, work-in-progress or final products. Besides, robotic systems can acquire precise information relating to the demand of customers, then transmit commands to robots and machines with the appropriate functions. Therefore, it can ensure the full capacity of all equipment, machine and integrate the whole processes within the operation (Görçün 2018). Moreover, the machine-to-machine (M2M) helps develop the communication and interaction between machines. Besides, the technological revolution, particularly mobile devices also encourage the application of robotics in facilities. More importantly, the identification system, RFID, is the main component of the M2M. With the RFID, objects are automatically identified, and machines can realize their functions using a remote sensing system. Nevertheless, this factory-centered approach is simplistic, compared to outrageous advancements in technology (Bertola and Teunissen 2018).

2.5 Automation

Another beneficial digital technology that can be applied to attain lean manufacturing objective is automation. Even though the application of mechanical devices has long been adopted since the very first days of manufacturing, it was not until the twentieth century that the wave of automated production actually levitated (Endsley 2018) when manufacturing became the engine in economic growth in most nations. Because of that, automation was put into priority in order for firms to gain low production costs and increase competitiveness. One of the emblematic examples of such momentous innovation in the automation area must be named is the "Assembly Line" invented by Henry Ford in around 1913 that brought a solid foundation for the Ford company to thrive afterwards (O'Neill and McGinley 2014). Especially during the 1980s, as more interest was concentrated on the MRP (Material Requirements Planning) and ERP (Enterprise Resource Planning) systems, there was a huge outbreak of numerous control machines that were adopted, and a fully automated production line was officially introduced. Also, in the same period of around 1980s, although automation had been widely put into practice, it was still categorized as a subsidiary tool to support the production processes, for example, the transportation, storage or material handling stage (Park et al. 2002). Hence, most of the manufacturing processes at this time were semi-automated in which the system was a combination of automatic and manual operations (Boothroyd 2005).

As the demand for product quality increased (which also consists of the growing sophistication and customization of the product), the level of automation kept being developed and fully automated systems were brought into practice with the presence of "lights-out" factories. With the Industry 4.0, there has a huge improvement on the worldwide manufacturing industry. Furthermore, automation has also been exploited with many new advanced machines and technology to improve productivity. Endsley and Kiris (1995), Connors (1998), and Onnasch et al. (2014), claim the inefficiency of automation, if not being applied correctly could lead to degradation, disturbance, or even slackening performance due to downgrade in workers' skills. Depending on the appropriate level of automation for certain manufacturing processes, it is highly suggested that the harmony between automation and human activities (human–automation labor systems) is the key for successive manufacturing (Frohm et al. 2008).

Hedelind and Jackson (2011), reclaim the definition of "Automation" as a modern approach of manufacturing that involves less human effort and emphasizes the crucial role of mechanical devices. Indeed, the concept of utilizing machines to replace the traditional human labor in stages within the production process has long been the agreeable depiction of automation among other various academics. Nevertheless, when the concept called "knowledge worker" came into existence, the role of "cognitive labor" also emerged in which workers may not be replaced but will work along with the adoption of machinery and technology (Frohm et al. 2008). In that notion, automation is considered as an effective tool to support and improve productivity rather than to replace human labor. More particularly, while Satchell (1998) highlighted the surge of mechanical and information technology, Lee and See (2004) defined automation is an inclination towards its data utility function. They asserted automation as the application of technology to control the processes by collecting, transforming and managing data.

According to Parasuraman et al. (2000), automation systems can be categorized into four main types: acquiring information, analyzing information, making decisions and conducting actions. More specifically, one automation system can function to achieve more than one purpose listed above. The authors also denote another way to classify automation which is by determining how much involvement does humans impose in the process. Parasuraman et al. (2000) demonstrated the scale with ten levels of automation in which one is the point where humans work closely with the system to mutually support each other and 10 is the level that the automated system is left working completely independently.

The goal of automation is to create automatic control over some stages or the whole manufacturing process. Womack et al. (1990), one of the pioneers in defining automation, mentioned that the usage of automation not only helps reducing human effort but also maneuvers less space, less lead time, fewer defects and higher productivity. Hoff and Bashir (2015) acknowledge in their publication the values of automation. In specific, they believe that the sole value of automation lies in its capability to perform complicated and repetitive tasks with higher efficiency and, assures the quality of products by avoiding any possible defects generated. Automation is very helpful to replace complex human activities, due to its ability in performing such activities, which gives people the opportunity to focus on other important tasks. Automation can also protect humans by replacing them to work in hazardous working conditions (Hoff and Bashir 2015).

In Lean Manufacturing, the ultimate objective is to upgrade the traditional manufacturing and logistics processes to be more agile and efficient, mainly via reducing waste and boosting productivity. Following that direction, automation can contribute for firms to reach those targets by eliminating defects in production and curtailing the lead time that used to be longer if done mainly through manual ways. Supporting the relationship between Lean and Automation, Dulchlnos and Massaro (2005), mentioned that the application of automation to make business production more efficient, raises the concept of "Lean automation". In simple terms, it is a set of techniques that assures a sufficient degree of automation for a specific task to build an optimal manufacturing process. Nicoletti (2013) specified in the chapter "Lean and Automate Manufacturing and Logistics" that automation can be included in Lean Manufacturing to create a perfect production streamline.

Taking the Six Sigma methodology into consideration, Nicoletti (2013) suggested a 6-step framework of incorporating automation to achieve Lean Manufacturing. The concept of Six Sigma is widely used as a set of principles and practices to accomplish Lean Manufacturing (Tjahjono et al. 2010). Its fundamental is to detect and get rid of production wates, eliminate defects and output volatility, thus guaranteeing the overall quality and lowering the cost of production. Based on that cornerstone, the 6-step of automation as mentioned by Nicoletti includes:

- **Step 1: Identify and measure**
  This is the phase when the project is launched, and all relevant stakeholders are identified. Firms should then define the roles and responsibilities of in-charge persons and start determining the objectives of the works as well as metrics and/or methods for quality control. To diagnose the objectives, it is necessary for firms to research their customers or target stakeholders first.
- **Step 2: Analyze and design production process**
  The processing map should be sketched in this phase. From that, firms are expected to find methods to improve or achieve the noted objectives in the most efficient way. Firms could use Kaizen principles to enhance its implementation plan.
- **Step 3: Automation-inclusive architecture**
  After having the overall scheme of how the production process should be conducted, this phase is spent on evaluating and designing the adoption of automation into the construction.
- **Step 4: Construct and testing**
  When finishing with planning the manufacturing process and choosing the automation technique that is most theoretically suitable, that architecture should be built up for implementation. The role of management needs to be taken in carefulness to conclude whether the solution is working as expected or not. Monitoring process and updates should be recorded continuously for the best outcome of evaluation.

- **Step 5: Double-check**
  A thorough verification should be operated to not only control the implementation process but also improve its performance if necessary. A list of envisioned benefits will be needed to compare and contrast with the performance of the running processes. At this phase, observation and development is crucial; improvements are likely to be made for both the system and human labor to assure the harmony between them.
- **Step 6: Reduplicate the process in different situations (Optional)**
  For some certain circumstance that the solution is anticipated to work under different environments or with different conditions (e.g., applied in other departments), firms should take those variabilities into account and re-operate the process in different contexts. Thereby, firms can increase the percentage of success and ensure the same qualified outcomes even in dissimilar situations when it comes to real life implementation.

The above 6-step framework is one way to integrate automation with business processes improvement. In general, the framework can be referred at any production stages and by any organizations. As mentioned above, automation techniques could be separated into two main branches: data-related and action-related automation. To pick the most prominent technologies that well-represent the two branches, Internet of Things (IoT) and RFID could be the perfect examples for data-related and action-related automation, respectively. RFID has unique functionality in bolstering Lean Manufacturing by accurate and on time information availability. The concept of IoT is briefly introduced in the following section.

Internet of Things (IoT) is the advanced technology that was first named by Kevin Ashton in 1999 when he worked at Procter & Gamble. In the recent decades (after the Internet revolution), with the rise of smart sensor adoption besides the continuing improvement in network connections and wireless communication, IoT is emerging as the modern technology that is believed to create a huge impact on all kinds of industries and businesses (Premkumar and Roberts 1999). IoT is also called "Machine to Machine" or M2M that aims to connect both wired and wireless devices and allow ubiquitous communication. Its operating principle is to have a central hub that would receive, analyze and transmit data through a remote network of multiple devices.

In Lean Manufacturing, applying IoT could save a lot of time in sharing and updating information within departments, stakeholders and even at the macro level outside the factory. Information such as physical pressure, light intensity, position, and temperature, can be collected so fast through direct input from a wide range of connected devices. For instance, information exchange is just a low-level function of IoT, but it has been employed into the MRP and ERP systems with Internet and cloud computing to improve flexibility and enhance collaboration in an incredible way (Vermesan and Friess 2014). The purpose of IoT, according to Atzori et al. (2017), is to connect objects and showcase the information to any person anytime and anywhere via the Internet. The outstanding point is that those objects and people can work in real-time to complete and accomplish a common task together. With the potential to be more advanced in the near future, IoT is believed to help achieve the objectives of Lean Manufacturing in the Industry 4.0 era in almost every industry (Atzori et al. 2010).

Apart from the benefits that automation can bring to the manufacturing sector, there are some associated risks. Parasuraman and Riley (1997) discussed some risks associated with automation implementation. In specific, the authors emphasized the necessity of system design in relation to the external factors such as surrounding interference, operator exploitation, or software-hardware mismatch that could lead to failures in the automation system. Therefore, proper monitoring coordination should be ensured to create trust in automation and protect human safety as well (Hoff and Bashir 2015).

2.6 IT Applications

The era of Industry 4.0 includes the production attributes of cyber-physical systems (CPS), combined with the knowledge integration and utilization of heterogeneous data (Ahmad et al. 2020). The CPS in this industrial revolution can help manufacturers improve their production and efficiency with the application of some information technologies (IT) (Fig. 4). There are various technologies that contribute to Industry 4.0 such as big data, the internet of things (IoT), and enterprise resource planning (ERP), (Lu 2017).

2.6.1 Business Intelligence System

Business intelligence system (BIS) is known as an umbrella term when enterprises combine various technology applications, software systems, and tool strategies to

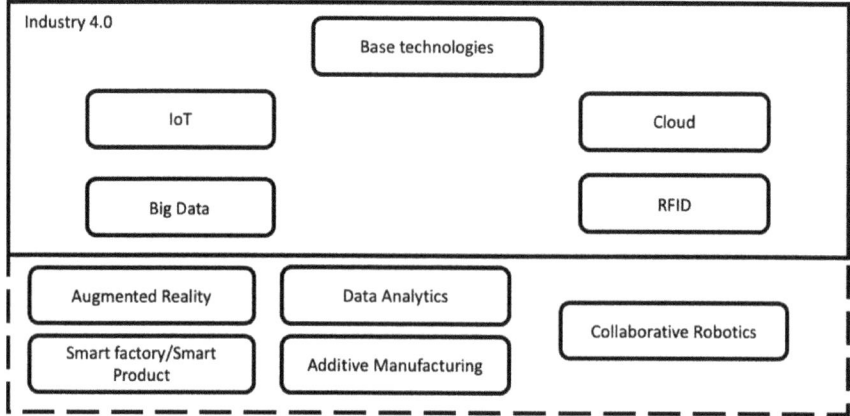

Fig. 4 Technologies applied in Industry 4.0 framework

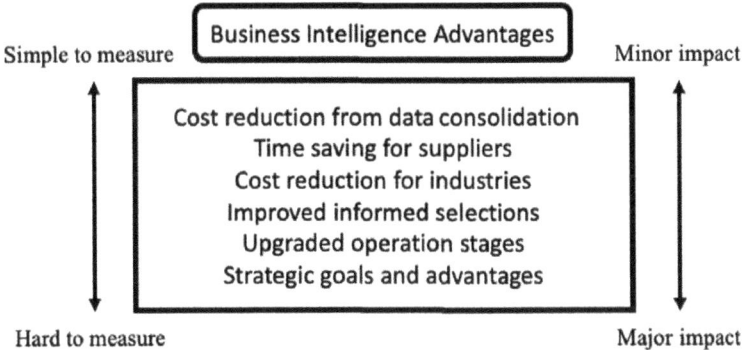

Fig. 5 Advantages of BIS implementation

support their business operation in terms of decision making and utilizing data (Muntean and Muntean 2012). Through BIS, businesses can gain the advantage with the possession of data, both internal and external, through multiple platforms and channels, and increase efficiency in the operation. In the competitive picture of the T&A industry at this current time, entrepreneurs can use the essential information to sustain businesses' competitiveness by interpreting and understanding their capabilities (Ahmad et al. 2020). Even though the integration of different technologies was not widely accepted by SMEs (Small and medium-sized enterprises) in the fashion and textile industry due to the barriers of cost (Chong and Chan 2012), the agile progress in the development of IT has facilitated the adaptation of BIS in Europe and the U.S. for years (Gartner Research 2009). Moreover, giant famous enterprises such as Nike or Zara have adopted the advanced BIS, which sustain their position in the global marketplace (Ahmad et al. 2020).

To thoroughly describe the reason for BIS adoption in the T&A industry, Fig. 5 lists the main advantages. BIS can be used in any department of a company, such as logistics, production, inventory management, finance, marketing, and retailing (Ahmad et al. 2020). By collecting and analysing the data from all departments, an efficient process is created, thus minimizing the inappropriate cost, forecasting business strategy, gaining customer loyalty, and identifying frauds. There are some popular BI solutions that are used by many T&A companies, such as Oracle, Tableau, and Dematic.

2.6.2 Big Data

Big data consists of extremely large data volume which cannot be stored and analysed using the conventional database (Hashem et al. 2015). Specifically, big data relates to large volume, high-variety, and high-speed data with the requirements of a modern form of information processing. However, the definition of big data at the current time may change in the future due to the accumulation of storage capacity. This refers to the utilization of very advanced techniques of analytic to examine, process and extract information from an excessive amount of diverse data. The outcome is believed to deliver faster data analytic results, have higher quality of data visualization, disclose hidden insights and thus support companies in making more effective decisions (Vaidya et al. 2018). In the era of the Industry 4.0, enterprises use various types of data, from unstructured, semi-structured, to structured data (Ahmad et al. 2020). Data is collected through multiple channels, both internal and external sources, using the internet to transmit and store in a cloud base (Chen et al. 2016). With the integration of various information technologies, T&A companies with the application of big data are allowed to have flexibility in decision-making, especially in financing and marketing strategy, and the capability to exploit advantageous knowledge to upscale their businesses.

2.6.3 Internet of Things (IoT)

The introduction of IoT creates a possibility to integrate various advanced technologies to support monitoring, traceability, coordination, and collaboration between enterprises (Ahmad et al. 2020). IoT represents a network of multiple devices that can generate information and provide information on replacement, maintenance, repair, and recalling of things or products. IoT transforms daily physical objects into a world of information and makes them more interactive, intelligent, and trackable (Papahristou et al. 2017). IoT can create some added value for the T&A industry, in several areas such as retail environment and manufacturing floor. T&A manufacturers can attach IoT sensors to their machines and equipment to track their efficiency. Besides, it enables development of products and operations, as well as facilitates the modern infrastructure where data can be collected, analysed and shared. Moving to the retail environment, IoT helps the application of augmented reality, which can fulfil the demand of modern customers, who demand the virtual dressing room to find the appropriateness of a fashion item (McKinsey & Company 2015). Nevertheless, some barriers still prevent the T&A industry from exploiting the benefits of IoT, such as the security issue, affordability, availability of infrastructure, and lack of knowledge and understanding (Papahristou et al. 2017).

2.6.4 Augmented Reality

Augmented reality (AR) is a component of additive manufacturing with the feature of enumerating sustainability (Ahmad et al. 2020). AR technology aims to complement the existing physical world with digital integration by exploiting technological devices (screen, camera, headset, etc.) to embellish with digital components such as virtual sound, visualization, or sensory stimulators. In the past, manufacturers applied augmented reality to fix and adjust tasks within the industry. Nowadays, as AR's dynamics is to enhance the overlapping between virtual objects and the real world, AR can help highlight the physical world's features with the assistance of technology, thus drawing out the insights that can be used to apply or improve in real world circumstances. Particularly, in contemporary businesses, it is exploited as a 3D application that can enhance the experience for customers by creating a virtual world. With a simple device such as a smartphone, tablet, or smart glasses, customers are provided with the full apparel collection or an overlay of that product through the cloud service (Zhou et al. 2008). With these attributes, enterprises can increase the efficiency of their services significantly. Moreover, they can also minimize the cost of workers since a service technician can proceed complex maintenance with augmented reality and a device, regardless of insufficient training and skills. More importantly, AR can be an effective tool for firms not only in insights and understanding of the process but also in their decision making. Specifically in the field of manufacturing, AR can help bolster the interaction between workers and their working-on equipment or any other digital products (Mayr et al. 2018).

2.6.5 Additive Manufacturing

Additive manufacturing (AM) or so-called 3D printing technology can be described as a process that is based on a computer-controlled system of sequential layering used to generate physical products with three dimensions. Products can be constructed with various layers created and placed one after the other. A three-dimensional 3D design must be provided in advance that can be created in Computer-Aided Design (CAD). AM technology is considered as a breakthrough innovation that can support in making of prototypes using bottom-up approach.

Until now, AM has been utilized in manufacturing processes that need industrial components created at a fast pace, in sophistication, and in only one step. Additionally, AM can not only reduce the stages and effort in the production process but also can improve the products' customization attributes (Bahrin et al. 2016). AM is applied to minimize input cost, operational cost, or energy consumption (Ahmad et al. 2020). This also helps producers to accelerate the product development process, manufacture customized, innovative, or complex products (Despeisse and Ford 2015), as well as creating sustainability within the production by minimizing the emission of waste (Niaki et al. 2019). AM consists of three main features namely: complexity for free, tool-less manufacturing, and less resource-intensive. For complexity for free, instead of using traditional machine systems, manufacturers are allowed to produce any geometrical complicated design at a lower cost. With tool-less manufacturing, a flexible system is created for producers to create customized products through the 3D CAD models. Then, less resource-intensive feature helps in reducing any possible input of a production system, such as time, workers, raw materials, or energy. Nevertheless, AM has not been widely adopted due to the technical aspects in application cost and reliability in production, such as time to produce and quality of final products (Niaki et al. 2019).

2.6.6 Mobile Computing

Mobile computing promotes ubiquitous access to various resources of ICT (Information and communications technology) (Ahmad et al. 2020). Mobile devices facilitate the business of retailers and fashion brands as well as the operation of firms in the T&A industry, especially in the contemporary trade. It helps producers to manage and track Near Field Communication (NFC) and RFID through these devices. Specifically, with the NFC tag, manufacturers can figure out the replicate of their products, track loyal customers, and minimize the waste by removing the price tag (Kishore and Rajeshwari 2017). Besides, it also enhances the speed for BIS users of mobile devices, thus fulfilling the collaboration and convenience. More importantly, with the mobile application, users have an interface which provides dressing advice, expedites the dry-cleaning process, and track the frequency of usage (Kishore and Rajeshwari 2017).

2.6.7 Cloud Computing

Cloud computing is one the most emerging technologies in the Industry 4.0 since producers apply it to enumerate operational, management processes, and other purposes. Through cloud computing, a wide range of computing services such as data storage, communication, networking, and data analytics, are available to use with less effort than ever before. Specifically, cloud computing facilitates access to the shared pool of data and information through any device. Therefore, it helps industrial manufacturers to minimize the maintenance cost of IT infrastructure and innovate the operational strategies (Ahmad et al. 2020). T&A industries utilize cloud computing to transform the conventional business model to an agile model that enhance productivity, increase flexibility, and induce collaboration (Mell and Grance 2011). Consequently, this can help to improve communication and information sharing between stakeholders in an organization in addition to enhancing the whole system's performance (Vaidya et al. 2018). Furthermore, the complexity and accumulation of data and information is easily stored in a virtual storage, thus creating a convenient way to convert them into advantageous knowledge. Besides, producers can manage the production plans every time and everywhere through cloud computing technology. It also eliminates the impediment of humans, input costs, and environmental status. As mentioned earlier, decision making has been a barrier in this industry due to seasonal attributes and variation in customers' demand. Therefore, cloud computing can promote transparency in decision-making by integrating the distributed resources (Ahmad et al. 2020).

2.7 Other Technologies

Above sections discuss some of the digital technologies that are believed to contribute a leading role in delivering and achieving the objectives of Lean Manufacturing. The list is conducted based on the unique functions each of those technologies owns that either is a groundbreaking innovation or consist of beneficial characteristics to be exploited in the future vision of manufacturing.

Nevertheless, the list of discovered digital technology is conjoined with numerous different kinds of technology. Hence, besides the aforementioned 5 typical technologies, there are still many promising technologies for Lean Manufacturing. This section will introduce some of the upcoming technologies that are believed to be propitious to make use of, especially in this fourth industrial revolution era, or the Industry 4.0 era.

The list of additional digital technology-related recommendations is based upon the research of Valamede et al. (2020) in their journal article "Lean 4.0: A New Holistic Approach for the Integration of Lean Manufacturing Tools and Digital Technologies".

The authors constructed a thorough analysis of the benefits received from the combination of two emerging philosophies: Lean Manufacturing and Industry 4.0. In this research, they put firms into a current context where our world is witnessing a strong surge in the role of technology (or the Industry 4.0). Therefore, the authors took into consideration and demonstrated the advantages (or the opportunities) the companies, especially manufacturing companies, can gain by applying both Lean Manufacturing and new technologies, such as the Industry 4.0). In other words, advanced technologies, in integration with other Lean Manufacturing tools, should be considered as the core strategy for firms to achieve that goal of Lean Manufacturing in a more efficient way to cope with the new digital era.

Following the scientific research of Kolberg and Zühlke (2015), Sanders et al. (2016), Mayr et al. (2018), and Pereira et al. (2019), the two authors Valamede and Akkari (2020) came up with a list of digital technologies that are substantiated to be among the best applications to be exploited in such a digital time. Except the abovementioned 5 technologies that have already been interpreted, the other technologies include:

- **Automated Guided Vehicles (AGVs)**

    In the era of Industry 4.0, AGVs can be referred to as automatic, portable robots or devices resorted to working for a certain industrial task within a company. Those robots can follow markers or any kinds of noticing signals on the manufacturing floors to perform their programmed functions. The remarkable thing about AGVs is that it does not require any manual or external control while those robots are doing their jobs, hence, making the manufacturing process operating faster, more collaborative and more economical (Mayr et al. 2018). In point of fact, AGVs can be utilized in various areas in manufacturing processes such as transportation by loading industrial parts or accessories or in assembly lines by fabricating or moving products within the workstations.

- **Virtual Simulation (VS)**

  Simply explained, VS is usually referred to as the process of mirroring a particular facet of our real world (reality) in a virtual setting, by which people who participate in that virtual environment can interact with that environment and perform certain activities as in reality in that setting. VS is constructed based on a computational system, using real-time data to simulate physical reality through a virtual one with the ability to integrate anything, even humans, products or any other kinds of objects. Thus, Bahrin et al. (2016) claimed that this technology could be useful to the manufacturing industry by providing a simulation that can be used to analyse the methods or techniques which are going to be applied into the manufacturing process. By measuring the expected result, performance can be put into careful consideration for production process implementation.

- **Cybersecurity**

  Cybersecurity (or information technology security) relates to the use of technology for the purpose of securing software and hardware programs such as systems, data, networks, or even computers, protecting those from possible digital attacks. Those cyber attacks' dangerousness can vary, but some common intentions include accessing, sabotaging the information, hindering business operations and for some typical cases, hackers are likely to take advantage of those vulnerabilities to blackmailing. With cybersecurity, preservation, safety and reliability of crucial computer systems and communication with sensitive information will be assured. In such a world that high-tech devices are emerging in numerous forms and functions, effective cybersecurity is suggested. From that, business information and resources management can be well-defended. In the era of Industry 4.0 manufacturers with highly advanced communication standards and connected chains of information will certainly benefit from this technology (Ustundag and Cevikcan 2017).

3 Recent Lean Trends in Fashion and Future Directions

3.1 Sustainable Steps of a Fast-Fashion Brand—Uniqlo

Uniqlo, a Japanese casual wear brand has revolutionaised the world's fashion perspective with the minimalist style. The brand has successfully established over 2,000 stores in nearly 20 countries in Australia, Europe, North America and Asia. While people are getting more conscious about societal and environmental issues, a fast-fashion brand like Uniqlo definitely needs to have certain strategic actions to remain in its position in the market as a popular yet sustainably developed fashion brand. In 2021, three consecutive activities have been organized by the brand, including the Re-Uniqlo, Blue Cycle Jeans and Empowering Women events. Uniqlo has always focused on reducing waste, increasing productivity and efficiency by implementing lean tools and digital technologies. Their major drive is to reduce

Fig. 6 Re-Uniqlo event from the Uniqlo fashion brand

waste in lean manufacturing by adopting the 3R concept- Reduce, Recycle, and Reuse.

3.1.1 Re-Uniqlo—Reduce, Recycle, and Reuse

To compensate for the significant environmental effect on the earth, Uniqlo is making considerable effort to motivate people to take action with their old clothes. In the Re-Uniqlo event. Worldwide consumers are encouraged to take their used clothes for a qualified recycle process with the brand, in which Uniqlo will collect all of the garments and transform them into new products. Alongside that, the brand also partners with NGOs and other organizations to deliver wearable used clothes to people in need (e.g., refugees, and poor people), with the promise to utilize the power of clothing to create a better future (Fig. 6).

3.1.2 Blue Cycle Jeans—Eco-Friendly, and Exceptionally Made

Jeans are obviously indispensable fashion pieces in every wardrobe, and Uniqlo is paying its effort to reduce up to 99% of water usage in the production process of this popular bottom wear. The Blue Cycle process allows jeans to reach the finishing process with just about a teacup of water. With the application of durable and powderless eco-stones and ozone-mist washing process, the brand is able to reduce the large amount of water needed to wash the jeans to create a specific fading effect. The two advanced technologies used in the jeans-making process at Uniqlo, motivate other jeans manufacturers to take the same actions for a better living planet.

3.2 Le-Agile and Conscious Collection of H&M

H&M is a leading fast fashion brand with the headquarter in Stockholm, Sweden. H&M is famous for trendy clothes for adults, teenagers, and kids. The brand was started in 1947, and now it is operational in 74 countries with more than 5,000 stores to serve the customers. H&M is one of the successful brands due to its efficient supply chain management. H&M emphasizes short lead time, affordable prices, and quality products for customers, which is essential for the success of the brand.

3.2.1 Lean Manufacturing at H&M

H&M as a fast fashion brand, which needs to operate in the pull approach to respond to the customer's demand providing the right product, at the right time, right quality, and right cost. In order to meet these objectives, H&M has adopted the concepts of Le-agile (or lean and agile combined) to its manufacturing line. By adopting the concepts of Le-agile manufacturing, H&M takes lower lead time and produces good-quality products with very few defects and variability among the products. Lee-agile concept also focuses on waste elimination, cost optimization, high productivity, and efficiency.

3.2.2 H&M Conscious Collection

In response to the current high sustainability consciousness, H&M is also altering its action to introduce the H&M Conscious collection. The Conscious collection includes clothes made from green materials such as organic cotton and recycled polyester. Moreover, customers are encouraged to recycle unwanted garments at any store of the brand to receive a discount for future purchases. By applying the eco-friendly materials and sustainable production methods (Nayak et al. 2020a, b), H&M attempts to reduce its environmental footprint while targeting its goal with only sustainable material sources by 2030.

4 Conclusions

In conclusion, this chapter has discussed the use of various digital technologies such as Industry 4.0, Blockchain, RFID, AI & Robotics, AGV technology and Cyber-security, that can complement Lean Manufacturing to achieve its objectives. For example, Industry 4.0 has been implemented for continuous improvement, in operational processes by eliminating the traditional wastes present in it. The results of combining digital technologies with Lean Manufacturing yield the concept of "Lean Industry 4.0". Similarly, blockchain can be implemented across the supply chain network, to eliminate financial loss from the fake products and expedite various processes. RFID technology can provide results with high levels of accuracy, and the technology is less time-consuming which can help to save labor costs while increasing product visibility. AI and robotics can deal with complex processes, which can lead to an increase in productivity and efficiency with the automatic system and decrease production costs.

Various other technologies can also be implemented to support the Lean Manufacturing concepts, including automated guided vehicles (AGVs), virtual stimulation (VS), and cybersecurity. These technologies can also eliminate waste and optimize the runtime of various operations in fashion and textile supply chain. To consolidate our findings, two case studies focusing on global fashion brands such as Uniqlo and H&M are presented to give realistic viewpoints of digital technology adoption. Through the cases of two leading fashion brands, it is undeniable that the implementation of the digital technologies in addition to the Lean Manufacturing can lead to maximizing the productivity and achieve sustainability, which is the priority for many brands towards a better future. Hence, digital technologies aligning with lean manufacturing are certainly on the rise, bringing the world on more modern and sustainable pathways. The chapter also conducted a qualitative research study on the application of technologies such as RFID in fashion and textile sector. The findings of a survey based on Vietnam's fashion and textile industries on the use of RFID technology also showed the reduction of wastage and sustainable benefits to these industries by digital technologies.

In the global consumer society when sustainability values are at the top of people's concerns, digital technologies for Lean Manufacturing have a clean pathway to strive in the future with more adaptation from a variety of industries. Especially, the fashion and textile industries, with the support from a diversity of high-tech approaches for sustainable steps, can become promisingly competitive when being able to bring about a modern yet eco-friendly products to highly demanding customers, with the "clean" production process that also ensures the well-being of both internal and external stakeholders. Hence, it can be concluded that the integration of digital technologies can lead to reducing waste and bring sustainable practices in the fashion and textile manufacturing and supply chain practices.

References

Ahmad S, Miskon S, Alabdan R, Tlili I (2020) Towards sustainable textile and apparel industry: exploring the role of business intelligence systems in the era of Industry 4.0. Sustainability 12(7):2632. https://doi.org/10.3390/su12072632

Alcácer V, Cruz-Machado V (2019) Scanning the Industry 4.0: a literature review on technologies for manufacturing systems. Eng Sci Technol Int J 22(3):899–919. https://doi.org/10.1016/j.jestch.2019.01.006

Asif Z, Mandviwalla M (2005) Integrating the supply chain with RFID: a technical and business analysis. Commun Assoc Inform Syst 15(1):393–427

Atzori L, Iera A, Morabito G (2010) The internet of things: a survey. Comput Netw 54(15):2787–2805. https://doi.org/10.1016/j.comnet.2010.05.010

Atzori L, Iera A, Morabito G (2017) Understanding the Internet of Things: definition, potentials, and societal role of a fast evolving paradigm. Ad Hoc Netw 56:122–140. https://doi.org/10.1016/j.adhoc.2016.12.004

Aydos TF, Ferreira JC (2016) RFID-based system for lean manufacturing in the context of Internet of Things. In: 2016 IEEE international conference on automation science and engineering (CASE). IEEE, pp 1140–1145. https://doi.org/10.1109/COASE.2016.7743533

Azevedo S, Carvalho H, Cruz-Machado V (2012) RFID technology in the fashion supply chain: an exploratory analysis. Fashion supply chain management: industry and business analysis. https://doi.org/10.4018/978-1-60960-756-2-2ch017

Azouz N, Pierreval H (2019) Adaptive smart card-based pull control systems in context-aware manufacturing systems: training a neural network through multi-objective simulation optimization. Appl Soft Comput 75:46–57. https://doi.org/10.1016/j.asoc.2018.10.051

Baggchi U, Guiffrida A, O'Neill L, Zeng A, Hayya J (2007) The effect of RFID on inventory management and control. In: Jung H, Jeong B, Chen FF (eds) Trends in supply chain design and management. Springer series in advanced manufacturing. Springer, London

Bahrin MAK, Othman MF, Azli NHN, Talib MF (2016) Industry 4.0: a review on industrial automation and robotic. Jurnal Teknologi 78(6–13). https://doi.org/10.11113/jt.v78.9285

Bertola P, Teunissen J (2018) Fashion 4.0. Innovating fashion industry through digital transformation. Res J Text Appar 22(4):352–369. https://doi.org/10.1108/RJTA-03-2018-0023

Boothroyd G (2005) Assembly automation and product design, 2nd edn. CRC Press

Bozarth CC, Handfield RB, Weiss HJ (2008) Introduction to operations and supply chain management. Pearson Prentice Hall, Upper Saddle River, NJ

Brintrup A, Ranasinghe D, McFarlane D (2010) RFID opportunity analysis for leaner manufacturing. Int J Prod Res 48(9):2745–2764. https://doi.org/10.1080/00207540903156517

Chechi DP, Twinkle K, Kaur P (2005) The RFID technology and its applications: a review. Int J Electron Commun Instr Eng Res Dev (IJECIERD) 3(2):109–120. https://www.researchgate.net/publication/232575248_THE_RFID_TECHNOLOGY_AND_ITS_APPLICATIONS_A_REVIEW

Chen KM, Chen JC, Cox RA (2012) Real time facility performance monitoring system using RFID technology. Assem Autom 32(2):185–196. https://doi.org/10.1108/01445151211212334

Chen Y, Lee GM, Shu L, Crespi N (2016) Industrial internet of things-based collaborative sensing intelligence: framework and research challenges. Sensors (Switzerland) 16(2)

Chiarini A, Baccarani C, Mascherpa V (2018) Lean production, Toyota production system and Kaizen philosophy: a conceptual analysis from the perspective of Zen Buddhism. The TQM Journal 30(4):425–438. https://doi.org/10.1108/TQM-12-2017-0178

Chong AYL, Chan FT (2012) Structural equation modeling for multi-stage analysis on radio frequency identification (RFID) diffusion in the health care industry. Expert Syst Appl 39(10):8645–8654. https://doi.org/10.1016/j.eswa.2012.01.201

Chongwatpol J, Sharda R (2013) Achieving lean objectives through RFID: a simulation-based assessment. Decis Sci 44(2):239–266. https://doi.org/10.1111/deci.12007

Chowdhury B, Chowdhury MU, D'souza C (2008) Challenges relating to RFID implementation within the electronic supply chain management—a practical approach. Stud Comput Intell 149:49–59. https://doi.org/10.1007/978-3-540-70560-4_5

Christidis K, Devetsikiotis M (2016) Blockchains and smart contracts for the internet of things. IEEE Access 4:2292–2303. https://doi.org/10.1109/ACCESS.2016.2566339

Chu J (2015) Applications of RFID technology. IEEE Microwave Mag 16(6):64–65. https://doi.org/10.1109/MMM.2015.2419891

Conca J (2015) Making climate change fashionable—the garment industry takes on global warming. https://www.forbes.com/sites/jamesconca/2015/12/03/making-climate-change-fashionable-the-garment-industry-takes-on-global-warming/#44c047d479e4. Accessed 31 Aug 2021

Connors MM (1998) Teaming humans and automated systems in safely engineered environments. Life Support Biosph Sci 5(4):453–460

Demirkol İ, Al-Futaıh AA (2020) The relationship between Industry 4.0 and lean production: an empirical study on Bursa manufacturing industry. İşletme Araştırmaları Dergisi 12(2):1083–1097. https://doi.org/10.20491/isarder.2020.897

Denuwara N, Maijala J, Hakovirta M (2019a) Sustainability benefits of RFID technology in the apparel industry. Department of Forest Biomaterials, North Carolina State University, Raleigh, NC 27695, USA, 11, 1–14. https://doi.org/10.3390/su11226477

Denuwara N, Maijala J, Hakovirta M (2019b) Sustainability benefits of RFID technology in the apparel industry. Sustainability 11(22):6477. https://doi.org/10.3390/su11226477

Despeisse M, Ford S (2015) The role of additive manufacturing in improving resource efficiency and sustainability. In: IFIP international conference on advances in production management systems. Springer, Cham, pp 129–136

Dolgui A, Ivanov D, Potryasaev S, Sokolov B, Ivanova M, Werner F (2020) Blockchain-oriented dynamic modelling of smart contract design and execution in the supply chain. Int J Prod Res 58(7):2184–2199. https://doi.org/10.1080/00207543.2019.1627439

Domdouzisa K, Kumarb B, Anumbaa C (2007) Radio-frequency identification (RFID) applications: a brief introduction. Adv Eng Inform 21(4):350–355

Dong AH, Leung SYS (2009) A simulation-based replenishment model for the textile industry. Text Res J 79(13):1188–1201. https://doi.org/10.1177/0040517508096224

Dulchlnos J, Massaro P (2005) The time is right for labs to embrace the principles of industrial automation. Drug World Discov 2006:25–28

Ejsmont K, Gladysz B, Corti D, Castaño F, Mohammed WM, Martinez Lastra JL (2020) Towards 'Lean Industry 4.0'—current trends and future perspectives. Cogent Bus Manag 7(1):1781995. https://doi.org/10.1080/23311975.2020.1781995

Endsley MR, Kiris EO (1995) The out-of-the-loop performance problem and level of control in automation. Hum Factors 37(2):381–394. https://doi.org/10.1518/001872095779064555

Endsley MR (2018) Automation and situation awareness. In: Automation and human performance: theory and applications. CRC Press, pp 163–181

Fennani B, Hamam H, Dahmane AO (2011) RFID overview. In: ICM 2011 proceeding. IEEE, pp 1–5. https://doi.org/10.1109/ICM.2011.6177411

Fragapane G, Ivanov D, Peron M, Sgarbossa F, Strandhagen JO (2020) Increasing flexibility and productivity in Industry 4.0 production networks with autonomous mobile robots and smart intralogistics. Ann Oper Res 1–19

Frohm J, Lindström V, Winroth M, Stahre J (2008) Levels of automation in manufacturing. Ergonomia 30(3)

Fujimoto T (2012) Evolution of firms and industries. Evol Inst Econ Rev 9(1):1–10

Fullerton RR, Kennedy FA, Widener SK (2014) Lean manufacturing and firm performance: the incremental contribution of lean management accounting practices. J Oper Manag 32(7–8):414–428. https://doi.org/10.1016/j.jom.2014.09.002

Furman J, Seamans R (2019) AI and the economy. Innov Policy Econ 19(1):161–191. https://doi.org/10.1086/699936

Gaiardelli P, Resta B, Dotti S (2019) Exploring the role of human factors in lean management. Int J Lean Six Sigma 10(1):339–366. https://doi.org/10.1108/IJLSS-08-2017-0094

Gartner Research (2009) Gartner EXP worldwide survey of more than 1,500 CIOs shows IT spending to be flat in 2009. http://www.gartner.com/it/page.jsp?id=855612

Ghobakhloo M (2020) Industry 4.0, digitization, and opportunities for sustainability. J Clean Prod 252:119869. https://doi.org/10.1016/j.jclepro.2019.119869

Ghode DJ, Yadav V, Jain R, Soni G (2020) Blockchain adoption in the supply chain: an appraisal on challenges. J Manuf Technol Manag 42(1):42–62. https://doi.org/10.1108/JMTM-11-2019-0395

Gilchrist A (2016) Industry 4.0: the industrial internet of things. Apress

Görçün ÖF (2018) The rise of smart factories in the fourth industrial revolution and its impacts on the textile industry. Int J Mater Mech Manuf 6(2):136–141

Gräßler I, Yang X (2016) Interdisciplinary development of production systems using systems engineering. Proc CIRP 50:653–658. https://doi.org/10.1016/j.procir.2016.05.008

Guo ZX, Wong WK, Leung SYS, Li M (2011) Applications of artificial intelligence in the apparel industry: a review. Text Res J 81(18):1871–1892. https://doi.org/10.1177/0040517511411968

Guo ZX, Wong WK, Leung SYS, Fan JT, Chan SF (2008) A genetic-algorithm-based optimization model for solving the flexible assembly line balancing problem with work sharing and workstation revisiting. IEEE Trans Syst Man Cybern Part C (Appl Rev) 38(2):218–228. https://doi.org/10.1109/TSMCC.2007.913912

Haddud A (2011) A study of the relationship between radio frequency identification (RFID) technology and lean manufacturing. Master's Theses and Doctoral Dissertations. 329. https://commons.emich.edu/cgi/viewcontent.cgi?referer=https://scholar.google.com/&httpsredir=1&article=1329&context=theses

Hakeem A, Solyali D, Asmael M, Zeeshan Q (2020) Smart manufacturing for Industry 4.0 using radio frequency identification (RFID) technology. J. Kejuruter 32:31–38. https://doi.org/10.17576/jkukm-2020-32(1)-05

Hashem IAT, Yaqoob I, Anuar NB, Mokhtar S, Gani A, Khan SU (2015) The rise of big data on cloud computing: review and open research issues. Inf Syst 47:98–115

Hearnshaw EJ, Wilson MM (2013) A complex network approach to supply chain network theory. Int J Oper Prod Manag 33(4):442–469. https://doi.org/10.1108/01443571311307343

Heckmann R, Lengauer T (1998) Computing closely matching upper and lower bounds on textile nesting problems. Eur J Oper Res 108(3):473–489. https://doi.org/10.1016/S0377-2217(97)00049-0

Hedelind M, Jackson M (2011) How to improve the use of industrial robots in lean manufacturing systems. J Manuf Technol Manag 22(7):891–905. https://doi.org/10.1108/17410381111160951

Hoff KA, Bashir M (2015) Trust in automation: integrating empirical evidence on factors that influence trust. Human Factors 57(3):407–434. https://doi-org.ezproxy.lib.rmit.edu.au/10.1177/0018720814547570

Holweg M (2007) The genealogy of lean production. J Oper Manag 25(2):420–437

Hu ZH, Ding YS, Yu XK, Zhang WB, Yan Q (2009) A hybrid neural network and immune algorithm approach for fit garment design. Text Res J 79(14):1319–1330. https://doi.org/10.1177/0040517508100726

Ilie-Zudor, E., Kemeny, Z., Egri, P., & Monostori, L. (2006, November). The RFID technology and its current applications. In *Conference Proceedings of the Modern Information Technology in the Innovation Processes of the Industrial Enterprises (MITIP)* (pp. 29–36).

Inui S (1996) A combined system of computer aided design and genetic algorithm for apparel designing. Sen'i Gakkaishi 52(11):605–611. https://doi.org/10.2115/fiber.52.11_605

Jabłońska M, Stawska J, Czechowska DI (2020) Country-specific determinants of textile industry development in Poland: comparative analysis of the years 2007 and 2017. Autex Res J 20(2):186–193. https://doi.org/10.2478/aut-2019-0064

Kamble S, Gunasekaran A, Dhone NC (2020) Industry 4.0 and lean manufacturing practices for sustainable organisational performance in Indian manufacturing companies. Int J Prod Res 58(5):1319–1337. https://doi.org/10.1080/00207543.2019.1630772

Kampakaki E, Papahristou E (2020) Digital intelligence as prerequisite of artificial intelligence's integration in the clothing industry 4.0. In: SETN Workshops. pp 36–41

Karakasa Y, Suwa H, Ohta T (2007) Evaluating effects of RFID introduction based on CO_2 reduction. In: Proceedings of the 51st annual meeting of the ISSS - 2007, Tokyo, Japan, vol 51,no 2. https://journals.isss.org/index.php/proceedings51st/article/view/493

Kim HS, Cho SB (2000) Application of interactive genetic algorithm to fashion design. Eng Appl Artif Intell 13(6):635–644. https://doi.org/10.1016/S0952-1976(00)00045-2

Koehl L, Zeng X, Camargo M, Fonteix C, Delmotte F (2008) Analysis and identification of fashion oriented industrial products using fuzzy logic techniques. In: 2008 3rd international conference on intelligent system and knowledge engineering, vol 1. IEEE, pp 465–470. https://doi.org/10.1109/ISKE.2008.4730975

Kolberg D, Zühlke D (2015) Lean automation enabled by Industry 4.0 technologies. IFAC-PapersOnLine 48(3):1870–1875. https://doi.org/10.1016/j.ifacol.2015.06.359

Korpela K, Hallikas J, Dahlberg T (2017) Digital supply chain transformation toward blockchain integration. In: Proceedings of the 50th Hawaii international conference on system sciences. https://doi.org/10.24251/HICSS.2017.506

Landt J (2005) The History of RFID. IEEE Potentials 24(4):8–11. https://doi.org/10.1109/MP.2005.1549751

Laurence T (2019) Blockchain for dummies. Wiley

Lee JD, See KA (2004) Trust in automation: designing for appropriate reliance. Hum Factors 46(1):50–80. https://doi.org/10.1518/hfes.46.1.50_30392

Lee HL, Padmanabhan V, Whang S (1997) Information distortion in a supply chain: the bullwhip effect. Manag Sci 43(4):546–558

Lee M, Yun JJ, Pyka A, Won D, Kodama F, Schiuma G et al (2018) How to respond to the fourth industrial revolution, or the second information technology revolution? Dynamic new combinations between technology, market, and society through open innovation. J Open Innov Technol Market Complex 4(3):21. https://doi.org/10.3390/joitmc4030021

Leung SYS, Wong WK, Mok PY (2008) Multiple-objective genetic optimization of the spatial design for packing and distribution carton boxes. Comput Ind Eng 54(4):889–902. https://doi.org/10.1016/j.cie.2007.10.018

Liu JH, Gao WD, Wang HB, Jiang HX, Li ZX (2010) Development of Bobbin tracing system based on RFID technology. J Text Inst 101(10):925–930. https://doi.org/10.1080/00405000903028299

Liu S, Leat M, Moizer J, Megicks P, Kasturiratne D (2013b) A decision-focused knowledge management framework to support collaborative decision making for lean supply chain management. Int J Prod Res 51(7):2123–2137. https://doi.org/10.1080/00207543.2012.709646

Liu R, Koch A, Zell A (2013a) Mapping UHF RFID tags with a mobile robot using a 3D sensor model. In: Proceedings of the 2013a IEEE/RSJ international conference on intelligent robots and systems, Tokyo, Japan, pp 1589–1594

Lo WS, Hong TP, Jeng R (2008) A framework of E-SCM multi-agent systems in the fashion industry. Int J Prod Econ 114(2):594–614. https://doi.org/10.1016/j.ijpe.2007.09.010

Loebbecke C, Palmer J, Huyskens C (2006) RFID's potential in the fashion industry: a case analysis. In: BLED 2006 Proceedings 34

Lu Y (2017) Industry 4.0: a survey on technologies, applications and open research issues. J Ind Inf Integr 6:1–10. https://doi.org/10.1016/j.jii.2017.04.005

Lu H, Chen Y, Wang Y (2008) Knowledge-based customers' desirability acquisition of clothing. In: 2008 fourth international conference on natural computation, vol 7. IEEE, pp 94–98. https://doi.org/10.1109/ICNC.2008.909

Luo X, Hou W, Li Y, Wang Z (2007) A fuzzy neural network model for predicting clothing thermal comfort. Comput Math Appl 53(12):1840–1846. https://doi.org/10.1016/j.camwa.2006.10.035

Mani M, Wheeler D (1998) In search of pollution havens? Dirty industry in the world economy, 1960 to 1995. J Environ Dev 7(3):215–247

Marr B (2018) How fashion retailer H&M is betting on artificial intelligence and big data to regain profitability. www.forbes.com/sites/bernardmarr/2018/08/10/how-fashion-retailer-hm-is-betting-on-artificial-intelligence-and-big-data-to-regain-profitability/#4b5220095b00. Accessed 31 Aug 2021

Mayr A, Weigelt M, Kühl A, Grimm S, Erll A, Potzel M, Franke J (2018) Lean 4.0—a conceptual conjunction of lean management and Industry 4.0. Procedia Cirp 72:622–628. https://doi.org/10.1016/j.procir.2018.03.292

McKinsey & Company (2015) The internet of things: mapping the value beyond the hype. https://www.mckinsey.de/

McQuinn A, Castro D (2019) A policymaker's guide to blockchain. Information Technology and Innovation Foundation.

Medeiros CR, Costa JR, Fernandes CA (2008) RFID smart shelf with con-fined detection volume at UHF. IEEE Antennas Wirel Propag Lett 7:773–776. https://doi.org/10.1109/LAWP.2008.2008672

Mell P, Grance T (2011) The NIST definition of cloud computing. National Institute of Standards and Technology

Min H (2019) Blockchain technology for enhancing supply chain resilience. Bus Horiz 62(1):35–45. https://doi.org/10.1016/j.bushor.2018.08.012

Moon KL, Ngai EWT (2008) The adoption of RFID in fashion retailing: a business value-added framework. Ind Manag Data Syst 108(5):596–612. https://doi.org/10.1108/02635570810876732

Moroz R (2004) Understanding radio frequency identification (RFID). Understanding Radio

Muni DP, Pal NR, Das J (2006) Texture generation for fashion design using genetic programming. In: 2006 9th international conference on control, automation, robotics and vision. IEEE, pp 1–5. https://doi.org/10.1109/ICARCV.2006.345073

Muntean M, Muntean C (2012) Evaluating a business intelligence solution. Feasibility analysis based on Monte Carlo method. MPRA 47(2):85–102

Nakamba CC, Chan PW, Sharmina M (2017) How does social sustainability feature in studies of supply chain management? A review and research agenda. Supply Chain Manag Int J 22(6):522–541. https://doi.org/10.1108/SCM-12-2016-0436

Nayak R, Singh A, Padhye R, Wang L (2015) RFID in textile and clothing manufacturing: technology and challenges. Fash Text 2(1):1–16. https://doi.org/10.1186/s40691-015-0034-9

Nayak R, Akbari M, Far SM (2019) Recent sustainable trends in Vietnam's fashion supply chain. J Clean Prod 225:291–303

Nayak R, Padhye R (2017). Automation in garment manufacturing. Woodhead Publishing

Nayak R, Padhye R (2018a) Artificial intelligence and its application in the apparel industry. In: Automation in garment manufacturing. Elsevier.

Nayak R, Padhye R (2018b) Introduction to automation in garment manufacturing. In: Automation in garment manufacturing. Elsevier.

Nayak R, Nguyen LTV, Panwar T, George M, Ulhaq I (2020a) 'Sustainable supply chain management', Supply chain management and logistics in the global fashion sector: the sustainability challenge: 1

Nayak R, Nguyen LTV, Panwar T, George M, Ulhaq I (2020b) Sustainable supply chain management: challenges from a fashion perspective. Supply chain management and logistics in the global fashion sector. pp 3–32

Nayak R (2019) Radio frequency identification (RFID): technology and application in garment manufacturing and supply chain. CRC Press

Nayak R (2020) Supply chain management and logistics in the global fashion sector: the sustainability challenge. Routledge

Negara SD (2010) Fragmentation of electronics and textile industries from Indonesia to CLMV countries. In: Banomyong R, Ishida M (eds) A study on upgrading industrial structure of CLMV countries. ERIA Research Project Report, Jakarta, ERIA.

Ng KKR, Rajeshwari K (2017) Interactive clothes based on IOT using NFC and mobile application. In: 2017 IEEE 7th annual computing and communication workshop and conference (CCWC). IEEE, pp 1–4. https://doi.org/10.1109/CCWC.2017.7868339

Niaki MK, Torabi SA, Nonino F (2019) Why manufacturers adopt additive manufacturing technologies: the role of sustainability. J Clean Prod 222:381–392. https://doi.org/10.1016/j.jclepro.2019.03.019

Nicoletti B (2013) Lean and automate manufacturing and logistics. In: IFIP International Conference on Advances in Production Management Systems. Springer, Berlin, Heidelberg, pp 278–285

Nofer M, Gomber P, Hinz O, Schiereck D (2017) Blockchain. Bus Inf Syst Eng 59(3):183–187. https://doi.org/10.1007/s12599-017-0467-3

Noor A, Saeed MA, Ullah T, Uddin Z, Ullah Khan RMW (2021) A review of artificial intelligence applications in apparel industry. J Text Inst 1–10.https://doi.org/10.1080/00405000.2021.1880088

O'Neill JW, McGinley S (2014) Operations research from 1913 to 2013: the Ford assembly line to hospitality industry innovation. Int J Contemp Hosp Manag 26(5):663–678. https://doi.org/10.1108/IJCHM-08-2013-0331

Onnasch L, Wickens CD, Li H, Manzey D (2014) Human performance consequences of stages and levels of automation: an integrated meta-analysis. Hum Factors 56(3):476–488. https://doi.org/10.1177/0018720813501549

Ozel Y, Kayar M (2008) An application of neural network solution in the apparel industry for cutting time forecasting. In: 8th WSEAS international conference on simulation, modelling and optimization (SMO'08). Santander, Cantabria, Spain, pp 23–25

Papahristou E, Kyratsis P, Priniotakis G, Bilalis N (2017, October) The interconnected fashion industry—an integrated vision. In: IOP conference series: materials science and engineering, vol 254, no 17. IOP Publishing, p 172020

Parasuraman R, Riley V (1997) Humans and automation: use, misuse, disuse, abuse. Hum Factors 39(2):230–253. https://doi.org/10.1518/001872097778543886

Parasuraman R, Sheridan TB, Wickens CD (2000) A model for types and levels of human interaction with automation. IEEE Trans Syst Man Cybern Part A Syst Hum 30(3):286–297. https://doi.org/10.1109/3468.844354

Park J, Reveliotis SA, Bodner DA, McGinnis LF (2002) A distributed, event-driven control architecture for flexibly automated manufacturing systems. Int J Comput Integr Manuf 15(2):109–126. https://doi.org/10.1080/09511920110046083

Patti AL, Narsing A (2008) Lean and RFID: friends or foes? J Bus Econ Res (JBER) 6(2). https://doi.org/10.19030/jber.v6i2.2393

Pereira AC, Dinis-Carvalho J, Alves AC, Arezes P (2019) How Industry 4.0 can enhance lean practices. FME Trans 47(4):810–822. https://doi.org/10.5937/fmet1904810P

Premkumar G, Roberts M (1999) Adoption of new information technologies in rural small businesses. Omega 27(4):467–484. https://doi.org/10.1016/S0305-0483(98)00071-1

Prisecaru P (2016) Challenges of the fourth industrial revolution. Knowl Horizons Econ 8(1):57

Rafique MZ, Ab Rahman MN, Saibani N, Arsad N, Saadat W (2016) RFID impacts on barriers affecting lean manufacturing. Ind Manag Data Syst 116(8):1585–1616. https://doi.org/10.1108/IMDS-10-2015-0427

Rose ANM, Deros BM, Ab Rahman MN (2013) A study on lean manufacturing implementation in Malaysian automotive component industry. Int J Autom Mech Eng 8:1467–1476. https://doi.org/10.15282/ijame.8.2013.33.0121

Saberi S, Kouhizadeh M, Sarkis J, Shen L (2019) Blockchain technology and its relationships to sustainable supply chain management. Int J Prod Res 57(7):2117–2135. https://doi.org/10.1080/00207543.2018.1533261

Sanders A, Elangeswaran C, Wulfsberg JP (2016) Industry 4.0 implies lean manufacturing: research activities in Industry 4.0 function as enablers for lean manufacturing. J Indus Eng Manag (JIEM) 9(3):811–833. https://doi.org/10.3926/jiem.1940

Schindler R, Schmalbein N, Steltenkamp V, Cave J, Wens B, Anhalt A (2012) Smart trash study on RFID tags and the recycling industry. European Commission. 30-CE-0395435/00–31

Schmidt CG, Wagner SM (2019) Blockchain and supply chain relations: a transaction cost theory perspective. J Purch Supply Manag 25(4):100552. https://doi.org/10.1016/j.pursup.2019.100552

Scott AJ (2006) The changing global geography of low-technology, labor-intensive industry: clothing, footwear, and furniture. World Dev 34(9):1517–1536. https://doi.org/10.1016/j.worlddev.2006.01.003

Shishoo R (2012) The importance of innovation-driven textile research and development in the textile industry. In: The global textile and clothing industry. Woodhead Publishing, pp 55–76

Shuai W, Chongqi M, Hanming L (2010) Yarn quality tracking system based-on RFID. In: Proceedings of the 2010 international conference on computer and information application. IEEE, pp 3–5. https://ieeexplore.ieee.org/document/6141548

Su W, Ma L, Hu K, Zhang L (2009) A research on integrated application of RFID-based lean manufacturing. In: 2009 Chinese control and decision conference. IEEE, pp 5781–5784. https://doi.org/10.1109/CCDC.2009.5195231

Sun ZL, Choi TM, Au KF, Yu Y (2008) Sales forecasting using extreme learning machine with applications in fashion retailing. Decis Support Syst 46(1):411–419. https://doi.org/10.1016/j.dss.2008.07.009

Swan M (2015) Blockchain: blueprint for a new economy. O'Reilly Media, Inc.

Swanson T (2015) Consensus-as-a-service: a brief report on the emergence of permissioned, distributed ledger systems. Working paper

Tan CC, Sheng B, Li Q (2008) Secure and serverless RFID authentication and search protocols. IEEE Trans Wirel Commun 7(4):1400–1407. https://doi.org/10.1109/TWC.2008.061012

Thangarajoo Y, Smith A (2015) Lean thinking: an overview. Indus Eng Manag 4(2):2169–0316. https://doi.org/10.4172/2169-0316.1000159

Thoben KD, Wiesner S, Wuest T (2017) "Industrie 4.0" and smart manufacturing-a review of research issues and application examples. Int J Autom Technol 11(1):4–16. https://doi.org/10.20965/ijat.2017.p0004

Thomassey S, Happiette M (2007) A neural clustering and classification system for sales forecasting of new apparel items. Appl Soft Comput 7(4):1177–1187. https://doi.org/10.1016/j.asoc.2006.01.005

Thomassey S, Happiette M, Castelain JM (2005) A short and mean-term automatic forecasting system—application to textile logistics. Eur J Oper Res 161(1):275–284. https://doi.org/10.1016/j.ejor.2002.09.001

Tjahjono B, Ball P, Vitanov VI, Scorzafave C, Nogueira J, Calleja J et al (2010) Six Sigma: a literature review. International Journal of Lean Six Sigma 1(3):216–233. https://doi.org/10.1108/20401461011075017

Tokumaru M, Muranaka N (2008) An evolutionary fuzzy color emotion model for coloring support systems. In: 2008 IEEE international conference on fuzzy systems (IEEE world congress on computational intelligence). IEEE, pp 408–413. https://doi.org/10.1109/FUZZY.2008.4630400

Tong B, Yang DL, Pan X (2010) Research on effect of RFID on supply chain lead-time compression and coordination. Oper Res Manag Sci 19(5):52–58

Türkeş MC, Oncioiu I, Aslam HD, Marin-Pantelescu A, Topor DI, Căpușneanu S (2019) Drivers and barriers in using Industry 4.0: a perspective of SMEs in Romania. Processes 7(3):153. https://doi.org/10.3390/pr7030153

Ustundag A (2012) The value of RFID. Springer, London Limited

Ustundag A, Cevikcan E (2017) Industry 4.0: managing the digital transformation. Springer

Vaidya S, Ambad P, Bhosle S (2018) Industry 4.0—a glimpse. Proc Manuf 20:233–238. https://doi.org/10.1016/j.promfg.2018.02.034

Valamede LS, Akkari ACS, Cristina A (2020) Lean 4.0: a new holistic approach for the integration of lean manufacturing tools and digital technologies. Int J Math Eng Manag Sci 5(5):851–868. https://doi.org/10.33889/IJMEMS.2020.5.5.066

Varela L, Araújo A, Ávila P, Castro H, Putnik G (2019) Evaluation of the relation between lean manufacturing, Industry 4.0, and sustainability. Sustainability, 11(5):1439. https://doi.org/10.3390/su11051439

Vermesan O, Friess P (eds) (2014) Internet of things-from research and innovation to market deployment, vol 29. River Publishers, Aalborg

Wagner T, Herrmann C, Thiede S (2017) Industry 4.0 impacts on lean production systems. Procedia Cirp 63:125–131. https://doi.org/10.1016/j.procir.2017.02.041

Wang KS (2014) Intelligent and integrated RFID (II-RFID) system for improving traceability in manufacturing. Adv Manuf 2(2):106–120. https://doi.org/10.1007/s40436-014-0053-6

Wang S, Wan J, Zhang D, Li D, Zhang C (2016) Towards smart factory for Industry 4.0: a self-organized multi-agent system with big data based feedback and coordination. Comput Netw 101:158–168. https://doi.org/10.1016/j.comnet.2015.12.017

Wang Y, Han JH, Beynon-Davies P (2019) Understanding blockchain technology for future supply chains: a systematic literature review and research agenda. Supply Chain Manag Int J 24(1):62–84. https://doi.org/10.1108/SCM-03-2018-0148

Weinstein R (2005) RFID: a technical overview and its application to the enterprise. IT Prof 7(3):27–33. https://doi.org/10.1109/MITP.2005.69

Whitaker J, Mithas S, Krishnan MS (2007) A field study of RFID deployment and return expectations. Prod Oper Manag 16(5):599–612

Wolla S, Schug MC, Wood WC (2019) The economics of artificial intelligence and robotics. Soc Educ 83(2):84–88

Womack J, Jones D, Roos D (1990) The machine that changed the world. Rawson Associates

Wong W K, Guo ZX (2010) A hybrid intelligent model for medium-term sales forecasting in fashion retail supply chains using extreme learning machine and harmony search algorithm. Int J Prod Econ 128(2):614–624. https://doi.org/10.1016/j.ijpe.2010.07.008

Wüst K, Gervais A (2018) Do you need a blockchain? In: 2018 crypto valley conference on blockchain technology (CVCBT). IEEE, pp 45–54

Xu M, David JM, Kim SH (2018) The fourth industrial revolution: opportunities and challenges. Int J Financ Res 9(2):90–95. https://doi.org/10.5430/ijfr.v9n2p90

Zelbst PJ, Green KW, Sower VE, Bond PL (2019) The impact of RFID, IIoT, and blockchain technologies on supply chain transparency. J Manuf Technol Manag 31(3):441–457. https://doi.org/10.1108/JMTM-03-2019-0118

Zhou F, Duh HBL, Billinghurst M (2008) Trends in augmented reality tracking, interaction and display: a review of ten years of ISMAR. In: 2008 7th IEEE/ACM international symposium on mixed and augmented reality. IEEE, pp 193–202. https://doi.org/10.1109/ISMAR.2008.4637362. https://ieeexplore-ieee-org.ezproxy.lib.rmit.edu.au/abstract/document/7868339

Lean Manufacturing: Case Studies from Fashion and Textile Industries

Majo George, Nguyen Minh Ngoc, Le Khac Yen Nhi, Vuong Nguyen Dang Tung, Le Phan Thanh Truc, and Rajkishore Nayak

Abstract This chapter will focus on various case studies relating to the failure and success of various industries in the fashion and textile manufacturing sector. From the earlier chapters (Chaps. "Lean Manufacturing: Case Studies from Fashion and Textile Industries"–"Digital Technologies for Lean Manufacturing", it has been clearly mentioned that the implementation of various lean tools such as Kaizen, Muda, 5S and Kanban can help the traditional industries to improve their productivity and efficiency by reducing waste. This chapter will discuss the implementation of lean manufacturing by some of the fashion and textile industries and its impact on productivity and efficiency. The chapter has been divided into three sections for better flow and understanding. The first section will discuss some case studies focusing on the failure of some of the industries from Africa and India due to not implementing of lean manufacturing tools. The second section will focus on the case studies focusing on the success of some of the fashion and textile industries by adopting lean manufacturing tools. Finally, the third and last section will focus on the future directions and conclusions of this chapter.

Keywords Case studies · Fashion and textiles · Industry · Lean tools · Failure and success · Efficiency

1 Introduction

With the constant development of the fashion and textile industry, companies that insist on traditional methods and neglect the application of advantageous tools such as Lean manufacturing have to cope with inferior positions in their own growth as well as in the market. In specific, insisting on traditional methods expose them to competitive

---

M. George (✉) · N. M. Ngoc · L. K. Y. Nhi · V. N. D. Tung · L. P. T. Truc
School of Business and Management, RMIT Vietnam, Ho Chi Minh City, Vietnam
e-mail: majo.george@rmit.edu.vn

R. Nayak
School of Communication & Design, RMIT Vietnam, Ho Chi Minh City, Vietnam

© The Author(s), under exclusive license to Springer Nature Singapore Pte Ltd. 2022
R. Nayak (ed.), *Lean Supply Chain Management in Fashion and Textile Industry*,
Textile Science and Clothing Technology,
https://doi.org/10.1007/978-981-19-2108-7_11

disadvantages with lower production, lower level of customer satisfaction, an increase in the lead time, and so on. Furthermore, since the fashion and textile industries are running in steep competition and in a sustainable product-driven market, they need to be productive and efficient. Further, the diversity in customers' orders, differences in order size, and design requirements would expose the firms that choose not to adopt lean manufacturing, to a more vulnerable spot when they eventually cannot compete in both price and quantity because of their obsolescent and inflexible production tools. Therefore, these firms are unable to serve the customer demand and possibly face business distress.

In the light of the challenges in the fashion and textile industry, lean manufacturing and its tools can assist individual companies and the whole industry to overcome the aforementioned obstacles. In detail, lean practice, which includes tools such as Kanban, Kaizen, 5S system, and just-in-time (JIT), is capable of detecting nonvalue-added operations in the manufacturing process, which can be eliminated to improve the productivity and efficiency. The lean implementation can lead to a better production flow with value stream management, lower production lead time, better quality products and better inventory management. Hence, fashion and textile industries can achieve higher production and customer satisfaction, lower production costs, eliminate waste and unnecessary operation, thus resulting in higher profitability. The fundamental principles and concepts of lean manufacturing have been discussed in earlier chapters. This chapter will focus on various case studies from fashion and textile industries around the globe that are implementing the lean manufacturing concepts. This chapter is designed to understand the practical aspects of lean manufacturing and how to be successful in the implementation of lean manufacturing.

2 Case Studies

2.1 Case Study 1. The Adinkra Textile Sub-Sector in Ghana's Textile Industry

Since the 1980s when trade liberalization happened, the traditional fashion and textile manufacturing sector in many countries has been confronted with several challenges (Aboagyewaa-Ntiri and Mintah 2016). Especially, the conventional sub-sectors in the textile industry are recognized as the most vulnerable to those challenges which have the highlight being the onslaught of the rising digital age, hence changing the way textile production processes. The case study on the withering of Adinkra—a small-scaled textiles cloth printing sub-sector could be examined to demonstrate the bigger picture where there are knotty problems that the traditional fashion and textile supply chain is facing.

Quartey and Abor (2011) discovered in their research that there has been a consistent, continuing decline in the number of both large and small-medium textile and garment manufacturing companies in Ghana since the mid-1970s. This decline is seen as a result of both unfavorable government policies and changes in the market forces. From a wider point of view, by 2005, there were only four major companies working in the textile field. They include Ghana Textile Manufacturing Company, Akosombo Textile Limited, Ghana Textile Product, and Printex (Quartey and Abor 2011). This also dragged a severe downslide in the number of employees (roughly 8.4 times decreasing in less than three decades), as well as yards of fabric produced.

Yet, with the purpose to concentrate on the traditional textile industry, the case study chooses to centre on the Adinkra hand printing textile cloth sub-sector that belongs to the larger textile industry. Indeed, Adinkra has long attached itself to Ghanaian history and culture and Adinkra symbols are images used to visualize Ghana's aspects of aesthetics, history, identity, and daily life of the Ghanaian people (Quarcoo 1972). It led to the advent of the Adinkra textile cloth production sub-sector where Adinkra symbols are imprinted on fabric, producing symbolic textile cloth. With its deep meaning and cultural illustration, Adinkra textile cloth is considered a valuable and prestigious royal craft (Kuwornu-Adjaottor et al. 2016).

A field study conducted by Aboagyewaa-Ntiri and Mintah (2016) showed a comprehensive investigation of the reasons behind the drastic downward trend in the Adinkra cloth printing and producing sub-sector. According to their findings, there are four main reasons contributing to the gloomy performance of the sector which include (1) Noncompetitive quality of the Adinkra dye, color, and cloth; (2) Lack of financial support for Adinkra SMEs (Small and medium-sized enterprises); (3) Challenges on the supply chain; and (4) Outdated production technology.

Nevertheless, this case will focus on the third and fourth challenges to be consistent with the topic of the chapter. Firstly, the lack of collaboration between key players in the supply chain of producing Adinkra cloth appears to be the top reason leading to the unproductivity of the products. In reality, the practice to fulfil customer satisfaction in this sub-sector has failed due to the interrupted joint effort among the stakeholders such as fabric suppliers, dye suppliers, retailers, governmental agencies, and manufacturers. Thus, due to that absence of collaboration, the information flow is disrupted, and the producers of textile cloths only merely based on their assumptions as a result of missing out on customer insight, hence failing to meet the needs of clients (up to 88.5% of user respondents in Aboagyewaa-Ntiri and Mintah's (2016) field study expressed their dissatisfaction with the end products from artisans) and contributing to the downturn of the whole sub-sector. The inability to capture customers' expectations also led to failure in determining the right price of Adinkra textile cloth, making it too high to attract customers. When compared to other manufactured textile cloths, Adinkra clearly has no competitive advantage.

Additionally, the obsolescence of technology and machinery usage is another main reason causing the decline in productivity. Aboagyewaa-Ntiri and Mintah (2016) stated in their research that the technology used in producing Adinkra textile cloths was "laborious and time-consuming" and that since 1927, the technology had not been improved. While the process of finding and disposing of raw materials is of utmost difficulty and sophisticated, no adapted technology was adopted to facilitate the process and the outcome quality. Consequently, the artisans have to acquire the dyes from established manufacturing firms, but the dependence upon them also creates certain challenges related to price and quality.

Until 2014, vast productions of Adinkra textile were abandoned (Aboagyewaa-Ntiri and Mintah 2016), alerting the real state of affairs that Adinkra textile cloth production is on a sharp decline. Thereby, Adinkra is considered as a "dying art" both inside and outside of the country (Kquofi et al. 2013). The failure can be attributed to the use of traditional technologies and not being willing to adopt the lean manufacturing concepts in the manufacturing industries.

2.2 Case Study 2. Ethiopian Cotton Spinning Industry

In the light of the Ethiopian textile industry, it is considered as one of the oldest manufacturing sectors in the country and undoubtedly contributes a great deal to the national economy in regard to both industrial production, employment provision, and profit generation (Shiferaw 2017). In spite of the Ethiopian government's effort in providing favorable policies for the industry's production and export, the national textile industry also witnessed stagnation. Specifically, according to the Ethiopian Textile Industry Development Institute, the country's textile exports from 2009 to 2016 were falling apart from the plan, only covering around 50–70% of the plan (Syduzzaman et al. 2016). From a wider point of view, the global competitiveness in this textile industry in general and in the spinning industry, in particular, has arisen as an indispensable factor affecting the performance of the industry. The need for textile companies and spinning manufacturers, as asserted by Ahmmed and Ayele (2020), is to create higher product quality in addition to improving reputation for reliability, manufacturing capability, cost-effectiveness, and timely delivery in response to both local and international competition.

Nevertheless, those companies working in the cotton spinning industry in Ethiopia are still facing problems in reaching the optimal point of production capability. In other words, Ahmmed and Ayele (2020) clearly stated in their research the shortcomings of the traditional total quality management (TQM) methods that have been applied for a long time and certainly are no longer the best approach in such a current time. To be specific, most of the Ethiopian textile industries including cotton spinning are now enduring problems of quality assurance, which is believed to derive from high variations in the production process—the process that is contrary to lean manufacturing. The consequences happen all over the supply chain's different aspects.

On the supply side, there are insufficient raw materials. The operation side suffers from the poor performance of product manufacturing, low productivity, and incapability to utilize the production resources. Lastly, on the demand side, customer dissatisfaction is the clearest and most direct evidence showing the lack of an efficient production chain. As a result, more time is needed for the whole production process and more waste is generated during the mentioned process has constrained the industry from developing and gaining a competitive edge as well as sustainability in the export market as it used to have in the previous time. In the end, these problems led to low-quality products, high cost, and put most of the Ethiopian textile and spinning companies in the situation of being in inferior positions when it comes to competitiveness and profitability.

In their research, Ahmmed and Ayele (2020) find that spinning manufacturers in Ethiopia only perform using the traditional quality control system. The statement is finalized in light of the lack of new and more efficient quality control systems such as the Six Sigma method or other advanced technology, programs, and tools. Therefore, the authors choose to take those approaches as recommendations for the current Ethiopian spinning and textile industries. If continuing to adopt those traditional quality management systems, it is expected that those systems will no longer be suitable for the emerging challenges of the production detriments and the high cost associated with production waste.

2.3 Case Study 3. Traditional Hand-Loom Sector of Varanasi, Uttar Pradesh in India

Handloom weaving has always been a crucial, traditional industry of India that not only creates millions of employments for the country but also contributes to the economic as well as cultural aspects of Indian people (Tanusree 2015). Indeed, it is considered to play an essential role in the national economy, being the second largest sector that acquires a huge amount of labor recruitment in comparison to the agriculture industry and accounts for roughly 14% of total cloth produced by the country. Yet, when it comes to the light of industrialization around the world and in the country, the handloom industry has confronted a significant downturn (Tanusree 2015). This sector appears to be a brilliant case study when we take into consideration how much the rise of industrialization can affect such a purely traditional industry of a nation.

Before the industrial revolution, the handloom appeared to be a strong economic sector with the advantages of having diverse material resources, cheap labor, growing domestic demands, and relatively simple technology application (Niranjana 2004). However, since the 19th century with the powerful penetration of the industrial and technological waves, the sector has been placed in jeopardy and the risk of being suppressed in both domestic and international markets has quickly become visible when the competition started to be extremely intense due to globalization, market liberalization (Ganguly-Scrase 2003) and the rise of mechanized manufacturers (Roy 2002). Thus, it can be said that the handloom industry, represented by the handloom weavers, emerges as an illustration of deindustrialization (Tanusree 2015). Noticeably, the advent of more and more advanced technology, especially the power-loom machine (mostly from foreign manufacturers) that attracted more preference from the customers due to its competitiveness in the price and also consistent quality (Goswami 1990). Consequently, power-loom is getting predominant in the market, jumping from 37 to 68% in one and a half decades (1980–1995) to cover the overall Indian textile products (Mizuno 1996; Srinivasulu 1994).

The traditional way of management and resistance to adapt to new changes, as well as new technologies in the current Industry 4.0 era, appears to be the main causes for this decline of the handloom industry in India. In fact, keeping the outdated production system with a capitalist management style not only worsens the weaknesses of the industry's nature but also prevents the drivers to gain a better adjustment in the production system. More importantly, this seems to be the root cause of other problems that contribute to the overall disadvantage of the handloom sector, namely labor problems, a detrimental wage system, and a lack of connection in the supply chain.

In short, the traditional fashion and textile industry (i.e., the handloom sector) has been confronting numerous challenges when it comes to the fragile competitiveness of the industry is holding. Noticeably, there are two main reasons causing such a downward trend. Firstly, the companies or manufacturers that have long been following the traditional practices of textile production are unable to compete with the rising in both quantity and quality of their rivals. This is due to the waves of industrialization and technological revolution; thus, when globalization and market liberalization occurred, the traditional manufacturers were quickly put into a disadvantageous position.

The second reason derives from the stagnation of the sector in adopting new technologies as well as applying more advanced quality management philosophy and methodology. Their slow response and resistance to the stream of technological development have contributed to the downfall in all aspects of manufacturing. As a result, not only the productivity and quality decline for those companies that pursue the traditional manufacturing process, but they also encounter a huge risk of being eliminated from the market due to low customer satisfaction, high cost, and intricate production mechanism.

3 The Application of the 5S System

3.1 Case Study 1. Sample Section of the Apparel Industry in Bangladesh

The sample section plays a crucial role in the efficiency and productivity of every enterprise and factory in the apparel industry of Bangladesh. This section creates impressive communication and maintains a firm relationship with customers by proving quality and quick sample submissions (Khan and Islam 2013). The idea and design of the product are necessary to be displayed with the proper sample, thus avoiding sample rejection, order cancellation, and hesitation in proceeding with future orders for any vendors. The sample section is said to be immensely workload-loaded. Limited time and high expectations are given from the customers to dispatch a sample with adequate quality. Therefore, discipline and efficient management are vital.

However, there have been many issues occurring within the management of this section. According to Khan and Islam (2013), these problems can derive from insufficient file management, weak inventory management, sanitation standard in the working environment, and incapable workers with deficient skill and knowledge. Thus, it can lead to several impacts such as late submission, sample rejection, waste of time, demotivation in the working environment, or quality reduction. As a result, it brings about dissatisfaction from customers, work procrastination, and a higher workload.

Accordingly, the 5S methodology has been utilized to improve and optimize the current operation of the sample section in Bangladesh. The current working structure is adjusted with the removal of non-added value items and the restructuring of the working space.

- *Seiri—sort*
  In this phase, unskilled workers are pointed out to have better training as well as creating a team with experienced employees. Each team is divided into specific sub-section based on their expertise with a sufficient number of workers and distinct inventory storage. Unnecessary machines, materials, and documents are removed and replaced with the current order with proper labels.
- *Seiton—set in order*
  In this phase, workers are provided with the systematic arrangement of machinery and items to optimize working time and motions. Wooden or Partex board wall cabinets are used within the department to separate the operation. The lighting system and ventilation are given in suitable partitions.
- *Seiso—shine*
  Proper cleaning and arrangement are proceeded after every shift to optimize time management in the long run. The cleaning team is well-managed with proper items and separated into specific tasks.
- *Seiketsu—standardize*
  In Seiketsu, an efficient workflow is generated. The routine of machinery and utility maintenance, and cleaning basis are provided. Each team will have a captain to take charge of storing inventory, sample submission quality, deadline, and delivering instructions to their workers.
- *Shitsuke—sustain*
  In the final phase of the 5S philosophy, the proper procedure is established to sustain the habit of the 5S methodology in the first four steps. Regular training and practice are processed to enhance workers' skills, motivation, and discipline within the section. Moreover, assessing the implementation in the previous phases can help make appropriate adjustments that suit the current status of the enterprise.

After implementing the 5S methodology, sample sections in Bangladesh apparel firms could improve the inventory and file management with a better arrangement of important machinery, material, and the elimination of non-added value items. Besides, sanitation was maintained with a proper cleanliness routine. Moreover, personnel's skills, knowledge, and discipline were improved with numerous training and practice from the Japanese philosophy. Therefore, production order has been sustainable along with better satisfaction from customers.

3.2 Case Study 2. The Improvement of the Italian Luxury Fashion Brand

The name of the company is kept anonymous as Italian Fashion Company (IFC) due to confidentiality. With the primary operation in fur coats production, since the 1950s, IFC has been remarkably growing to become a multi-national brand from a quasi-artisan family company (Carmignani and Zammori 2015). Besides its main market in Europe, Eastern countries (Russia and China), North America, and Mexico are contributing a large portion of IFC's total sales volume. The company produces bags, clothing, accessories, and so on, in limited quantities, and all products are made by artisans. The whole operation, including design and production, is processed in Italy with 12 manufacturing factories located near the headquarters. IFC also has a global network of warehouses in Italy, Asia, the USA, and Europe.

Nevertheless, its business has been suffering from a downside lately as the contraction in margins, inventory shrinkage, higher transportation costs, and varied market trends. To be more specific, the out-of-date machinery and materials, as well as the increase in unsold products, were the main incidents.

Therefore, lean projects have been developed to improve the productivity and efficiency of the business. IFC's objective was to cut its operation cost, inventory levels, lead time and increase flexibility. Therefore, 5S techniques have been used as a "visual control" tool to identify wastes in operation and eliminate it to achieve the primary improvement. Numerous minor problems, adequate documentation, and consistent procedures have been solved and improved. Accordingly, different sections within the manufacturing facilities are separated with different colored marks to optimize the time and motion of workers. Moreover, the sanitation in the workplace, specifically the equipment, is improved, thus motivating employees with better working conditions. Standard chambers and KLT containers are utilized to separate materials, semi-finished products, final products, and spare parts. Dedicated allocation for each stocking criteria with smart tags can minimize the chance of damage during picking, and shipping as well as optimally identifying stock-keeping units (SKUs) and minute accessories. Besides, supervised training, workshops, and single-point lessons are provided to optimize the utilization rate of workers since the under-skilled workforce is minimized.

Thanks to the application of 5S techniques, the efficiency of the manufacturing process has been primarily improved. Therefore, in the combination of other lean methodologies and techniques, IFC can improve its profitability, optimal cost management, and inventory management. Nevertheless, the luxury fashion sector is different from other sectors due to demand seasonality and variability, product customization, and craftsmanship. Therefore, the application of this practice is considered to be rarely possible.

4 The Implementation of Value Strem Mapping

4.1 Case Study 1. Higher Customer Satisfaction as an Achievement in ABC Clothing Ltd. In India

Hitherto, the textile industry in India has grown significantly with the increasing production volume. According to Kumar and Thavaraj (2015), due to the dissimilar demand, the demanded quantity and design style varies for each customer, thus requiring higher productivity in each firm to serve the demand punctually. Another difficulty that incurs is the high Work-in-process (WIP) in the conventional batch production. This is the manufacturing inventory that is especially wasteful and should be eliminated to avoid obstructing the production rate. Altogether, these impediments obstruct the company from achieving higher customer satisfaction, hence a potential customer risk that would severely impact the business operation.

In the case of an Indian garment manufacturing company—ABC Clothing Ltd. with a 100% export-oriented business model, it has to cope with numerous different orders from the UK, USA, and Germany, indicating relevant issues from the industry. With the goal of better productivity, lean manufacturing is a recommended tool that ABC tends to implement in its business. This method aims to detect the major source of operational underperformance and offers proper tools to address mentioned problems.

Starting with the value stream mapping (VSM), it can reduce wastes in inventory and defects by visualizing the production process to detect different types of wastes and suggest appropriate methods to mitigate or eliminate them. The application of VSM goes through the process of selecting a product line as an improvement target and the comparison between current and future state maps to manage the implementation process.

In this case, the chosen product was men's innerwear with a focus on the sewing sector (Kumar and Thavaraj 2015). From the current VSM, the cycle time is calculated to set the foundation for improvement in the future VSM. Specifically, before each process, specific statistics of the labor force, inventory level, cycle, and change over time are collected under the approach of Rother and Shook (1999) to estimate the cycle time and highlight Muda (process waste) in the current operation. As a result, various things are highlighted, such as (1) large inventories, (2) the difference between the lead-time and value-added time of the production process, (3) the on-schedule operating process, and (4) process ratio at an extremely low level. Accordingly, there are some recommended lean tools that are incorporated while planning for the future VSM, including single-piece flow, Kaizen, and 5 s methodology, to minimize or eliminate stated problems. The cellular layout is also under recommendation to achieve for the operations the least WIP level, cost-efficiency, labor force's skill advancement, and production lead-time minimization. Furthermore, after the non value-added components in the production progress are detected, some of them were extracted from the current VSM, while the others were integrated together to lower cycle time and operator association.

Eventually, this whole lean implementation yields the achievement of the optimal goal. In detail, the process ratio has increased significantly from the minimization of unnecessary process delay time. Simultaneously, the cycle time, WIP level, and a number of operations have been successfully reduced, which lowers production costs in short-run orders, and cuts down operator allocation and space within the existing VSM. Hence, ABC Clothing Ltd. is capable of being more responsive to customer demand whilst having high-quality manufacturing in the most efficient and economical way with waste reduction or elimination in human resources, inventory, time to market, and production space.

5 Application of Muda, Six Sigma and Statistical Process Control

5.1 Case Study 1. Applying Muda Concept to Natural Fiber Clothing Manufacturing

As more customers go against the cost of waste, businesses are searching for grand tools and techniques for the identification of the Muda in their operations (Rahani and Ashraf 2012). The Muda refers to the wastes (or non-value-added activities) that need to be eliminated in the business process for an organization's sustainable growth. The case study of a manufacturing clothing company specialized in natural fibre applying the Muda concept from the Journal of Textile Institute will be presented in this section.

To identify the Muda, a value stream map helping people visualize the whole production process from the method creation stage to the delivery to end consumers stage is constructed (Rother and Shook 1998). The research process is split into two categories. Firstly, the effects of lean manufacturing will be identified; and secondly, a suitable mapping tool's value flow to implement predetermined lean manufacturing will be built to critically find out the Mudas and in each step with an appropriate solution (Table 1). Specifically, through the company process (Table 2), seven Mudas are correspondingly identified (Table 3).

A value stream map has been illustrated to identify the Mudas in the whole production process (Rother and Shook, 1998). With this done, various wastes hindering the constant operation of production will be minimized (Teichgräber and de Bucourt, 2012). Afterwards, an analytical network process (ANP) is applied and ranking the Mudas respectively based on cost, quality, and delivery via Super Decision software. Hence, most Mudas are found in the sewing process, specifically, in the overload of

Table 1 Mapping tools to implement predetermined lean manufacturing to find out the Mudas in manufacturing

Tools	Research stages
Study	Studying the literature of leaning and the methods of value stream mapping dearing
Observation	Drawing the value stream map
Interview and consulting with the focus group	Identifying the Mudas in the value stream map
ANP technique	Prioritising the Mudas
Casual inferences	Presenting general solutions

Table 2 Production process of this company

Stage	Activity	Description
1	Cutting	Prepare feeding for the sewing stage: deliver the supplied fabric to the cutting shop The fabric can be cut in traditional way (hand cutting) or industrial way (utilizing a CNC machine)
2	Sewing	The sewing process takes place at three different shops, with ready-to-be-used materials such as wards and knots
3	Enzyme	Clothes are dyed at an enzyme shop and remain in the dyeing pots to achieve the intended colors in a certain period
4	Quality control and ironing	The counted returned clothes from the enzyme stage are highly monitored and will be ironed if no enzyme is defected
5	Barcoding, folding, and transporting to the storehouse	The finished clothes are barcoded, folded, and transported to the storehouse

transported parcels to each location. With this result, the value stream map is developed to mitigate the waste by increasing transportation frequency and decreasing transported volume, which considerably minimizes the process and waiting time while fulfilling the final expense.

5.2 Case Study 2. Applying Lean Six Sigma to a Spinning Mill In India

The revolution of quality standards in the textiles and clothing industry has pushed these involving industries to implement modern manufacturing models such as lean tools in their execution. This case study from the Journal of Textile Institute depicts

Table 3 Seven Mudas identified in the production process

Waiting	The waiting time in the sewing stage, including delivering activities and other processes in the cutting and enzyme stage
Motion	Improper design of workplaces in each station leading to operators' uneconomic movements and motions
Defect	Parts of the defected shirts are reprocessed
Over-processing	Defective products are rechecked Redo the enzyme process for the returned products
Overproduction	This Muda stem from the malpractice in forecasting, inventories, and final product planning, especially during the sale season
Transportation	The flow of materials and work-in-process are interrupted due to illogical equipment arrangement The shop location and route planning is not optimized
Inventory	The full utilization of push production result in accumulated WIP inventories

an identical implementation of the Lean Six Sigma and ISO 9001:2008 standard-based Quality Management System (QMS) in a textile mill in India and supports the industry to save up to 2 million INR annually.

As from the name, Lean Six Sigma is the combination of lean manufacturing and Six Sigma (Karthi et al. 2011). The rising concern towards quality, cost, and delivery warrants scientific systems that require the textile industry to apply ISO 9001 for a sustained growth rate (Calisir 2007). From that, the integrated model of Lean Six Sigma and ISO 9001-L6QMS-2008 was created with DMAIC methodology, belt-based training infrastructure, and Lean Six Sigma in ISO 9001:2008 standard-based QMS (Table 4) to mitigate global challenges while boosting the quality degree of the production process. Based on that, twenty hypothetical steps have been built to fully adopt the model in the spinning mill in Andhra Pradesh, India (Table 5).

Table 4 The application of Lean Six Sigma QMS to the production process

1	Scope	Lean Six Sigma projects, belt personnel, belt training and applications
2	Normative references	Devane (2004), Harry and Schroeder (2005), George (2003), George et al. (2005) and Taghizadegan (2006)
3	Terms and definition	Lean and Six Sigma terms and definitions
4	Quality management system	General Lean Six Sigma requirements
5	Management responsibility	Responsibilities of champion, master black belt, black belts and green belts
6	Resource management	Lean Six Sigma belt based training infrastructure requirements
7	Product realisation	Define phase requirements
8	Measurement, analysis and Improvement	MAIC phases' requirements

Table 5 The implementation of L6QMS-2008 model divided in 20 steps

Step	Activity
1	Coordinator's meeting
2	Conduction of cognizance programme
3	Modalities discussion meeting
4	Head of departments meeting
5	Management representative interview
6	Other elements' determination
7	Meeting with the technical manager
8	Other elements' preparation
9	Elements submission to top management
10	Quality manual amendments
11	Actions planning
12	Belt-based training infrastructure enhancement
13	Instruction classes provision
14	L6QMS-2008 project begun
15	L6QMS-2008 project in execution
16	L6QMS-2008 project in review
17	Digital related documents of the project
18	Quantitative performance of the project
19	Achievements compliment to the top management
20	Project reviewed by the top management

In the execution, two projects aiming to reduce sliver waste generation and ring frame tenter training lead time are carried out respectively to confirm the credibility of the model. Particularly, the journey of this case study is illustrated below.

Step 1. Coordinator's meeting

Structuring the board of executives began the journey of all, in which the Project Coordinator, Technical Manager, and Coordinator were nominated to implement the L6QMS-2008. The coordinator will be assisted by the other two to be directly in charge of the process.

Step 2. Conduction of cognizance programme

Specific reasons for the integration of Lean Six Sigma and ISO 9001:2008 is pinpointed in this programme.

Step 3. Modalities discussion meeting

Senior executives will discuss the most optimized strategy for the implementation. After that, a pilot project is decided to be executed before continuing with the whole project.

Step 4. Head of Departments meeting

There will be a meeting between heads of departments to carry out the brainstorming work. This aims to identify the detailed tasks involved in this project to define ways for a profitable execution.

Step 5. Management Representative interview

An interview between the coordinator and management representative will be conducted to collect the necessary information for the implementation.

Step 6. Other elements' determination

The comparison of the present ISO 9001:2008 standard and L6QMS-2008 model has been made by the coordinator to determine additional QMS elements to be appended in the existing IMS for L6QMS-2008 implementation.

Step 7. Meeting with the Technical Manager

As an outcome of the meeting, new elements of the L6QMS-2008 model were appended to the IMS of Unit A.

Step 8. Other elements' preparation

The second level documentation procedures including other quality and elements to implement the L6QMS-2008 model were created. The Vice President of Technical Affairs (VP-Tech) was allocated as the Champion and the Deputy General Manager (DGM) as the Master Black Belt.

Step 9. Elements submission to Top Management

The L6QMS-2008 model's quality manual was submitted to the VP-Tech from the Coordinator and Technical Manager.

Step 10. Quality Manual amendments

From the received recommendations by the DGM, the Coordinator revised the quality manual. This document was approved by the VP-Tech.

Step 11. Actions planning

As depicted in Fig. 1, the "L6QMS 2008 Unit A" folder was logically organized with sub-folders for instruction classes, standard and manual, procedures and records (project and training).

Step 12. Belt-based training infrastructure enhancement

Production Manager and HR Manager were assigned as the Black Belts. Production Supervisors and one HR Assistant, who took the responsibility of each project respectively, were chosen as the Green Belts. These employers give detailed descriptions of the methods with appropriate tools and techniques after the completion of every phase, with recorded meeting minutes.

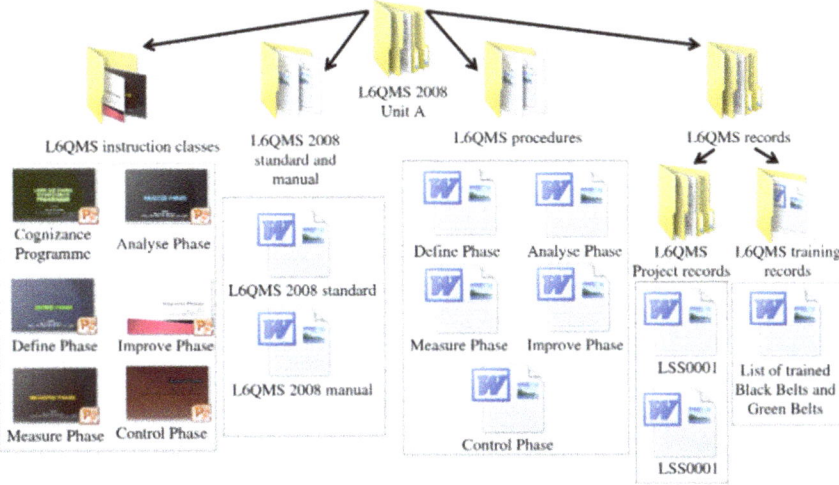

Fig. 1 The "L6QMS 2008 Unit A" folder

Step 13. Instruction classes provision

For the Define phase, instruction class about Project Charter, SIPOC diagram, Responsibility-Accountability-Consult-Inform (RACI) chart and cost-benefit analysis was conducted.

For the Analysis phase, Value Stream Map, CTQ tree diagram, Data Collection and Analysis Plan, Measurement System Analysis and Gauge Repeatability and Reproducibility analysis were taught in the Measure Phase session. To balance between each class, the Improve phase and Control phase were trained in another session.

Both training sessions are carried out by the coordinator.

Step 14. L6QMS-2008 project begun

The projects are started by the trainees and supervised by the Coordinator and Technical Manager. Besides, the whole project is monitored by the Master Black Belt and the MR of Unit A every day.

Step 15. L6QMS-2008 project in execution

The two projects were consecutively denoted as LSS0001 and LSS0002, each of which execution steps were highly monitored. For the LSS0001 project, it was aimed to reduce 0.35% of waste (from 1.85 to 1.5%). For the other, it was aimed to reduce the training lead time by 3 months (from 9 to 6 months).

Step 16. L6QMS-2008 project in review

The coordinator continually scrutinized the projects in terms of their high-level process map (Table 6), cause and effect (Fig. 2), rectification actions and results.

Table 6 High-level process maps of both projects

Silver waste	
1	Variables in carding
2	Variables in SL/RL
3	Variables in the Comber
4	Variables in the Drawing
5	Variables in Simplex
Training lead time	
1	Induction program
2	Disciplinary training
3	Safety training
4	Off the job training
5	Piecing training
6	Gaiting training
7	Doffing training

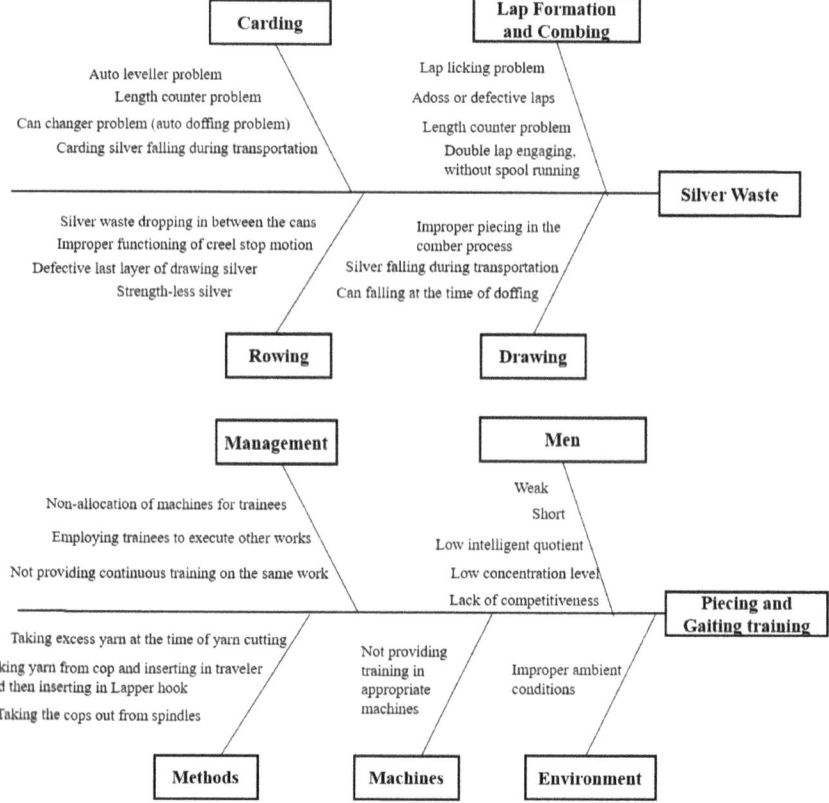

Fig. 2 Cause and effect of both projects

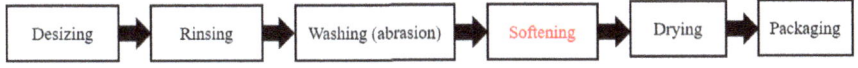

Fig. 3 Denim washing process

Step 17. Digital related documents of the project

All documents from the two projects are electronically recorded in the "L6RP_Lean Six Sigma record for projects" subfolder, which is illustrated in Fig. 3.

Step 18. Quantitative performance of the project

Through applying tools and techniques and organizing training, the implementation of two projects achieved one million INR for each project, resulting in a total volume of two million INR annually This proves an outstanding improvement of a textile mill.

Step 19. Achievements compliment to the Top Management

The achievements are proudly presented to the VP, HR, and Administration with an impressive reduction in lead time in Unit A. The L6QMS-2008 project is then requested to be implemented in other spinning units.

Step 20. The project reviewed by the Top Management

In appreciation of the coordinator and belt members, the top management meeting was conducted at the central office to review the accomplishment.

In conclusion, the L6QMS-2008 project has been successfully completed for a spinning mill in India. The combination of Lean and Six Sigma has never been tried in the textile unit, however, it has then proved to be successful at a full cooperation level without any difficulties. The twenty hypothetical steps have successfully led the employees through an understandable and practicable process to achieve an annual cost reduction of two million INR without any significant problems in the given timeframe. This means that no modifications are necessary for the project to be largely launched for the whole business.

5.3 Case Study 3. Applying Statistical Process Control (SPC) to a Denim Garment in Turkey

This case study pays particular attention to the application of SPC to determine washing defects in the industrial washing processes of the denim garment industry.

Denim garments are considered the most important part of the textile industry, since this type of fabric, despite being the oldest one in the world, can always remain popular with a wide range of consumers thanks to its family-wearable characteristic. Throughout the production process, the most important step lies in the washing operation, regarded as the final and a base step to complete any denim product by strengthening a tensile in the fabric and removing hard attitude in the raw fabric. The denim washing technology creates new color tones that suit customer preferences. Specifically, the denim is dyed with indigo dye, which will be discolored after being washed. The SPC method is applied particularly in this process to determine the most common defect and its impact on the overall product quality. The denim washing process will be evaluated in a general framework, utilizing statistical process control methods, namely Pareto analysis, cause and effect diagram, and P control graphs.

As illustrated in Fig. 3, the denim washing process includes six stages. The first stage—desizing refers to using amylase enzymes to break down long molecular (water-insoluble) starch chains into shorter (water-soluble) molecules that can be washed away more quickly. After that, the garment is made to become wearable in the rinsing part. At this step, the de-sized garment will be softened to improve the hand feel of the garment. The washing stage ensures the finished products acquire high color-contrast, color pull, and low back-staining degree with enzymes. Next, the softening stage, as it is called, provides a soft touch to denim products. Specifically, industrial denim washing includes three types of softeners, namely cationic (for all soft products except white-colored ones), non-ionic (only for white and light-colored products), and silicones (provided softness as well as lubricity). The fifth stage of drying the garment hydro-extract denim garments to remove 80% of water. Since too many garments in the dryer lead to a longer drying time which makes the denim garment back staining, this stage can affect the softening stage result to some extent.

After going into details of each step, certain defects have been identified, Specifically, the resin spray, permanganate spray, pigment spray, desizing, stone washing, crinkle, enzyme washing, bleaching, tint, softening processes in the wet process, alongside the sanding, whisker effect, damage, swift, and laser operations in the dry process has resulted in tons of defects (Fig. 4). Those lie in every stage of the production and even in the textile sector, which means that the formation of a defect in the denim washing process is also inevitable.

That given, with the Pareto analysis technique, many defect types have been identified with different levels of importance (Fig. 5), main problems leading to these defects are also clearly defined through the cause and effect chart (Fig. 6).

Eventually, after applying the Pareto analysis and cause and effect diagrams, it is depicted that 80% of the defects stem from the chemical and color-based ground beside other causes. Hence, proper solutions have also been raised to improve the overall process for a more sustainable textile industry (Table 7).

In conclusion, certain defect problems have been addressed to better control the washing processes and take every stage to minimize industry wastes and environmental pollution. The application of SPC is one of the identical examples of a method to achieve the aforementioned objectives.

Lean Manufacturing: Case Studies from Fashion ...

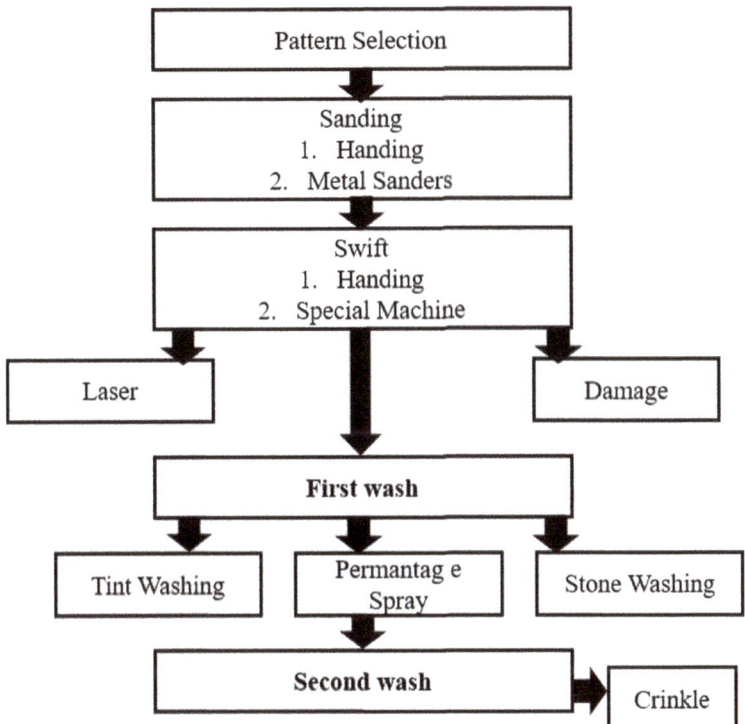

Fig. 4 An alternative flowchart for denim washing process

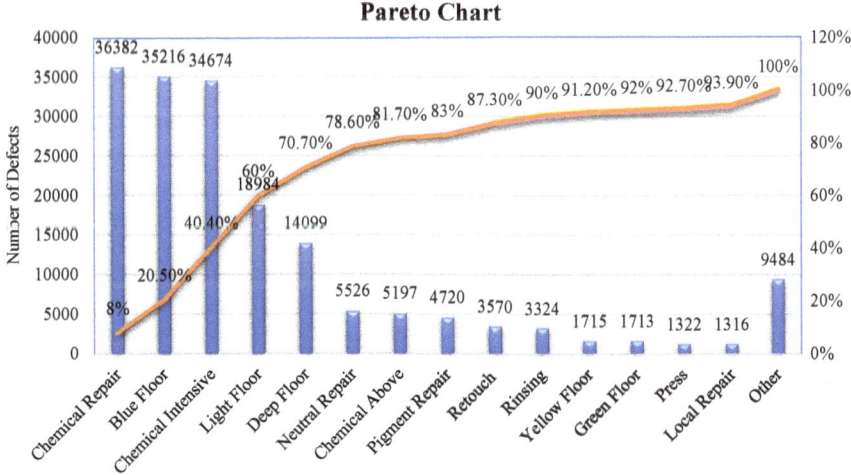

Fig. 5 Pareto chart for denim washing defects

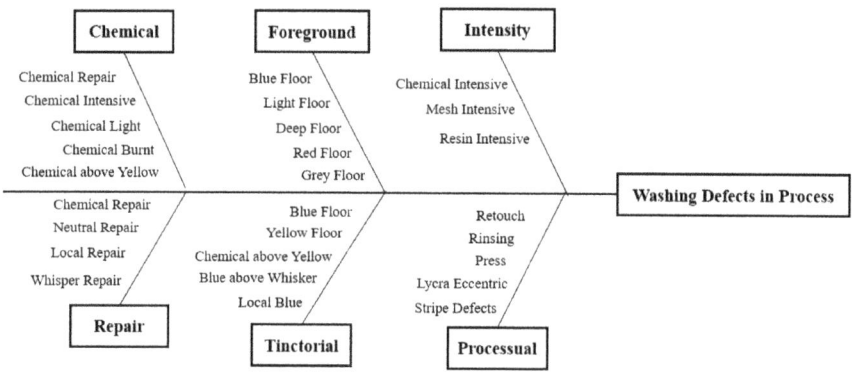

Fig. 6 Cause and effect diagram

Table 7 Recommended corrective actions for washing defects in process

No	Defect type	Causes	Corrective actions
1	Chemical defects	Desizing enzymes, homogeneity (exceeding Ph 6–7)	Dispersing Indigos Using cationic or silicone for softening
2	Ground defects	Insufficient or intensive local permanganate and pigment	Back staining by ozone
3	Intensity problem	After the type of washing, as a result of the narrowing in the width and neck of the fabric and the increase of weft and warp frequencies	Investigations should be made on rins, enzymes, and stone and stony bleach wash recipes. Time, temperature and stone used in washing recipes, enzyme and bleach, weight, dimensional stability
4		Reasons for repairs are wet processes and operational errors due to spray	Faulty products with tone difference are light or dark color on the floor, permanganate at local insufficient or dense
5	Tinctorial	Parameters such as temperature, duration and concentration	Sodium hypochlorite or potassium permanganate bleaching process, neutralisation and rinsing
6	Processual	Failure to rinse with soft water. The problem is that the rinse water is not cold. Failure to remove peroxide. Hard water causes problems in dyeing due to the high rate of lime it contains	Acetic acid and peroxide killer are added. In addition to these, soft water is added and the temperature is raised to 50 degrees and kept for 15 min. If peroxide is 0 and ph = 7, it means that suitable environment is provided for dyeing

6 Conclusions

The traditional textile and fashion industries have been proven to struggle with difficulties by following conventional methods. To be more specific, globalization and market liberalization have induced competitiveness between corporations, thus emphasizing high productivity and effectiveness. Nevertheless, each section of the industry is dealing with distinct barriers around the globe. For instance, the shortcomings of the traditional management system in the apparel industry of Bangladesh or in the cotton spinning industry of Ethiopia have led to declined demand. Moreover, the obsolescence of production technology and techniques in the Adinkra textile sub-sector in Ghana or the Italian luxury fashion industry has led to the loss of productivity. Additionally, the resistance to the adaptation of new changes in the hand-loom sector of Ind, has made this sector sick. Altogether, each sector is facing analogous impediments, which result in underproduction, inefficiency, and low-quality products.

To overcome this problem, many industries have adopted different strategies, techniques, and methodologies that are derived from the lean philosophy. The apparel industry of Bangladesh adopted the 5S system to get rid of the insufficient management, low sanitation standard, and incapabilities of workers. Similarly, the Italian giant of the fashion industry utilized the 5S methodology to eliminate non-added value items and waste within its operation. Furthermore, the ABC industry in India applied the value stream mapping and 5S method to improve inventory management and reduce defects within its production system. Besides, the concept of Muda, Six Sigma, and Statistical Process Control (SPC) is implemented respectively in natural fibre clothing manufacturing, Indian spinning mill, and denim garment manufacturing in Turkey to improve productivity and efficiency.

The improvement of Lean philosophy is distinguished between different implementations. Nonetheless, it yields identical results that create advantages for corporations in terms of production and competitiveness. In general, the lean philosophy can help reduce non-added value items, cut production costs, enumerate management systems and customer satisfaction, increase productivity with a lower lead time and profitability. As highlighted above, the textile and fashion industries have been attempting to adopt lean philosophy to increase its efficiency and competitiveness. In addition, with the 'waste' reduction, which is the most critical essence of lean, these industries can be able to achieve sustainability due to its major impact on global environmental, social, and economic problems.

References

Aboagyewaa-Ntiri J, Mintah K (2016) Challenges and opportunities for the textile industry in Ghana: a study of the adinkra textile sub-sector. Int Bus Res 9(2):127–136

Ahmmed AS, Ayele M (2020) In-depth analysis and defect reduction for ethiopian cotton spinning industry based on TQM approach. J Eng 2020:1–8. https://doi.org/10.1155/2020/5792434

Calisir F (2007) Factors affecting service companies' satisfaction with ISO 9000. Manag Serv Qual 17:579–593

Carmignani G, Zammori F (2015) Lean thinking in the luxury-fashion market: evidences from an extensive industrial project. Int J Retail Distrib Manag 43(10/11):988–1012. https://doi.org/10.1108/IJRDM-07-2014-0093

Ganguly-Scrase R (2003) Paradoxes of globalization, liberalization, and gender equality: the worldviews of the lower middle class in West Bengal, India gender and society. Sage Publ 17(4):544–566

Goswami O (1990) Sickness and growth of India's textile industry: analysis and policy option. Econ Pol Wkly 25(44):2429–2439

Karthi S, Devadasan SR, Murugesh R (2011) Lean Six Sigma through ISO 9001 standard-based quality management system: an investigation for research. Int J Prod Qual Manag 8:180–204

Khan AM, Islam M (2013) Application of 5S system in the sample section of an apparel industry for smooth sample dispatch. Res J Manag Sci 2(7):28–32. ISSN 2319–1171

Kquofi S, Amate P, Tabi-Agyei E (2013) Symbolic representation and socio-cultural significance of selected Akan proverbs in Ghana. Int Inst Sci Technol Educ 3(1):86–98

Kumar S, Thavaraj S (2015) Impact of lean manufacturing practices on clothing industry performance. Int J Text Fash Technol (IJTFT) 5(2):1–14

Kuwornu-Adjaottor JET, Appiah G, Nartey M (2016) The philosophy behind some Adinkra symbols and their communicative values in Akan. Philos Pap Rev 7(3):22–33. https://doi.org/10.5897/PPR2015.0117

Mizuno K (1996) Rural industrialization in indonesia: a case study of community-based weaving industry in West Java, 1st edn. Institute of Developing Economies

Niranjana S (2004) Thinking with handlooms: perspectives from Andhra Pradesh. Econ Pol Wkly 39(6):553–563

Quarcoo AK (1972) The language of the adinkra patterns, 1st edn. Institute of African Studies, University of Ghana, Legon

Quartey P, Abor J (2011) Do Ghanaians prefer imported textiles to locally manufactured ones? Mod Econ 2(1):54–61. https://doi.org/10.4236/me.2011.21009

Rahani AR, Al-Ashraf M (2012) Production flow analysis through value stream mapping: a lean manufacturing process case study. Int Symp Robot Intell Sens 2012:1727–1734

Rother M, Shook J (1998) Learning to see: value stream mapping to create value and eliminate Muda. The Lean Enterprise Institute, Brookline, MA

Roy T (2002) Acceptance of innovations in early twentieth century Indian weaving. Econ Hist Rev New Ser 55(3):507–532

Shiferaw A (2017) Productive capacity and economic growth in Ethiopia. CDP Background 34;1–24

Srinivasulu K (1994) Handloom weavers' struggle for survival. Econ Pol Wkly 29(36):2331–2333

Syduzzaman Md, Biswas MAS, Yeasmin D (2016) Developing a framework for implementing total quality management in the apparel industry: case study on a Bangladeshi apparel manufacturing factory. Int J Textile Sci 5(5):87–95

Tanusree S (2015) A Study of the present situation of the traditional handloom weavers of Varanasi, Uttar Pradesh, India. Int Res J Soc Sci 4(3):48–53

Teichgräber UK, de Bucourt M (2012) Applying value stream mapping techniques to eliminate non-value-added waste for the procurement of endovascular stents. Eur J Radiol 81:e47–e52. https://doi.org/10.1016/j.ejrad.2010.12.045

Benefits, Drawbacks, and Future Directions of Lean on the Fashion and Textile Industry

Hung Manh Nguyen, Scott McDonald, Bill Au, and Mohammadreza Akbari

Abstract Lean supply chain principles have been well applied to help Fashion and Textile (F&T) industry to become more competitive and sustainable. In recent times, the F&T industry has been facing the COVID-19 pandemic with many supply chain disruptions. Lean concepts have changed during the crisis to create a balanced and flexible flow of textile and garment products through lean with agility, supply chain visibility, and green supply chain practices. This chapter discusses the benefits and drawbacks of the lean manufacturing in F&T industries. Further, this chapter also recommends the future development in lean and agility, sourcing flexibility and green supply chain practices that can help the industry to recharge and adjust during such a difficult time. Information technologies such as big data and the Internet of Things play supporting roles in enhancing visibility and integration for a lean and sustainable F&T development, which will be future in the lean manufacturing.

Keywords Lean · Agility · Sourcing flexibility · Visibility · Supply chain green practices

1 Introduction

The rapid growth in the production of Fashion and Textile (F&T) products, associated with rising wealth and fast consumption in developing nations enrich consumers experience with "fast fashion". For decades, it has been defined as "a business model based on offering consumers frequent novelty in the form of low-priced, trend-led products" (Cleff et al. 2018; Niinimäki et al. 2020). The global fast fashion market is expected to grow from $25.09 billion in 2020 to $30.58 billion in 2021 at a compound annual growth rate of 21.9% (Research & Markets 2021). The forecast for the next

---

H. M. Nguyen · S. McDonald · B. Au
RMIT University Vietnam, Ho Chi Minh City, Vietnam

M. Akbari (✉)
James Cook University, Douglas, Australia
e-mail: reza.akbari@jcu.edu.au

© The Author(s), under exclusive license to Springer Nature Singapore Pte Ltd. 2022
R. Nayak (ed.), *Lean Supply Chain Management in Fashion and Textile Industry*,
Textile Science and Clothing Technology,
https://doi.org/10.1007/978-981-19-2108-7_12

10 years, based on this growth, projects that the fast fashion industry will be worth $44 billion by 2028 (Teodoro and Rodriguez 2020). Several studies investigated lean and global consumption surges (competitive price, good quality and delivery terms). Lean manufacturing has been cited as a key player in enhancing competitiveness and the growth of the industry (Amaro et al. 2019; Hodge et al. 2011; Purvis et al. 2014).

However, the other side of the global consumption, fast fashion trend, sustainable ethics create a big debate in the F&T industry (Shen 2014; Nayak 2020). Literature often questions the balance between lean operations for competitiveness and at the same time sustainable development in F&T supply chains (Anguelov 2015; Gardetti and Muthu 2020; Huo et al. 2019; Paksoy et al. 2019). Since the textile and fashion industry has a long and complex supply chain (Nayak et al. 2019), starting from agriculture and petrochemical production to manufacturing, logistics and retail (Niinimäki et al. 2020), managing the green and social development along with economic aspects become essential (Anguelov 2015; Huo et al. 2019; Martin 2016). Sustainability is easy to overlook, but it is an important value for development in the F&T industry, especially during pandemic outbreaks like Covid-19 (Lu 2021; Zhao and Kim 2021).

The COVID-19 pandemic has created enormous variabilities that push F&T industries to restructure their supply chain practices and make quick changes to adapt to the new environment. Supply chain risks have been reported as the second most important factor affecting the F&T industry, according to the 2020 fashion benchmarking study in the USA (Lu 2021). Covid-19 put more pressures on F&T systems to be lean and flexible to sourcing strategies when dealing with suppliers. Agility has become the forefront during such an epidemic like Covid-19 to handle supplier risks and minimize future wastes. Future development in F&T manufacturing continues to be agility and supply chain integration (SCI). As SCI has become the forefront in manufacturing for the last few decades, it would be essential to discuss the benefits of Lean and Green Strategies in the context of disruption such as Covid-19.

The purpose of this chapter is to develop a research framework and suggest future directions for lean F&T sustainable development. While previous chapters focus on the technical aspect of lean tools and techniques (Chapters "Fundamental Concepts of Lean and Agile Manufacturing", "Lean Concept in Fashion and Textile Manufacturing" and "Standardized Work in Fashion Industry"), this chapter focuses on structural design and supply chain network that facilitates both lean and green strategies in the F&T industry. Section 1 provides a literature review for lean and sustainable supply chains in the F&T industry. Section 2 describes sustainable benefits, challenges and expectations from Lean implementation. Section 3 also provides a research framework for future development of Lean and Green integration with the importance of supply chain agility, integration and information technology.

2 Lean Supply Chain in the Fashion and Textile Industry

Literature review has been conducted on case studies and articles describing lean production and lean thinking implementations in the F&T industry. Generally, lean production aims at reducing waste and variability in the processes, adding more value to customers and providing operational performance improvement (Stevenson 2005). Major literature on lean concepts focuses on product and process design, structural design, manufacturing control and planning. While standardization and module concepts can help garment manufacturers like Zara to excel in the market, the postponement process has shown the importance of lean theory in the F&T industry (Aftab et al. 2017; Venkatesh and Swaminathan 2004). The supporting goals for lean are to eliminate disruptions from yarns, fabrics to garment manufacturing (Manfredsson 2016; Stevenson 2005); create flexible systems to meet diverse customer's requirements; and eliminate wastes (Torres-Luna et al. 2020). These goals drive the system to have a balanced and flexible flow (Stevenson 2005). Lean concepts have been used for decades and in the F&T industry (Abernathy et al. 1999; Chakraborty and Biswas 2020; Prasad et al. 2020) for productivity improvement (Adikorley et al. 2017; Palange and Dhatrak 2021). Lean principles offer a process improvement mindset and engage everyone in the process (Alves et al. 2019; Amaro et al. 2019).

However, due to the complexity of the F&T supply chains, organizations are not yet fully aware of the Lean principles as they do not apply the approach to an entire value stream (i.e., to products or families of products) but only to parts of value streams (i.e., to garment manufacturing sectors rather than fabrics processes). The main issue that has been reported in the literature was the lack of integration among different supply chain members. The supply chain in the F&T industry is complex with several factors, namely a short lifecycle, high volatility, low predictability, and high impulse purchase (Bruce et al. 2004). Consequently, managing a good supplier–buyer relationship in the F&T supply chain is essential for operational improvement such as reducing lead times and achieving a quick response. These characteristics of the industry highlight the need to use an approach such as agility.

Agility is the ability to sense changes and to rapidly and flexibly respond to fluctuations (Christopher 2000; Li et al. 2008). As rapid and flexible responses may as well be considered elements of supply chain flexibility, which is an important dimension of supply chain agility. Competition intensity and diverse customer needs will continue to push these F&T supply chains to be more agile. In addition, the recent Covid-19 outbreaks further push F&T supply chains to be more agile as well.

H&M and Zara have built agility into every link of their supply chains to ensure responsiveness to fluctuating supply and demand through agile design processes, postponed manufacturing, and state-of-the-art sorting and material handling technologies (Lee 2004). Besides, pressures from diverse customer needs will foster fast fashion companies like Zara to restructure their supply chain to handle higher product complexity (i.e., a wider variety for customer selection). Through supply chain agility, manufacturers can deploy delayed configuration, standardization and modular design that enable firms to cost-efficiently deliver products characterized by numerous styles and sizes, and adjust production processes in response to change in supply and demand (Li et al. 2008). Standardized parts have a high commonality that can share product platforms and resources that enable firms to meet diversified customer requirements while being competitive in terms of cost and quality (Robertson and Ulrich 1998). Zara produces over 450 million items and launches around 12,000 new designs annually. Indeed, the company gained a high level of supply chain agility, exercised efficiently product standardization and postponement strategies to ensure that they deliver new products every fortnight with the constant refreshment of store-level collections goes off smoothly and efficiently (Roll 2020).

3 Lean and Sustainable Supply Chains: Benefits and Challenges

Several case studies show how lean principles can facilitate efficient F&T production and help the supply chain become more competitive. The following section discusses how lean strategy can align with green practices and directly impact responsible consumption, promoting sustainable industrialization (Shen et al. 2017). Different practices have been used in the F&T industry to pursue environmental sustainability objectives. Previous literature has revealed that Lean operations have attempted to veer in several directions, using generic models (Fig. 1) with three pillars: social, economic, and environmental aspects. Each has its challenges and benefits to the industry as well as the consumers alike and each has an impact on the other two.

Sustainability is something that should be easy to understand and something everyone should embrace wholeheartedly, but this has not always been the case

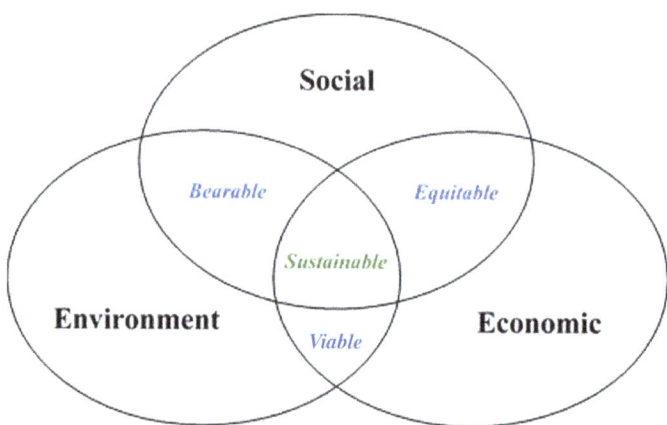

Fig. 1 Sustainable supply chain model in F&T industry

(Nayak et al. 2020). Sustainability in the F&T industry, and particularly the environmental concerns, has historically been ignored or at least was avoided due to a lack of understanding or a feeling that individual designers could not make a difference anyway (Beard 2008). Where should we start to make a difference? How could we possibly get the attention of industry players?

If the current leaders in the F&T industry choose not to participate in environmentally sustainable practices, then perhaps it might be better to turn our efforts towards the future and the industry players of the future (Shen et al. 2020). As with other industries, these efforts have started to become more prevalent in higher education by delivering the appropriately designed curricula to bring the underlying issues to the surface. The F&T industry influencers of tomorrow could be the key to building a sustainability mindset.

Vietnam is entrenched in the global fashion (apparel) industry to the point where it is one of the most important economic export elements in the country (Nayak et al. 2019). This sector has the second-highest number of exports just behind the electronics industry and employs one out of every five workers in the total labour force making it one of the key industries driving the nation's growth. Even during the COVID-19 pandemic, over the period from 2019–2021, the F&T industry continued to progress steadily as forecasted in 2017 to extend this trend well into 2025, as displayed in Table 1. The export market for Vietnamese products such as textiles and apparels, shoes, furniture, various electronic devices and spare parts shipped to the USA reaching an import–export turnover of $543.9 billion USD which was up by 5.1% in 2020 over 2019 and showed the highest trade surplus since 2016. In the first three months of 2021, the US market became the largest recipient of Vietnamese items trading $22.2 billion USD indicating that the US was moving away from their traditional sourcing nations to other emerging countries such as Vietnam.

When sustainability is mentioned anywhere in the media, it is the environment that takes center stage as the world appears to be in crisis mode when addressing issues like pollution causing global warming and the destruction of the shrinking natural habitats all over the planet from our oceans to forests and other grasslands (e.g. Akbari and Hopkins 2019). When addressing the issue of sustainability in the F&T industry, due to the labour intensive nature of the industry, there is a focus on reducing human rights, violations, and the ever-present corruption found in such places.

Table 1 US textile and apparel Units

	2005	2007	2009	2011	2013	2015	2017	2019	2021	2023	2025
Vietnam	2.8	4.7	5.2	7	8.2	9.9	11	12.1	12.8	14.9	16.1
India	4.5	5.1	4.5	6	6.2	6.6	7.1	7.9	8.2	8.8	9.2
Indonesia	3.8	4.2	4.1	5.1	5.3	5.7	5.9	6.1	6.3	6.5	7
Mexico	7.1	5.6	4	4.7	4.4	4.2	4.2	4.1	4	4	4
Bangladesh	2.4	5.2	3.7	4.4	5.1	5.9	6.3	5.2	8	8.3	9.2

Green supply chain management (GSCM) is certainly relevant to the F&T industry; however, achieving good environmental performance is rather difficult. Due to geographically complexity and regional differences within F&T supply chains, coordination and collaboration with supply chain members can be challenging, especially in monitoring sustainable practices such as environmental practices in the fabrics manufacturing. Lack of integration with textile suppliers such as yarns and fabrics can be detrimental to sustainable efforts. F&T sectors and their related industries (e.g., yarns and fabrics processing) have a significant impact on the environment due to resource and labor-intensive practices (Nayak 2019).

From the materials being extracted, sourced, and placed into F&T production, most of them are used only once, being disposed of within the first year of use (Anguelov 2015; Gardetti and Muthu 2020). Therefore, achieving sustainable production patterns, especially engaging the entire value chain, from raw material to consumer, both globally and locally become essential for the F&T industry (Gardetti and Muthu 2020). Many examples include loose relationships between garment manufacturers and textile producers on supplier environmental performance. Caniato et al. (2012) found that Indian suppliers were unwilling to engage in environmental performance assessment. Similarly, Vermeulen and Ras (2006) identified challenges among the Dutch fashion companies in greening their global fashion supply chain. The results showed an inconsistency between global environmental challenges and company environmental disclosure; however, a disconnection between the specific environmental indicators reported by textile companies and the garment producers remain the hotspots for the GSCM discussions.

3.1 Sustainable Benefits

Several benefits may be achieved by the F&T industry such as performing code of conduct and social audits, collaborating with multi-tier suppliers, as well as offering incentives to the suppliers for being more socially responsible. This aspect is usually neglected by many companies in place of a focus on environmental sustainability (Akbari and McClelland 2020). The common environmental "green" factors include Life Cycle Assessments (LCAs), CO_2 (carbon dioxide) Emissions, and the use of recycled water as a sustainable process that are prominently displayed on company websites and product labels.

Since the recent sentiment of consumers is focused on sustainability and the circular economy, it would greatly benefit fashion designers to implement practices that reflect these two factors. Consumers have recently been leaning towards products and production processes that follow stricter "green" models. Due to the media attention over the last several years on global population explosions, global warming, climate change, water scarcity, and dwindling farmlands, consumers have taken a position to make a difference by protesting through choices made at the cash register. Brands that have tried to reduce their waste water, reduce the use hazardous chemicals, and eliminate human rights' violations have shown a marked increase in consumers' decisions to buy their products.

Fast Fashion has been a thorn in the side of this progress as it has taken the low road and gotten worse regarding their waste water, use of chemicals, and human rights abuses. All these can be attributed in the name of meeting the speedier demand of consumers (Hines and Bruce 2017). With a global turnover of 1.3 trillion dollars and 300 million workers employed in the fashion industry, it is well worth the attention of industry players to understand the impact of consumer behaviour and provide the market with what it demands.

The best-performing companies focused on sustainability by changing their processes and practices in their company governance and manipulating their business models into ones that take advantage of the growth that sustainability can bring to the entire supply chain.

3.2 Sustainable Challenges

As much as 'sustainability' itself has become a mainstay in organizations around the globe, it is still very much misunderstood (Thomas 2008) and confusing to many who seek to incorporate it into their daily business operations or creative outputs. Unfortunately, there is no clear definition for what sustainability means compared to what it is not in terms of designing the fashion and the manufacturing of it leaving consumers to try to figure all this out for themselves (Beard 2008).

Due to various players in the industry from the designers to the material suppliers, to the manufacturers, to the distributors, to the retailers, and finally to the consumers, each views their responsibilities towards the environment and ethical practices differently making sustainability much more challenging to accomplish (Haug and Busch 2016). The biggest hurdle to each of the players understanding sustainability the same way is due to differences in their education, life experiences, involvement in social or cultural movements, as well as simply the way it is interpreted (Thomas 2020).

One of the many challenges regarding sustainability is the perception that it carries in the business world. Executives call it the "S-word" with a negative connotation because it goes against everything that they work so hard regarding "growing" the business and its market share. At higher level strategy meetings, things like maximizing the ROI (return on investment), squeezing profits and budgetary models for the company along with the many challenges of scarcity of resources make sustainability more of a burden for some business managers (Thomas 2020).

The negative customer perceptions of brand compliance as well as the incompatibility with the customers' value propositions and the lack of scalability of sustainable business models hinder the industry from making a full-fledged commitment to changing their ways and pushing their supply chains to make similar changes to their practices (Pal and Gander 2018).

Perhaps one solution might be to help the fashion industry move from the current practices of simply 'reporting' their sustainability efforts, their financial statements, and corporate statements to 'taking action', creating sustainability value for those beyond their own company to include their entire supply chain and making committed changes to their policies, processes and practices by developing an *Industry Scorecard* that would add an element of transparency thus applying the needed pressure to act on their claims of being a sustainable organization (Garcia-Torres et al. 2017).

Besides, several green supply chain practices could not approach the entire value stream (i.e. to products or families of products) but only to parts of value streams (i.e. to fabrics sectors). That disconnection among garment supply chain partners or lack of integration (Raut et al. 2019) influence how Lean manufacturing directly impacts responsible consumption and production. Lean suppliers and greening suppliers (Laari et al. 2017; Yu and Huo 2019) have been making major efforts in lean manufacturing that can potentially reduce waste generation through prevention, making industries sustainable, with increased resource-use efficiency and greater adoption of clean and environmentally sound technologies and industrial processes (Nath et al. 2019; Sancha et al. 2016).

Lean-Green indices are developed to measure the Lean-Green practices in companies. This is important to evaluate and compare green practices. However, literature indicated a mix of measures and inconsistencies among the usages in different industries (Alves et al. 2019). Since the process-oriented industry like textile and garment has highly inflexible automatic machinery with high volume/low product variety (Mohan Prasad et al. 2020). This complexity of the textile industry renders the introduction of lean manufacturing strategies in a textile market with some measures that represent indicators related to sustainability (key environmental performance indicators—KEPI) that, many times, are unknown. Sustainable Apparel Index was developed by the Sustainable Apparel Coalition (SAC) to measure and promote sustainability built upon a common approach for evaluating sustainability performance (Radhakrishnan 2015). The Higg index, announced by the SAC, enables apparel companies to measure the environmental and social impact of apparel production throughout the product lifecycle, from design to end of use or recycling of the product. Although the Higg index can be applicable for many organizations in an apparel supply chain, implementation of a common tool and technique can be challenged as different size, companies' visions, and cultures. Figure 2 provides an overview of the sustainable associates of fashion brands in their supply chains.

4 Future Directions Lean and Sustainable Supply Chains

The textile sector continues to experience significant environmental issues related to the production process, including the overuse of chemical products and natural resources, resulting in a high environmental impact (Bubicz et al. 2021; Gardetti and Muthu 2020; Radhakrishnan 2020). Moreover, due to geographically complexity and regional differences within F&T supply chains, coordination and collaboration

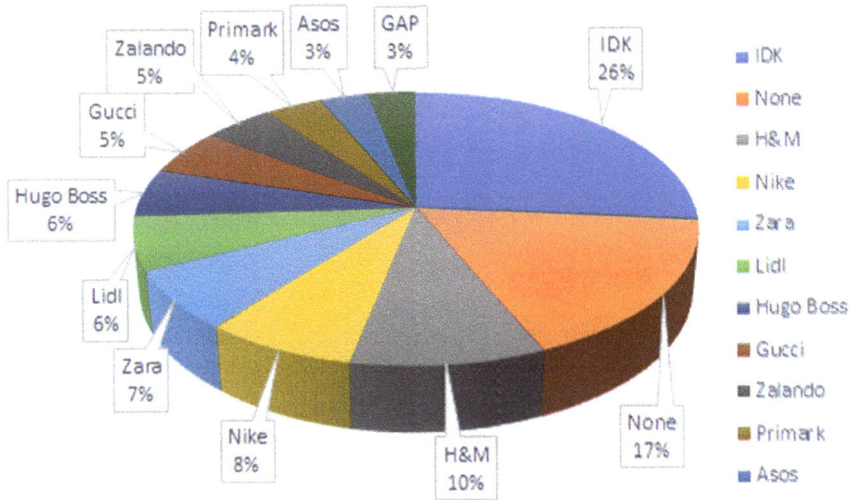

Fig. 2 Which, if any, of the following fashion brands or shops do you associate with having a sustainable supply chain? *Source* Adapted from Statista (2021)

with supply chain members can be challenged, especially in monitoring sustainable practices such as environmental practices in the fabrics manufacturing. This section proposes a framework where lean strategy can align to green supply chain practices to ensure a balanced flow of textiles and garments. The discussions from previous sections provide a foundation for integrating agility, supply chain integration and technology to lean and green approaches. These approaches can facilitate F&T companies to direct their efforts during situations such a Covid-19 pandemic.

Figure 3 proposes a framework to elaborate linkages between lean and green strategies in F&T operations with agility, integration, and supply chain technology (Shen and Li 2019).

Covid-19 puts more pressure on F&T systems to be flexible in sourcing strategies when dealing with suppliers. Supplier risks have been reported as the major source of supply chain disruptions, especially in the F&T industry (Butt 2021; Hoek 2020). Some strategies such as supply reduction base or centralization might not be effective. During the Covid-19 pandemic, we can see the trend of increasing buffering and switching suppliers (Espitia et al, 2021; Hoek 2020). When looking at the scale of supply/value chain disruptions, Covid-19 has hit hard in many F&T manufacturing nations including Bangladesh, China, Mexico, Turkey, and Vietnam.

Figure 4 indicates apparel production levels in these major F&T manufacturing countries reflecting the major supply chain disruptions. Literature mentioned a "triple hit" to global manufacturing due to the pandemic, referring to direct supply disruptions, demand disruptions and supply chain contagion. Supply-chain contagion can be seen in F&T supply chains when Chinese textile materials and fabrics become more expensive or impossible for importing due to cross border trade suspension during Covid-19. These amplifications created direct supply shocks in F&T sectors

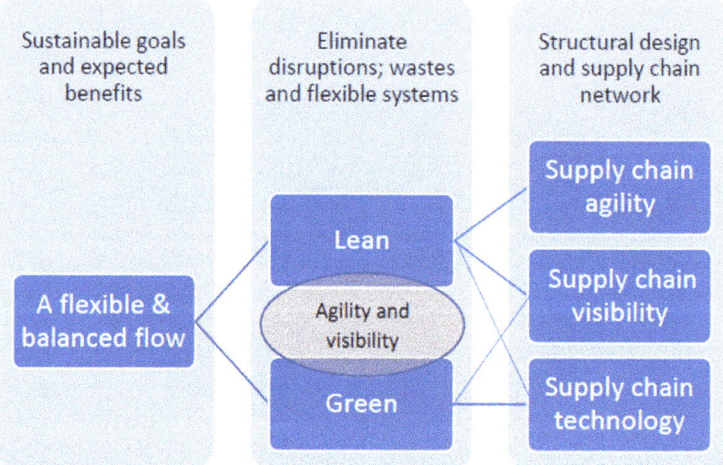

Fig. 3 Future lean supply chains and agility—Covid 19 pressures. *Source* Adapted from Butt (2021) and Hoek (2020)

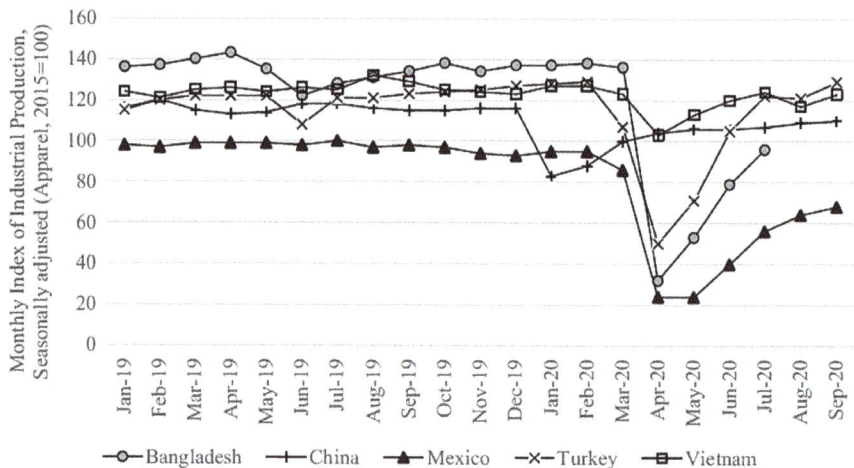

Fig. 4 Apparel production in major fashion and textile countries. *Source* UNIDO (2020)

in less-affected nations (e.g. Vietnam) that find it harder and/or more expensive to acquire the necessary imported fabrics and garment inputs from China (Dao et al. 2020). Therefore, achieving more sustainable production patterns, especially engaging supplier environmental monitoring and involving the entire value chain from a holistic perspective, from raw material to consumer, both globally and locally become essential for the F&T industry (Gardetti and Muthu 2020).

Agility has become the forefront during such an epidemic like Covid-19 to handle supplier risks and minimize future wastes. The benchmark study among USA textile manufacturers indicated that U.S. fashion companies are more actively exploring "Made in the USA" and nearshoring opportunities to improve agility and flexibility and reduce sourcing risks (Lu 2021).

Expected in the aftermath of the Covid-19 outbreaks, F&T supply chains are increasingly required to balance the management of key lean capabilities with resilience, responsiveness and growth (Pettit et al. 2019). The lean, agile and Le-agile approaches supply all effective sequences and manage the manufacturing process to reduce lead times (Bruce et al. 2004; Purvis et al. 2014). This is particularly relevant to the textiles and clothing industry, characterized by highly volatile markets, short product lifecycles and high product variety. Besides, due to the small size of these garment and textile firms, there are increased competitive forces compounded with small businesses where the problem becomes more challenging with less resources available (Bruce et al. 2004).

Agility plays an important role as the F&T sector has extremely low-profit margins so that producing and even holding small quantities of stock is not commonly a viable option. The incidents from Covid-19 supply shortage in F&T industry, many firms could not store large quantities of materials, and as fashion products have a very short life cycle and are seasonal. Zara observed global store closures amid the COVID-19 crisis in 2020, with sales falling 44% year-on-year in Q1 2020 and the company reporting a net loss of $482 million USD (Roll 2020). COVID-19 has resulted in a widespread sales decline and order cancellation among U.S. fashion companies. Almost all respondents (96%) from the recent benchmark expect their companies' sales revenue to decrease in 2020 (Lu 2021).

Therefore, the literature suggests that the textiles and clothing industry should try to combine lean and agility, which is often referred to as 'leagility' which handles the two drivers of low margins and volatility of demand. This would ensure fast product replenishment, the building and maintaining of supply chain partnerships and flexibility in response to the volatility of demand from retailers (Lu 2021; Purvis et al. 2014; Zhao and Kim 2021). These authors suggest increasing sourcing diversification and reshoring manufacturing and increasing flexible capacity plans. It was difficult to redirect to other sourcing destinations to China was hit hard by Covid-19. The "increasing production and sourcing cost" and associated logistics scandals such as Suez's canal put more pressures on "shipping and logistics" cost in 2020 and potentially in the next few quarters (Lu 2021; Zhao and Kim 2021). These authors observed that during COVID-19 and the USA-China trade war, more sourcing diversification and reshoring manufacturing become necessary, without any alternate choice. Around 29% of respondents indicated that they source more from Vietnam than from China in 2020, up further from 25% in 2019 (Zhao and Kim 2021). The trend continues to diversify their sourcing destinations, with 42.1% currently sourcing from more than 10 different countries or regions (Zhao and Kim 2021).

Nearshoring and reshoring manufacturing capabilities are emerging topics in the fashion industry. Recent research indicates the need for flexible supply source planning can help F&T companies avoid global trade disruption (Castañeda-Navarrete et al. 2021; Xu et al. 2020). Deploying nearshoring opportunities supports flexibility and better autonomy for production. This is extremely important for the textile and garment industry where the life cycle becomes shorter due to customer diverse experience. In the recent fashion benchmarking report 2020, Lu (2021) indicates that U.S. fashion companies now are more actively exploring "Made in the USA" sourcing opportunities to improve agility and flexibility and reduce sourcing risks. Approximately 25% of the respondents expect to increase sourcing locally in the United States in the next 2 years, which would be the highest level since 2016.

5 Future Lean and Supply Chain Visibility

Future development in F&T manufacturing continues to be agility and supply chain visibility (SCV). Zara has claimed a super-efficient and super-flexible supply chain. Zara demonstrates the ability to overcome the vagaries of the fast fashion industry by its agility and is considered legendary (Zhelyazkov 2011), launching thousands of fashion models in a fortnight. Besides, the pressure from e-commerce and omnichannel marketing increases the need for speed, quick and visible inventory turnover. SCV has become the forefront in manufacturing for the last decade, it would be essential to discuss the benefits of Lean and green strategies in the context of disruption of Covid-19.

To be "truly agile" research indicates the need to enhance market-sensitive capabilities together with abilities on sharing information across all supply chain partners (Bruce et al. 2004; Brusset 2016). Visibility capabilities provide companies tracking and tracing of goods flows, and reporting tools. In a supply chain, visibility capabilities often referred to as boundary-spanning or inter-organizational technologies that link supply chain members, as well as other procedures, processes and routines which enable distinct firms to work together towards a common goal. Research indicated that visibility capabilities can bring not just information but know-how and other intellectual capital, which enables firms to make informed decisions (Christopher 2000). These techniques can be integrated with ERP (enterprise resource planning) systems, and with other supply chain management software to enhance agility in a supply chain (Brusset 2016).

Visibility brings many influences on lean and supply chain integration. During the Covid-19 pandemic, manufacturers have demanded greater visibility into the supply chains of their suppliers so that the brand or retailers can be more dynamic in their planning. In the fashion and textile industry, this capability remains the essential partnerships between brands, retailers, and suppliers around the world. On the other hand, it pushes up the level of confidence in the long-term outlook for the fashion industry (Schattem et al. 2020a, b). So, it is more important than ever for fashion companies to work with their suppliers, to develop joint collaborative activities, enhance information and linkage to meet the key sourcing goals of speed-to-market, flexibility and agility, sourcing cost control and low compliance risks.

COVID-19 has caused severe supply chain disruptions to global fashion companies, reflecting the increasing number of order cancellations and supply shortages. COVID-19 has resulted in a widespread sales decline and order cancellation among

U.S. fashion companies. Almost all respondents (96%) expect their companies' sales revenue to decrease in 2020 (Lu 2021). The fragility of the clothing supply chain operating in South Asian countries indicating by a lack of detailed information on supply disruption and sourcing contracts (Majumdar and Sinha 2020). Nearly half of self-identified retailers said the sourcing orders they cancelled or postponed without much notice. Another 40% expected order cancellation and postponement could extend further to the fourth quarter of 2020 or even beyond. The order cancellation or postponement has affected vendors in China, Bangladesh, and India the most. These discussions raise the importance of the information infrastructure and alignment to the company objectives.

6 Future Lean Supply Chains Are Digital

To connect different Lean tools, supply chain technologies are essential. The exchange of information and data between F&T buyers and suppliers is crucial for agile supply chains (Christopher 2000). Firms need to have the capability to react to possible volatile fluctuations in demand. Improving supply chain visibility among members can help companies reduce the amount of inventory held in anticipation of predicted and often distorted demand (Akbari and Ha 2020). For many industries, using different information technologies become essential to sustain the extended enterprise where collaborative alliances support the exchange of information to enable such activities as joint product development and common systems (Bruce et al. 2004). The real strength, however, does not lie in the individual information technology tools but in the integration of these tools with the redesigned production processes.

Tools embedded with the synchronized information and production flow for their smooth functioning would bring about the reduction of nonvalue-added activities and increase customer satisfaction (Mahmood 2020). Chapter "Digital Technologies for Lean Manufacturing" has discussed some of the important technologies and their fundamental principles in the implementation of lean manufacturing in F&T industries. This chapter briefly highlights how different technologies such as blockchains, big data, RFID (Radio-frequency identification), ERP systems, and the Internet of things (IoT) will influence the future of supply chain in the F&T sector. The next section discusses some of the recent technologies' potential impact on lean and green partnerships (Kumar et al. 2012).

A study by Wang and Ha-Brookshire (2018), demonstrated the growing demand for digital competency amongst common fashion supply chain functions (Forecasting, consumer research, design, product development, merchandising, sourcing/production and retailing/distribution). Digital competency, which can be defined as possessing abilities and skills, or competency, to deal with digital resources and environments (Prifti et al. 2017) has been noted to be of particular importance in the initial stages of the fashion business cycle (Forecasting, consumer research and design). Further competencies in the area of social media and social networks were deemed important, being able to identify the DNA of a specific brand on online searches and analytical capability leveraging a variety of channels such as magazines, the internet and retail shopping to be able to identify and present market trends.

Wang and Ha-Brookshire (2018), identified downstream functions required less but more focused digital competency, for example, sourcing required strong competency in product lifecycle management applications and distribution required skills that focused on ERP system knowledge and skills.

This trend of increasing digital competency is driven by the continuing evolution of supply chains from Industry 3.0 to Industry 4.0 and studies have indicated that the future of the F&T industry involves adopting the major characteristics of Industry 4.0 as the focus shifts towards customer-oriented products, rapid response, and customer service, all of which are key goals of Industry 4.0 (Chen and Xing 2016).

Organizations have and are continuing to realize the need to remain competitive through the use of digital technology (Chen and Xing 2016). A supply chain that is powered by emerging technologies such as blockchain and the IoT will nurture collaboration, drive visibility and allow for Chen and Xing better decision making and, risk identification and management. Integration through technology will allow for organizations to engage suppliers with common visions, develop symbiotic relationships across functions and allow for better measurement of performance and accountability of responsibilities.

When looking towards the future, emerging markets also need to be considered along with their impacts on the global economy. Economic growth and the increasing appetite of China and other developing countries for resources is advancing global prices, which in turn makes it more complex to plan and configure supply chain assets (Ghoshray and Pundit 2021). This in combination with the growing corporate awareness of sustainability and greater government regulation adds to the complexity and uncertainty faced by those in supply chain management (Le et al. 2019; Kane et al. 2021). The digitization of supply chains has been a strategic evolution that has been ongoing for years as technologies mature and mainstream acceptance increases (Akbari and Do 2021). However, the COVID-19 pandemic acted as a catalyst for digital transformation and a sharp reminder of the costly impacts of disruption on supply chains not geared towards flexibility, reshaping how people in the industry are thinking about the future of the supply chain management (Schatteman et al. 2020a, b).

Digital transformation of supply chains may also allow for more decentralized supply chain models. Schmitt et al. (2015) suggested that using a decentralized design reduces cost variance through the risk diversification effect, thus, splintering traditional monolithic supply chains into smaller ones providing more resilience and flexibility. While decentralized, splintered supply chains utilize the same assets and resources, the key to its success is the utilization of information through technology, which assist in the management of complexity. This approach may suit organizations that are risk-averse looking to minimize disruption.

6.1 Blockchain and the Internet of Things

Two technologies that have promising potential to revolutionize the F&T supply chain are blockchain and IoT. Blockchain is considered to be a very disruptive technology, having the potential to impact and affect material change in most if not all industries. While research and application of blockchain started with a focus on financial transactions and distributed ledgers, it has major future implications on the financial sector, providing alternative structures and models which promote financial activities without the need of trusted and centralized intermediaries such as banks (Tapscott and Tapscott 2017).

Wang and Ha-Brookshire (2018), put forward that blockchain has the potential to disrupt many other domains, with the nature of blockchain allowing for the secure exchange of data using a decentralized model. It will, in turn, change the way an organization is governed, how relationships across the supply chain are structured and how transactions are handled. Lean supply chain management in the F&T industry also stands to gain the same benefits. Traceability has been a persisting problem in the F&T supply chain, while transparency is arguably a more impactful issue that is also not well addressed (ElMessiry 2018). These issues have an exacerbated impact across participants of the supply chain and pressures entities to create local nodes to use localized information.

When blockchain is integrated with other maturing technologies like IoT, it is possible to document and share actionable records related to a products, end-to-end journey through the supply chain. Wang and Ha-Brookshire (2018) asserted that creating this level of visibility where each step and action can be recorded and shared securely affords a level of traceability, authenticity and legitimacy that current systems are unable to attain. In addition, other benefits can be gained, such as compliance, error reduction, & payment processing (Tapscott and Tapscott 2017), while IoT allows for the generation of Big Data (Wang et al. 2019).

While the possible benefits of blockchain and IoT are widely related to traceability and transparency being an obvious benefit, the current application of supply chains and in particular within the F&T industry is still maturing, to realize these benefits (Sunny et al. 2020).

6.2 Cloud Computing

Cloud computing technologies offer several benefits which range from scalability to reduction of cost. However, the biggest benefit cloud computing has on LSCM (Logistics & Supply Chain Management) is the ability to integrate a variety of different platforms, enabling synchronization and collaboration across the supply chain (Dave et al. 2016). It is possible due to the integration of cloud computing and mobile technology that has evolved to modern standards, allowing for a real-time flow of information, available for consumption anywhere (Sanders et al. 2016).

Due to the aforementioned benefits, supply chains can react and make decisions in real-time in turn reducing, mitigating and potentially preventing the bullwhip effect.

6.3 Artificial Intelligence (AI)

Decision-making has always been difficult in supply chain management due to the complex nature and uncertainty of the environment. Inaccurate demand forecast and decision making can often lead to undesirable situations such as overproduction, logistical wastage and delays leading to financial costs and opportunity costs. The application of AI in F&T industry is broad and yield many benefits. For example, Torres-Luna et al. (2020) demonstrated benefits such as the elimination of waste in production, reduction of inventory surplus, reduction of lead and waiting times, improved efficiency in the movement of materials and the overall increase in productivity in the entire lean supply chain.

AI has also been applied to analyze and traffic to optimize delivery routes, improving Just-In-Time (JIT) models, reducing delivery times, and increasing customer satisfaction according to Kang et al. (2019). To further improve profitability, AI was applied to schedule product deliveries, optimize a fleet with multiple delivery vehicles, and delivery time requirements, resulting in more efficient fuel consumption and maximizing profitability (Kang et al. 2019). There has been a wide range of applications of AI in F&T industry that can help to achieve the objective of lean manufacturing (Nayak and Padhye 2018a, b).

6.4 Big Data

The application of the above technologies would lead supply chains to generate large amounts of data. These datasets would be critical in evolving LSCM, being able to mine useful information and knowledge from various data points across the supply chain (Bevilacqua et al. 2019), allowing for data visualization, analysis of key metrics and the enablement of semi-automized or automated decision making throughout the supply chain (Roy and Roy 2019). As LSCM adopts more digital technology, the generation of data will continue to proliferate. LSCM leaders will increasingly rely on data from their digital ecosystem to allow them to measure efficiency from various functions in the supply chain ranging from their carbon footprint, performance of supply communities, customer experience and satisfaction (Nguyen et al. 2015; Plambeck 2012).

6.5 Virtual Reality/Augmented Reality (VR/AR)

VR and AR have been applied more widely in various industries such as tourism, entertainment, and education to name a few, however, in the F&T industry there are limited applications (Walter et al. 2009). AR has been applied more widely due to the technology being available on mobile devices and has resulted in AR-based fashion applications, allowing customers to view products within their environments (Walter et al. 2009). However, De Silva et al. (2019) suggested that new product development in the apparel industry will be collaborative and demand virtual simulation technologies. The two technologies will provide benefits to different stages, with VR technologies providing more benefit to collaboration on new product design in the development stage such as prototyping, while AR tools being complimentary for consumer integration stages such as concept tests and fit assessments. It should be noted that while the possibilities of VR/AR technology are evident, there is very little literature, indicating the immaturity of the technology.

7 Conclusions

This chapter provides a comprehensive review of literature and cases describing Lean supply chains and sustainable development in the fashion and textile industry. While previous chapters deal with different tools and techniques, especially Chapter "Digital Technologies for Lean Manufacturing" discussing various technologies that can be integrated with the lean manufacturing, this chapter discusses several approaches to enhance lean and green supply chains to ensure smooth production and distribution of fashion and textiles products. The chapter places fashion and textile positioning in the volatile and disruptive environment like Covid-19. Lean manufacturing has been used for decades and continues to be important for the fashion and textile in reducing wastes, minimizing disruptions, and creating a balanced flow of products. Agility and visibility have been at the forefront in supporting lean manufacturing and supply chain development. This chapter confirms the importance of lean thinking, which could be applied in alignment with solutions to the sustainability paradox in the fashion and textile industry.

This chapter also discusses the future development of sustainable fashion and textile industry in alignment between lean, agility and visibility. Information technology plays a crucial role in linking lean, agility, and visibility through promoting a culture of continuous improvement, engaging everyone in the process. This chapter has briefly explained how the digital technologies such as blockchain, big data, RFID, ERP systems, and the Internet of things (IoT) will influence the future of supply chain in the fashion and textile industries. The implementation of some of the recent technologies has the potential on lean and green partnerships to make the fashion and textile sector more productive as well as to meet the sustainable standards in product manufacturing.

References

Abernathy FH, Dunlop JT, Hammond JH, Weil D (1999) A stitch in time: Lean retailing and the transformation of manufacturing–lessons from the apparel and textile industries. Oxford University Press

Adikorley RD, Rothenberg L, Guillory A (2017) Lean Six Sigma applications in the textile industry: a case study. Int J Lean Six Sigma 8(2):210–224. https://doi.org/10.1108/IJLSS-03-2016-0014

Aftab MA, Yuanjian Q, Kabir N, Aftab MA (2017) Postponement application in the fast fashion supply chain: A. Int J Bus Manag 12(7):1–14. https://doi.org/10.5539/ijbm.v12n7p115

Akbari M, Do NAT (2021) A systematic review of machine learning in logistics and supply chain management: current trends and future directions. Benchmarking: Int J (In-Press). https://doi.org/10.1108/BIJ-10-2020-0514

Akbari M, McClelland R (2020) Corporate social responsibility and corporate citizenship in sustainable supply chain: a structured literature review. Benchmarking: Int J 27(6):1799–1841. https://doi.org/10.1108/BIJ-11-2019-0509

Akbari M, Ha N (2020) Impact of additive manufacturing on the Vietnamese transportation industry: an exploratory study. Asian J. Shipp. Logist. 36(2):78–88. https://doi.org/10.1016/j.ajsl.2019.11.001

Akbari M, Hopkins J (2019) An investigation into anywhere working as a system for accelerating the transition of Ho Chi Minh city into a more livable city. J Clean Prod 209:665–679. https://doi.org/10.1016/j.jclepro.2018.10.262

Alves AC, Kahlen F-J, Flumerfelt S, Siriban-Manalang AB (2019) Lean engineering for global development. Springer

Amaro P, Alves AC, Sousa RM (2019) Lean thinking: a transversal and global management philosophy to achieve sustainability benefits. In: Alves A, Kahlen FJ, Flumerfelt S, Siriban-Manalang A (eds) Lean engineering for global development. Springer, Cham. https://doi.org/10.1007/978-3-030-13515-7_1

Anguelov N (2015) The dirty side of the garment industry: fast fashion and its negative impact on environment and society. CRC Press

Beard ND (2008) The branding of ethical fashion and the consumer: a luxury niche or mass-market reality? Fash Theory 12(4):447–467. https://doi.org/10.2752/175174108X346931

Bevilacqua M, Ciarapica FE, Antomarioni S (2019) Lean principles for organizing items in an automated storage and retrieval system: an association rule mining–based approach. Manag Prod Eng Rev 10(1):29–36. https://doi.org/10.24425/MPER.2019.128241

Bruce M, Daly L, Towers N (2004) Lean or agile: a solution for supply chain management in the textiles and clothing industry? Int J Oper Prod Manag 24(2):151–170. https://doi.org/10.1108/01443570410514867

Brusset X (2016) Does supply chain visibility enhance agility? Int J Prod Econ 171(1):46–59. https://doi.org/10.1016/j.ijpe.2015.10.005

Bubicz ME, Dias Barbosa-Póvoa APF, Carvalho A (2021) Social sustainability management in the apparel supply chains. J Clean Prod 280:124214. https://doi.org/10.1016/j.jclepro.2020.124214

Butt AS (2021) Strategies to mitigate the impact of COVID-19 on supply chain disruptions: a multiple case analysis of buyers and distributors. Int J Logist Manag. (In-Press). https://doi.org/10.1108/IJLM-11-2020-0455

Caniato F, Caridi M, Crippa L, Moretto A (2012) Environmental sustainability in fashion supply chains: an exploratory case based research. Int J Prod Econ 135(2):659–670. https://doi.org/10.1016/j.ijpe.2011.06.001

Castañeda-Navarrete J, Hauge J, López-Gómez C (2021) COVID-19's impacts on global value chains, as seen in the apparel industry. Dev Policy Rev. (In-Press). https://doi.org/10.1111/dpr.12539

Chakraborty S, Biswas MC (2020) Impact of COVID-19 on the textile, apparel and fashion manufacturing industry supply chain: case study on a ready-made garment manufacturing industry. J Supply Chain Manag Logist Procure 3(2):1–19. https://doi.org/10.2139/ssrn.3762220

Chen Z, Xing MJ (2016) Brief analysis of textile manufacturing industry status in our country and industry updating suggestion. Cotton Text Technol 44(4):80–83

Christopher M (2000) The agile supply chain: competing in volatile markets. Ind Mark Manag 29(1):37–44. https://doi.org/10.1016/S0019-8501(99)00110-8

Cleff T, van Driel G, Mildner L-M, Walter N (2018) Corporate social responsibility in the fashion industry: how eco-innovations can lead to a (more) sustainable business model in the fashion industry. In: New developments in eco-innovation research. Springer, pp 257–275

Dao TB, Barysheva GA, Bich Ngoc Tran T (2020) The impact of the COVID-19 pandemic on socio-economic development: a case study of tourism services, textile and garment industry in Vietnam. In: Proceedings of the Research Technologies of Pandemic Coronavirus Impact (RTCOV 2020). https://doi.org/10.2991/assehr.k.201105.079

Dave B, Kubler S, Främling K, Koskela L (2016) Opportunities for enhanced lean construction management using Internet of Things standards. Autom Constr 61:86–97. https://doi.org/10.1016/j.autcon.2015.10.009

De Silva RKJ, Rupasinghe TD, Apeagyei P (2019) A collaborative apparel new product development process model using virtual reality and augmented reality technologies as enablers. Int J Fash Des Technol Educ 12(1):1–11. https://doi.org/10.1080/17543266.2018.1462858

ElMessiry M, ElMessiry A (2018) Blockchain framework for textile supply chain management. In: Proceedings of international conference on blockchain (ICBC 2018). Springer, Cham, pp 213–227. https://doi.org/10.1007/978-3-319-94478-4_15

Espitia A, Mattoo A, Rocha N, Ruta M, Winkler D (2021) Pandemic trade: COVID-19, remote work and global value chains. World Econ. (In-Press). https://doi.org/10.1111/twec.13117

Garcia-Torres S, Rey-Garcia M, Albareda-Vivo L (2017) Effective disclosure in the fast-fashion industry: from sustainability reporting to action. Sustainability 9(12):2256. https://doi.org/10.3390/su9122256

Gardetti MA, Muthu SS (2020) The UN sustainable development goals for the textile and fashion industry. Springer

Ghoshray A, Pundit M (2021) Economic growth in China and its impact on international commodity prices. Int J Financ Econ 26(2):2776–2789. https://doi.org/10.1002/ijfe.1933

Haug A, Busch J (2016) Towards an ethical fashion framework. Fash Theory 20(3):317–339. https://doi.org/10.1080/1362704X.2015.1082295

Hodge GL, Goforth Ross K, Joines JA, Thoney K (2011) Adapting lean manufacturing principles to the textile industry. Prod Plan Control 22(3):237–247. https://doi.org/10.1080/09537287.2010.498577

Hoek RV (2020) Responding to COVID-19 supply chain risks—Insights from supply chain change management, total cost of ownership and supplier segmentation theory. Logistics 4(4):23. https://doi.org/10.3390/logistics4040023

Huo B, Gu M, Wang Z (2019) Green or lean? A supply chain approach to sustainable performance. J Clean Prod 216:152–166. https://doi.org/10.1016/j.jclepro.2019.01.141

Kane V, Akbari M, Nguyen L, Nguyen T (2021) Corporate social responsibility in Vietnam: views from corporate and NGO executives. Soc Responsib J. (In-Press). https://doi.org/10.1108/SRJ-10-2020-0434

Kang Y, Lee S, Do Chung B (2019) Learning-based logistics planning and scheduling for crowdsourced parcel delivery. Comput Ind Eng 132:271–279. https://doi.org/10.1016/j.cie.2019.04.044

Kumar S, Teichman S, Timpernagel T (2012) A green supply chain is a requirement for profitability. Int J Prod Res 50(5):1278–1296. https://doi.org/10.1080/00207543.2011.571924

Laari S, Töyli J, Ojala L (2017) Supply chain perspective on competitive strategies and green supply chain management strategies. J Clean Prod 141:1303–1315. https://doi.org/10.1016/j.jclepro.2016.09.114

Le T, Nguyen T, Phan T, Tran M, Phung X, Tran T, Giao K (2019) Impact of corporate social responsibility on supply chain management and financial performance in Vietnamese garment

and textile firms. Uncertain Supply Chain Manag 7(4):679–690. https://doi.org/10.5267/j.uscm.2019.4.002
Lee HL (2004) The triple-A supply chain. Harv Bus Rev 82(10):102–113
Li X, Chung C, Goldsby TJ, Holsapple CW (2008) A unified model of supply chain agility: the work-design perspective. Int J Logist Manag 19(3):408–435. https://doi.org/10.1108/09574090810919224
Lu S (2021) 2020 fashion industry benchmarking study. United States Fashion Industry Association. http://www.usfashionindustry.com/pdf_files/20200731-fashion-industry-benchmarking-study.pdf
Mahmood A (2020) Smart lean in ring spinning—a case study to improve performance of yarn manufacturing process. J Text Inst 111(11):1681–1696. https://doi.org/10.1080/00405000.2020.1724461
Majumdar A, Sinha SK (2019) Analyzing the barriers of green textile supply chain management in Southeast Asia using interpretive structural modeling. Sustain Prod Consum 17:176–187. https://doi.org/10.1016/j.spc.2018.10.005
Majumdar A, Shaw M, Sinha SK (2020) COVID-19 debunks the myth of socially sustainable supply chain: a case of the clothing industry in South Asian countries. Sustain Prod Consum 24:150–155. https://doi.org/10.1016/j.spc.2020.07.001
Manfredsson P (2016) Textile management enabled by lean thinking: a case study of textile SMEs. Prod Plan Control 27(7–8):541–549. https://doi.org/10.1080/09537287.2016.1165299
Martin M (2016) Fix textile and garment supply chains. In: Building the Impact Economy. Springer, Cham, pp 43–60. https://doi.org/10.1007/978-3-319-25604-7_4
Mohan Prasad M, Dhiyaneswari JM, Ridzwanul Jamaan J, Mythreyan S, Sutharsan SM (2020) A framework for lean manufacturing implementation in Indian textile industry. Mater Today: Proc 33:2986–2995. https://doi.org/10.1016/j.matpr.2020.02.979
Nath SD, Eweje G, Bathurst R (2019) The invisible side of managing sustainability in global supply chains: evidence from multitier apparel suppliers. J Bus Logist. (In-Press). https://doi.org/10.1111/jbl.12230
Nayak R, Akbari M, Maleki Far S (2019) Recent sustainable trends in Vietnam's fashion supply chain. J Clean Prod 225:291–303. https://doi.org/10.1016/j.jclepro.2019.03.239
Nayak R, Padhye R (2018a) Artificial intelligence and its application in the apparel industry. In: Automation in garment manufacturing. Elsevier
Nayak R, Padhye R (2018b) Introduction to automation in garment manufacturing. In: Automation in garment manufacturing. Elsevier
Nayak R, Nguyen LTV, Panwar T, George M, Ulhaq I (2020) Sustainable supply chain management', Supply chain management and logistics in the global fashion sector: the sustainability challenge, pp 1–29
Nayak R (ed.) (2019) Sustainable technologies for fashion and textiles. Woodhead Publishing
Nayak R (2020) Supply chain management and logistics in the global fashion sector: the sustainability challenge. Routledge
Nguyen J, Donohue KL, Mehrotra M (2015) The buyer's role in improving supply chain energy efficiency. SSRN. https://ssrn.com/abstract=2564287
Niinimäki K, Peters G, Dahlbo H, Perry P, Rissanen T, Gwilt A (2020) The environmental price of fast fashion. Nat Rev Earth Environ 1(4):189–200. https://doi.org/10.1038/s43017-020-0039-9
Paksoy T, Weber G-W, Huber S (2019) Lean and green supply chain management. Optimization models and algorithms. Springer International Publishing, Cham
Pal R, Gander J (2018) Modelling environmental value: an examination of sustainable business models within the fashion industry. J Clean Prod 184:251–263. https://doi.org/10.1016/j.jclepro.2018.02.001
Palange A, Dhatrak P (2021) Lean manufacturing a vital tool to enhance productivity in manufacturing. Mater Today: Proc 46(1):729–736. https://doi.org/10.1016/j.matpr.2020.12.193

Pettit TJ, Croxton KL, Fiksel J (2019) The evolution of resilience in supply chain management: a retrospective on ensuring supply chain resilience. J Bus Logist 40(1):56–65. https://doi.org/10.1111/jbl.12202

Plambeck EL (2012) Reducing greenhouse gas emissions through operations and supply chain management. Energy Econ 34(1):64–74. https://doi.org/10.1016/j.eneco.2012.08.031

Prasad MM, Dhiyaneswari J, Jamaan JR, Mythreyan S, Sutharsan S (2020) A framework for lean manufacturing implementation in Indian textile industry. Mater Today: Proc 33:2986–2995. https://doi.org/10.1016/j.matpr.2020.02.979

Prifti L, Knigge M, Kienegger H, Krcmar H (2017) A competency model for "Industrie 4.0" employees. In: Leimeister JM, Brenner W (eds.) (Hrsg.): Proceedings der 13. Internationalen Tagung Wirtschaftsinformatik (WI 2017), St. Gallen, S. 46–60. https://aisel.aisnet.org/wi2017/track01/paper/4/

Purvis L, Gosling J, Naim MM (2014) The development of a lean, agile and leagile supply network taxonomy based on differing types of flexibility. Int J Prod Econ 151:100–111. https://doi.org/10.1016/j.ijpe.2014.02.002

Radhakrishnan S (2015). The sustainable apparel coalition and the higg index. In: Roadmap to sustainable textiles and clothing. Springer, pp 23–57

Radhakrishnan S (2020) Sustainable consumption and production patterns in fashion. In: The UN sustainable development goals for the textile and fashion industry, pp 59–75. https://doi.org/10.1007/978-981-13-8787-6_4

Raut R, Gardas BB, Narkhede B (2019) Ranking the barriers of sustainable textile and apparel supply chains: an interpretive structural modelling methodology. Benchmarking: Int J 26(2):371–394. https://doi.org/10.1108/BIJ-12-2017-0340

Research & Markets (2021) Fast fashion global market report 2021: COVID-19 growth and change to 2030. https://www.researchandmarkets.com/reports/5321430/fast-fashion-global-market-report-2021-covid-19

Robertson D, Ulrich K (1998) Planning for product platforms. Sloan Manag Rev 39(4):19

Roll M (2020) The secret of zara's success: a culture of customer co-creation. https://martinroll.com/resources/articles/strategy/the-secret-of-zaras-success-a-culture-of-customer-co-creation/

Roy M, Roy A (2019) Nexus of Internet of Things (IoT) and big data: roadmap for smart management systems (SMgS). IEEE Eng Manage Rev 47(2):53–65. https://doi.org/10.1109/EMR.2019.2915961

Sancha C, Wong CWY, Gimenez Thomsen C (2016) Buyer–supplier relationships on environmental issues: a contingency perspective. J Clean Prod 112:1849–1860. https://doi.org/10.1016/j.jclepro.2014.09.026

Sanders A, Elangeswaran C, Wulfsberg JP (2016) Industry 4.0 implies lean manufacturing: research activities in industry 4.0 function as enablers for lean manufacturing. J Ind Eng Manag 9(3):811–833. https://doi.org/10.3926/jiem.1940

Schatteman O, Woodhouse D, Terino J (2020b) Supply chain lessons from Covid-19: time to refocus on resilience. Bain & Company Inc., Boston, MA, pp 1–12

Schatteman O, Woodhouse D, Terino J (2020a) Supply chain lessons from Covid-19: time to refocus on resilience. Bain & Company, Inc., Boston, MA, pp 1–12. https://www.bain.com/contentassets/d6b7a373b2b242d2be114cb46345fce2/bain-brief-supply-chain-lessons-from-covid-19.pdf

Schmitt AJ, Sun SA, Snyder LV, Shen ZJM (2015) Centralization versus decentralization: risk pooling, risk diversification, and supply chain disruptions. Omega 52:201–212. https://doi.org/10.1016/j.omega.2014.06.002

Shen B (2014) Sustainable fashion supply chain: lessons from H&M. Sustainability 6(9):6236–6249. https://doi.org/10.3390/su6096236

Shen B, Li Q (2019) Green technology adoption in textile supply chains with environmental taxes: production, pricing, and competition. IFAC-PapersOnLine 52(13):379–384. https://doi.org/10.1016/j.ifacol.2019.11.153

Shen B, Li Q, Dong C, Perry P (2017) Sustainability issues in textile and apparel supply chains. Sustainability 9(9):1592. https://doi.org/10.3390/su9091592

Shen B, Zhu C, Li Q, Wang X (2020) Green technology adoption in textiles and apparel supply chains with environmental taxes. Int J Prod Res 59(14):4157–4174. https://doi.org/10.1080/00207543.2020.1758354

Statista (2021) Which, if any, of the following fashion brands or shops do you associate with having a sustainable supply chain? https://www.statista.com/statistics/1099579/consumer-opinions-on-fashion-brands-supply-chain-sustainability-europe/

Stevenson WJ (2005) Operations management. McGraw-hill

Sunny J, Undralla N, Pillai VM (2020) Supply chain transparency through blockchain-based traceability: an overview with demonstration. Comput Ind Eng 150:106895. https://doi.org/10.1016/j.cie.2020.106895

Tapscott D, Tapscott A (2017) How blockchain will change organizations. MIT Sloan Manag Rev 58(2):10

Teodoro A, Rodriguez L (2020) Textile and garment supply chains in times of COVID-19: challenges for developing countries. 53 (Second Quarter 2020). https://unctad.org/news/textile-and-garment-supply-chains-times-covid-19-challenges-developing-countries

Thomas S (2008) From "green blur" to eco-fashion: fashioning an eco-lexicon. Fash Theory 12(4):525–539. https://doi.org/10.2752/175174108X346977

Thomas K (2020) Cultures of sustainability in the fashion industry. Fash Theory 24(5):715–742. https://doi.org/10.1080/1362704X.2018.1532737

Torres-Luna S, Valdivia-Ríos J, Macassi-Jáuregui I, Palomino ER, Viacava-Campos G, León-Chavarri C (2020) Waste reduction model design in textile industry. In: Proceedings of the 6th Brazilian technology symposium (BTSym'20), pp 427–434. https://doi.org/10.1007/978-3-030-75680-2_48

Venkatesh, S., & Swaminathan, J. M. (2004). Managing product variety through postponement: concept and applications. In The Practice of Supply Chain Management: Where Theory and Application Converge (pp. 139–155). Springer.

Vermeulen WJ, Ras PJ (2006) The challenge of greening global product chains: meeting both ends. Sustain Dev 14(4):245–256. https://doi.org/10.1002/sd.270

Walter L, Kartsounis GA, Carosio S (2009) Transforming clothing production into a demand-driven, knowledge-based, high-tech Industry. In: Transforming clothing production into a demand-driven, knowledge-based, high-tech industry: the leapfrog paradigm, Springer, London, pp 1–2

Wang Y, Han JH, Beynon-Davies P (2019) Understanding blockchain technology for future supply chains: a systematic literature review and research agenda. Supply Chain Manag: Int J 24(1):62–84. https://doi.org/10.1108/SCM-03-2018-0148

Wang B, Ha-Brookshire JE (2018) Exploration of digital competency requirements within the fashion supply chain with an anticipation of industry 4.0. Int J Fash Des Technol Educ 11(3):333–342. https://doi.org/10.1080/17543266.2018.1448459

Xu Z, Elomri A, Kerbache L, El Omri A (2020) Impacts of COVID-19 on global supply chains: facts and perspectives. IEEE Eng Manage Rev 48(3):153–166. https://doi.org/10.1109/EMR.2020.3018420

Yu Y, Huo B (2019) The impact of environmental orientation on supplier green management and financial performance: the moderating role of relational capital. J Clean Prod 211:628–639. https://doi.org/10.1016/j.jclepro.2018.11.198

Zhao L, Kim K (2021) Responding to the COVID-19 pandemic: practices and strategies of the global clothing and textile value chain. Cloth TextEs Res J 39(2):57–172. https://doi.org/10.1177/0887302X21994207

Zhelyazkov G (2011). Agile supply chain: zara's case study analysis. In: Design, manufacture & engineering management, Strathclyde University Glasgow, Velika Britanija, pp 2–11. http://idea-space.eu:19001/up/319cc877736053b53fe46efc79a29be6.pdf

Milton Keynes UK
Ingram Content Group UK Ltd.
UKHW021816190923
428961UK00003B/3